Colloids and Interfaces in Oil Recovery

Colloids and Interfaces in Oil Recovery

Special Issue Editor

Spencer Taylor

MDPI • Basel • Beijing • Wuhan • Barcelona • Belgrade

MDPI

Special Issue Editor
Spencer Taylor
University of Surrey
UK

Editorial Office
MDPI
St. Alban-Anlage 66
4052 Basel, Switzerland

This is a reprint of articles from the Special Issue published online in the open access journal *Colloids Interfaces* (ISSN 2504-5377) from 2018 to 2019 (available at: https://www.mdpi.com/journal/colloids/special_issues/Oil_Recovery)

For citation purposes, cite each article independently as indicated on the article page online and as indicated below:

LastName, A.A.; LastName, B.B.; LastName, C.C. Article Title. *Journal Name* **Year**, *Article Number*, Page Range.

ISBN 978-3-03921-106-7 (Pbk)
ISBN 978-3-03921-107-4 (PDF)

Cover image courtesy of Spencer Taylor.

Contents

About the Special Issue Editor

Spencer Taylor studied chemistry at the University of Surrey, where he gained both BSc and PhD degrees. He completed his doctorate under the supervision of Professor John Richards Jones in physical organic chemistry. After postdoctoral studies at in the Chemistry department at Queen's University, Kingston, Canada with Professors Erwin Buncel and Albert Norris, and the Inorganic Chemistry Laboratory at Oxford University with Professor Allen Hill, he joined BP Research at Sunbury-on-Thames to work on surface chemistry specifically related to crude oil. This built on earlier interests in surface chemistry gained when working at the Unilever Research Laboratory in Welwyn in the 1970s and, moreover, it renewed a working relationship with Dr David Graham. His career in BP lasted almost 22 years and almost entirely involved research into colloidal and interfacial aspects of the wide-ranging processes and products related to BP's diversification in the 1980s. However, the latter part of his time spent in BP was concerned with surface and colloidal aspects of aviation jet fuels. Since leaving BP, he returned to the Department of Chemistry at the University of Surrey as an academic, and in 2012, was given the opportunity by BP America to set up a facility to study steam-assisted gravity drainage used to recover bitumen from oilsands. In addition to petroleum-related research, his current interests continue to be in applying colloid and interface science principles to relevant processes and products. He has authored more than 100 research articles and patents, and is the recipient of the Institution of Chemical Engineers' Hutchison medal (1996) and Moulton medal (2010).

*colloids
and interfaces*

MDPI

Editorial

Colloids and Interfaces in Oil Recovery

Spencer E. Taylor

Centre for Petroleum and Surface Chemistry, Department of Chemistry, University of Surrey, Guildford, Surrey GU2 7XH, UK; s.taylor@surrey.ac.uk

Received: 30 May 2019; Accepted: 31 May 2019; Published: 31 May 2019

The role of surface and colloid chemistry in the petroleum industry is of great importance to the many current and future challenges confronting this sector. Even though according the latest BP Energy Outlook (https://www.bp.com/en/global/corporate/news-and-insights/press-releases/bp-energy-outlook-2019.html) alternative, more sustainable energy sources are making increasing contributions to global energy needs, crude oil output will nevertheless continue at least at the present level for decades to come. This increases the pressure on producers to recover crude oil more economically and with increasing environmental consideration. Therefore, the oil industry continues to work on many fronts towards improving procedures for extracting and processing these valuable resources.

The oil price has always been a key determinant for technological developments in oil recovery. For example, the high price of a barrel of crude oil in the 1970s and early-1980s stimulated the development of surfactant-based enhanced oil recovery (EOR) technology, as well as encouraging the pursuit of heavy oils and natural bitumen.

However, the fall in the oil price in the mid-1980s saw research into new technology decline, as extraction of incremental oil from conventional reservoirs, as well as heavy oil requiring unconventional methods, became uneconomic. Not until the early-2000s did the prospect of $100/barrel oil prices encourage oil producers to revisit the development of new technologies, based on improved physicochemical and geological understandings of the nature of crude oil reservoirs. One particular feature of this was that heavy oil and bitumen exploitation using mining and energy-intensive steam-based technologies increased in importance. In parallel with research into heavy oils, further research continued into conventional oil production. Thus, the past two decades have seen the development of more sophisticated approaches directed at improving the incremental recovery of conventional oil, with emphasis on oil/water/rock interfacial chemistry. Applied to crude oil reservoirs, studies have shown that rock wettability can be favorably influenced to aid oil recovery during waterflooding through the application of nanoparticles or by controlling the ionic composition of the water.

Therefore, the contents of this volume, contributed by an international group of researchers, address different aspects of oil recovery, emphasizing colloidal and interfacial aspects relevant to contemporary issues. Original research articles and reviews are included, with a focus on improving oil recovery, considering physical, chemical and microbial approaches as applied to conventional and heavy oils, as well as to natural bitumen.

Low salinity waterflooding (LSW), applicable to conventional oil recovery, was critically reviewed by Derkani et al. [1], focusing specifically on carbonate reservoirs. These authors provide evidence which confirms the benefits of LSW in improving recovery. They conclude that it is possible that no one mechanism is exclusively responsible for the effects seen, but rather, several of the reviewed mechanisms could be involved. In focusing on one specific mechanism, Taylor and Chu [2] explored the interfacial role of Ca^{2+} ions in LSW. Although a part of this investigation was in error, due to an inappropriate source of Ca^{2+} ions being used (Taylor et al. [3]), it nevertheless highlights that the coordination modes of Ca^{2+} ions are potentially relevant to crude oils, thereby warranting further research.

The roles of interfacial properties, and wettability in particular, on droplet displacement from surfaces were reviewed by Tangparitkul et al. [4] with reference to the action of different interfacially-active materials, including surfactants and nanoparticles (and derived nanofluids). Oil displacement mechanisms were considered in the presence of the different systems. Martins et al. [5] considered the specific effect of *n*-alcohols on the properties of asphaltene films at air/water and toluene/water interfaces, as these relate to crude oil emulsions formed during recovery.

Telles da Cruz et al. [6] demonstrated the possibility of modifying solid wettability through the use of cyclodextrins (CDs). The toroidal shapes of CDs create the ability for these supramolecular compounds to form inclusion complexes with linear hydrocarbons, producing what the authors term "pseudo-surfactants" and increasing the water-wettability of an initially hydrophobic C_{18}-modified quartz surface.

Two contributions consider sweep efficiency during waterflooding in porous media. In their study, Skauge et al. [7] firstly reviewed polymer flow in porous media, and followed that by introducing new rheological data from in-situ linear core-flood experiments. Significantly, it was found that the presence of oil reduced the apparent viscosity of the polymer solution. In the second contribution on this topic, Yeh and Juárez [8] demonstrated the use of a randomly close-packed porous micromodel, pre-saturated with silicone oil (viscosity = 5 cSt) to simulate the flow of polymer (polyvinylpyrrolidone) and surfactant (sodium dodecyl sulfate) solutions.

The remaining contributions focus on specific aspects of heavy oil and bitumen recovery. At the present time, heavy oil and natural bitumen amount to approximately 70% of global oil reserves. The physical properties of these resources, as well as the geology of the reservoirs, make the cost of recovery higher than for conventional oils. Particular emphasis is centered on cost and emissions reduction associated with recovery. Extending the requirement to improve sweep efficiency, in this case for the unfavorable mobility ratios associated with high viscosity heavy oils, Telmadarreie and Trivedi [9] investigated the application of surfactant and surfactant/polymer foams, in which they identified advantages in the presence of polymer. Various aspects of the interfacial chemistry involved in steam-based recovery methods were considered by Taylor [10], and how these might be improved with regard to efficiency of oil displacement under the high temperature conditions required to mobilize high viscosity oils. In a further development, Shibulal et al. [11] demonstrated the application of microbial EOR using two indigenous strains of *Bacillus* spp. isolated from soil contaminated with heavy crude oil from an Omani oilfield. In core tests, additional recoveries of around 8–10% were obtained.

At the outset, the aim of this Special Issue was to highlight the breadth of colloidal and interfacial applications in oil recovery, acknowledging the inevitability that the contributions could only be a snapshot of contemporary research. It is hoped that this has been achieved, at least in some small measure.

Conflicts of Interest: The author declares no conflict of interest.

References

1. Derkani, M.H.; Fletcher, A.J.; Abdallah, W.; Sauerer, B.; Anderson, J.; Zhang, Z.J. Low Salinity Waterflooding in Carbonate Reservoirs: Review of Interfacial Mechanisms. *Colloids Interfaces* **2018**, *2*, 20. [CrossRef]
2. Taylor, S.E.; Chu, H.T. Metal Ion Interactions with Crude Oil Components: Specificity of Ca2+ Binding to Naphthenic Acid at an Oil/Water Interface. *Colloids Interfaces* **2018**, *2*, 40. [CrossRef]
3. Taylor, S.E.; Chu, H.T.; Isiocha, U.I. Addendum: Taylor, S.E. Metal Ion Interactions with Crude Oil Components: Specificity of Ca2+ Binding to Naphthenic Acid at an Oil/Water Interface. *Colloids Interfaces* **2018**, *2*, 54. [CrossRef]
4. Tangparitkul, S.; Charpentier, T.V.J.; Pradilla, D.; Harbottle, D. Interfacial and Colloidal Forces Governing Oil Droplet Displacement: Implications for Enhanced Oil Recovery. *Colloids Interfaces* **2018**, *2*, 30. [CrossRef]
5. Martins, R.G.; Martins, L.S.; Santos, R.G. Effects of Short-Chain n-Alcohols on the Properties of Asphaltenes at Toluene/Air and Toluene/Water Interfaces. *Colloids Interfaces* **2018**, *2*, 13. [CrossRef]

6. Da Cruz, A.F.T.; Sanches, R.D.; Miranda, C.R.; Brochsztain, S. Evaluation of Cyclodextrins as Environmentally Friendly Wettability Modifiers for Enhanced Oil Recovery. *Colloids Interfaces* **2018**, *2*, 10. [CrossRef]

7. Skauge, A.; Zamani, N.; Jacobsen, J.G.; Shiran, B.S.; Al-Shakry, B.; Skauge, T. Polymer Flow in Porous Media: Relevance to Enhanced Oil Recovery. *Colloids Interfaces* **2018**, *2*, 27. [CrossRef]

8. Yeh, H.-L.; Juárez, J.J. Waterflooding of Surfactant and Polymer Solutions in a Porous Media Micromodel. *Colloids Interfaces* **2018**, *2*, 23. [CrossRef]

9. Telmadarreie, A.; Trivedi, J.J. Static and Dynamic Performance of Wet Foam and Polymer-Enhanced Foam in the Presence of Heavy Oil. *Colloids Interfaces* **2018**, *2*, 38. [CrossRef]

10. Taylor, S.E. Interfacial Chemistry in Steam-Based Thermal Recovery of Oil Sands Bitumen with Emphasis on Steam-Assisted Gravity Drainage and the Role of Chemical Additives. *Colloids Interfaces* **2018**, *2*, 16. [CrossRef]

11. Shibulal, B.; Al-Bahry, S.N.; Al-Wahaibi, Y.M.; Elshafie, A.E.; Al-Bemani, A.S.; Joshi, S.J. Microbial-Enhanced Heavy Oil Recovery under Laboratory Conditions by Bacillus firmus BG4 and Bacillus halodurans BG5 Isolated from Heavy Oil Fields. *Colloids Interfaces* **2018**, *2*, 1. [CrossRef]

colloids and interfaces

MDPI

Review

Low Salinity Waterflooding in Carbonate Reservoirs: Review of Interfacial Mechanisms

Maryam H. Derkani [1], Ashleigh J. Fletcher [1,*], Wael Abdallah [2], Bastian Sauerer [2], James Anderson [3] and Zhenyu J. Zhang [4,*]

[1] Department of Chemical and Process Engineering, University of Strathclyde, Glasgow G1 1XJ, UK; maryam.derkani@strath.ac.uk

[2] Schlumberger Middle East, S.A., Schlumberger Dhahran Carbonate Research Center, Dhahran Techno Valley, P.O. Box 39011, Dammam 31942, Saudi Arabia; WAbdallah@slb.com (W.A.); BSauerer@slb.com (B.S.)

[3] Chemical and Materials Engineering Group, School of Engineering, University of Aberdeen, Aberdeen AB24 3UE, UK; j.anderson@abdn.ac.uk

[4] School of Chemical Engineering, University of Birmingham, Birmingham B15 2TT, UK

* Correspondence: ashleigh.fletcher@strath.ac.uk (A.J.F.); z.j.zhang@bham.ac.uk (Z.J.Z.)

Received: 15 March 2018; Accepted: 8 May 2018; Published: 18 May 2018

Abstract: Carbonate rock reservoirs comprise approximately 60% of the world's oil and gas reserves. Complex flow mechanisms and strong adsorption of crude oil on carbonate formation surfaces can reduce hydrocarbon recovery of an oil-wet carbonate reservoir to as low as 10%. Low salinity waterflooding (LSW) has been confirmed as a promising technique to improve the oil recovery factor. However, the principal mechanism underpinning this recovery method is not fully understood, which poses a challenge toward designing the optimal salinity and ionic composition of any injection solution. In general, it is believed that there is more than one mechanism involved in LSW of carbonates; even though wettability alteration toward a more desirable state for oil to be recovered could be the main cause during LSW, how this alteration happens is still the subject of debate. This paper reviews different working conditions of LSW, previous studies, and field observations, alongside the proposed interfacial mechanisms which affect the colloidal interactions at oil–rock–brine interfaces. This paper provides a comprehensive review of studies on LSW in carbonate formation and further analyzes the latest achievements of LSW application in carbonates, which helps to better understand the challenges involved in these complicated multicomponent systems and potentially benefits the oil production industry.

Keywords: enhanced oil recovery; wettability alteration; recovery factor; surface charge; electric double layer; multicomponent ion exchange; fluid–fluid interactions

1. Introduction

Maximizing the amount of crude oil extracted from existing reservoirs is vital for the oil and gas industry to increase its profitability and sustainability. However, studies [1,2] have shown that about 70% of global oil reserves cannot be extracted using conventional oil recovery techniques. It is also suggested that improving the global oil recovery factor by only 1% has the potential to produce an extra 88 billion barrels of oil, equal to three years of annual oil production, at current rates [3]. In principle, all hydrocarbon recovery operations involve three phases. During the primary oil recovery phase, the natural pressure within the reservoir pushes the oil out from the oil well. This stage can generally extract only 10% of available oil reserves in the reservoir. During the secondary oil recovery phase, also known as improved oil recovery (IOR), waterflooding is commonly used to provide pressure support and improve the sweep efficiency [2]. Conventional oil recovery phases, including both the primary and secondary phases, can only extract a maximum of 30–35% of the available hydrocarbon from oil reservoirs [1]. During the tertiary oil recovery phase, also known as enhanced oil recovery (EOR), chemicals flooding (polymers, alkaline, and surfactants), miscible flooding (carbon dioxide, liquefied petroleum gases, methane, and nitrogen), thermal flooding (steam), microbial flooding (microorganisms), or a combination of them, are introduced, which helps the flow of the oil trapped in the reservoir rock by decreasing both the surface tension and viscosity of the crude oil [1,2,4,5]. This could be achieved by single or multiple changes of reservoir fluids properties and its interfacial energy. The application of each technique depends on the reservoir conditions, such as viscosity of the crude oil, brine salinity, rock permeability, as well as reservoir depth (temperature). EOR processes can recover an additional 5–20% of the original oil in place (OOIP), so that the total oil recovery can be improved potentially up to 50–70% depending on specific reservoir conditions [3].

During the secondary oil recovery phase, waterflooding (formation water) is conventionally used to maintain reservoir pressure above the bubble point pressure of the oil, improve sweep efficiency, and displace oil by water by taking advantage of viscous force [2,3]. It was first shown that either altering the brine composition or reducing the salinity of injected brine below that of the initial formation water can lead to additional oil recovery for Berea sandstone [6–12]. Such results attracted many oil and gas companies, such as British Petroleum [13–19], Shell [20–26], ExxonMobil [27,28], Schlumberger [29–31], TOTAL [32,33], and Statoil [34,35] to investigate and further explore the potential and applicability of low salinity waterflooding (LSW) for improved oil recovery. LSW, also known as designer waterflood, advanced ion management, and smart waterflooding, injects brine with controlled ionic concentration and composition (also known as smart water or dynamic water) into the well [17,20,27]. The devised formulation destabilizes the equilibrium of the initial oil–rock–brine system, which results in alteration of initial wettability conditions, and has a positive influence on the capillary pressure and relative permeability [3]. LSW can produce up to 10% extra crude oil compared to simple waterflooding methods [1]. During LSW, no expensive chemicals are added; therefore, this technique is cheap and environmentally friendly, and has no associated injection issues. Additionally, it is economically efficient to use LSW improved recovery from the start of the waterflooding process [3].

Water treatment processes for LSW are performed over two stages: nanofiltration and reverse osmosis. In the nanofiltration process, contaminations, such as sulphate, as well as other divalent ions, are removed to decrease the hardness of the brine and the probability of membrane blockage in the reverse osmosis process [3]. During the reverse osmosis process, salinity is reduced by removing salts from the injected brine. LSW permits variations in the operation window of key parameters allowing customization of the ionic concentration and composition of the injected brine, making it suitable for a particular reservoir condition, with the consideration of the clay swelling and reservoir souring, also to prevent corrosion and aerobic bacterial issues. The advantage of this technology is its lower operating and capital costs compared to most EOR processes [16,36–38]. Additionally, LSW is not only applicable from the early stages of the oil recovery process (unlike EOR techniques), but can also be applied during the late life cycle of the reservoir [36,37]. LSW techniques can also be used alongside chemical and thermal EOR processes. Studies show that using low salinity water, instead of sea water, in polymer flooding processes can decrease polymer consumption considerably (5–10 times), which is additional to the potential benefits of LSW itself [39]. Appropriate implementation of LSW could potentially improve hydrocarbon recovery efficiencies up to 40% of OOIP, corresponding to a reduction in residual oil saturation of up to 20% pore-volume [36,40–43]. LSW has been demonstrated as a promising approach to improve the oil recovery factor and it can be used both onshore and offshore. However, inconsistent and conflicting results about efficiency of this technique have been reported in literature. Additionally, the principal mechanism underpinning this technique is not yet fully understood and poses a challenge in designing and optimizing the salinity and ionic composition of the injected brine.

Approximately 60% of the world's oil reserves are found in carbonate fields (limestone, chalks, and dolomites) with a large portion located in the Middle East; these comprise 75% of oil and 90% of gas reserves for the region [44,45]. Carbonates are sedimentary rocks formed of minerals, including those predominately containing carbonate ions, e.g., calcite and dolomite [46]. Carbonate rock may also contain aragonite, anhydrite, phosphate, glauconite, chert, quartz, pyrite, ankerite, clay minerals, and siderite [44]. It is likely that carbonate rocks undergo dissolution, recrystallization, and mineralogical replacement by varying the temperature, pressure, and pore-fluid chemistry [44]. Limited knowledge with regards to the petrophysical properties of carbonate reservoir, such as heterogeneity, porosity, permeability, and wettability, is one of the greatest challenges that the oil industry is facing to manage such reservoirs and maximize its oil recovery factor [44]. Carbonate surfaces are originally water-wet, containing positively charged surface electrostatics over a wide range of pH [47]. However, adsorption of negatively charged carboxylic materials ($-COO^-$), present in the heavy end fractions of crude oil such as resin and asphaltene fractions, onto positively charged carbonate rock surfaces, results in large crude oil particles covering the carbonate surface and could promote mixed-wet or oil-wet characteristics [45,47–50]. Carbonate reservoir rocks have inherently higher chemical activities compared to minerals in sandstone reservoirs (quartz); additionally, there have been difficulties in modelling the distribution of permeability in carbonates and predicting reservoir behaviour, due to poor correlation between permeability and porosity. Researchers have also reported the presence of fractures, and large-scale heterogeneity, which generates complex paths available for fluid flow [51]; thus, the combination of these factors, as well as reduced water wetness, results in low oil recovery (30–10%) from oil-wet carbonate reservoirs [52,53].

Extensive studies on the application of LSW in both sandstone [9–13,18,54–57] and carbonate [58–62] reservoirs have been reported in the literature. However, the effect of LSW at field scale was not observed in carbonates until the recent work by Yousef and colleagues [37], where they conducted coreflooding experiments under reservoir conditions and executed successful field trails. Implementation of LSW in carbonate reservoirs has received inconsistent feedback compared to sandstone reservoirs. This is most likely due to the carbonate heterogeneity and considerably high bonding energy between carbonate surface and polar components in crude oil, in comparison with more homogeneous sandstone formations, as well as a deficiency of clay and certain minerals in carbonates [55,60]. Although improvement of oil recovery in carbonates was reported by using LSW injection [15,20,38,52,61,63–68], a few studies [69,70] reported failure in improving oil recovery efficiency when seawater or diluted seawater was used. Moreover, Fathi and coworkers [65] observed a reduction in oil recovery for outcrop chalk cores, from 60% to 15% of OOIP, by using diluted seawater compared to seawater, due to a decrease in active ions.

Improved oil recovery was observed in both secondary mode (at initial formation water saturation) and tertiary mode (after seawater residual saturation) [46,57,71,72], or solely in the secondary LSW mode [57,73–75]. No specific temperature range has been proposed for performing LSW, despite that the process is mostly carried out at temperatures of less than 100 °C [60,76]. Even though very low concentrations (up to 5 kppm) of total dissolved solids (TDS) have been proposed for LSW in sandstones [16,41,76,77], there are other works whereby improved oil recovery in carbonates was achieved by a considerably higher salinity of injected brine (up to 33 kppm) [60,65]. It is suggested that LSW in carbonates can improve oil recovery even at higher salinity of the injected brine, as long as it contains a different relative concentration of active ions compared to the formation water.

A survey of the literature shows that the wetting condition of a carbonate rock can be altered by increasing the concentration of divalent anion (e.g., SO_4^{2-}), decreasing the concentration of divalent cations (Ca^{2+} or Mg^{2+}), reducing the salinity of brine, or removing sodium chloride (NaCl) from seawater [62,78,79]. For example, it was reported that removing sodium chloride from the injection brine and increasing SO_4^{2-} concentration, at high temperature (above 90 °C), could modify the rock wettability significantly, improving oil recovery from 37% to 62% of OOIP from chalk cores, compared to original seawater flooding [79]. Depending on the characteristics of the crude oil, the reservoir condition, and the properties of the formation brine, several mechanisms have been suggested to interpret LSW in carbonate reservoirs. The primary mechanisms proposed include fines migration, rock dissolution, reduction of interfacial tension (IFT), fluid–fluid interaction and formation of microemulsions, multicomponent ionic exchange (MIE), and expansion of electric double layer (EDL) [46,80–84]. The desirable condition for enhanced oil recovery could be due to an alteration in the wettability of the rock, which is determined by the thickness and stability of a water film between the rock surface and crude oil. It is generally accepted that LSW disturbs the pre-established thermodynamic equilibrium between rock, formation water and oil interfaces and facilitates a new chemical equilibrium, which modifies the wetting properties towards a more desirable condition to improve oil recovery from reservoir rock pores. The wettability alteration is affected by temperature, pressure, pore water and crude oil chemistry [85]. In general, it is believed that there is more than one mechanism involved in LSW of carbonates; wettability alteration could be the main cause during LSW, although how this alteration happens is still the subject of debate. Understanding the colloidal interactions at oil–rock–brine interfaces, the underlying physiochemical mechanisms involved in LSW, as well as detecting its suitability to a specific type of reservoir, are essential before implementing this technique to a given carbonate reservoir.

This paper provides a critical review on LSW in carbonate reservoirs by discussing different variables involved in this technique, such as reservoir and the injected brine parameters, previous experimental and theoretical studies on surface wettability, interfacial tension and recovery factors, as well as field observations, and finally the proposed interfacial mechanisms. Our review summarizes and further analyzes the latest achievements of LSW in carbonates, which helps to better understand the challenges involved in these complicated multicomponent systems, and potentially benefits the oil industry.

2. Working Conditions

Based on previous studies, various conditions have been suggested to affect oil recovery improvement during LSW in carbonate formations, such as specific reservoir conditions, oil–rock–brine properties, and well pattern. Both reservoir and the injected brine parameters are summarized in this section.

2.1. Reservoir Parameters

2.1.1. Formation Water Composition and pH

The composition of the formation (connate) water in reservoirs and its pH can influence the efficiency of oil recovery by affecting the initial rock wettability and interactions with the injected brine. The charge of rock–brine interfaces is affected by potential determining ions (PDIs) present in the formation water, as well as the pH of the formation water [58]. Natural pH of formation water in carbonate reservoirs is slightly basic (7–8) [58,61]. However, due to the high buffer capacity of calcium carbonate from the formation water, the pH remains unchanged, due to chemical equilibrium at the oil–rock–brine interface. Thus, it is suggested that wettability alteration, as a result of pH variation is temporary [61]. Table 1 shows the variation in composition of brine (formation water and seawater) in different regions, such as the Persian Gulf, Ekofisk and common areas [3,30,38,86]. Formation water is usually extremely saline (up to 250 kppm), and contains a high concentration of Ca^{2+} ions, but much lower concentration Mg^{2+} ions. Additionally, SO_4^{2-} is known as the most active anion in altering wettability of carbonate rock. However, the concentration of SO_4^{2-} ions is very low, due to the high concentration of Ca^{2+} in the formation water, especially at high temperatures, which results in precipitation of anhydrite ($CaSO_4$) [87]. Romanuka et al. [88] suggested that injection of seawater with high concentrations of surface interacting ions (SO_4^{2-}, BO_3^{3-} or PO_4^{3-}), with the presence of divalent cations, e.g., barium (Ba^{2+}) and strontium (Sr^{2+}), in formation water, can increase the chance of scale formation in the production lines and plugging of reservoir rock in the production well. Additionally, injection of seawater with high concentration of SO_4^{2-} can turn sweet oil fields to sour oil fields [89]. The interactions between small amounts of SO_4^{2-} dissolved in formation water and the rock surface using chalk cores at 20–130 °C was studied by Shariatpanahi et al. [87]. Their experimental results showed that even small concentrations of SO_4^{2-} ions (up to 2 mmol/L) present in formation water can influence initial wetting conditions of carbonate rock reservoir considerably and alter it to a more water-wet state. Therefore, composition of the formation water in carbonate fields must be taken into account before designing the injected brine for LSW in order to prevent the chance of reservoir souring and plugging, and to be able to maximize oil production from a specific carbonate reservoir.

Table 1. Composition of the formation water and seawater compared in different regions (the Persian Gulf, Ekofisk, and common areas) [3,30,38,86].

Type	Seawater (ppm)			Formation Water (ppm)		
Ionic Composition	Persian Gulf	Ekofisk	Common	Persian Gulf	Ekofisk	Common
Na^+	18,040	10,345	10,890	59,491	15,745	31,275
K^+	0	390	460	0	0	654
Ba^{2+}	0	0	0	0	0	269
Ca^{2+}	650	521	428	19,040	9258	5038
Mg^{2+}	2160	1093	1368	2439	607	379
Sr^{2+}	10	0	0	0	0	771
SO_4^{2-}	4450	2305	2960	350	0	0
Cl^-	31,810	18,719	19,766	132,060	42,437	60,412
CO_3^{2-}	30	0	0	0	0	0
HCO_3^-	120	122	0	354	0	0
Total	57,270	33,497	35,872	213,734	68,050	98,798

2.1.2. Initial Water Saturation

Initial water saturation (S_{wi}) is determined as the amount of water adsorbed on the surface of rock porous channels per pore-volume and exist as free water in the pore space [90]. It is suggested that reducing initial water saturation results in reduction of the water-wet condition for sandstone cores [8,91]. The effect of initial water saturation on wettability alteration of reservoir limestone cores was studied by Strand et al. [59], where their experimental results showed an increase in oil recovery of reservoir limestone cores by increasing the initial water saturation from 9.1% to 14.8%. They showed that the core with higher initial water saturation produced an additional 15% compared with OOIP. Similarly, Puntervold et al. [92] examined a range of initial water saturations, such as 100% oil saturation ($S_{wi} = 0$), low initial water saturation ($S_{wi} = 10\%$), and high initial water saturation ($S_{wi} = 30$–50%), where they observed a decrease in water wetness of chalk cores by decreasing the initial water saturation. By contrast, in their later published work [93], two different residual water saturations (10% and 22%) and a crude oil were used, where the experimental results showed no correlation between initial water saturation and improved oil recovery from chalk cores, as the authors observed 50% recovery for both cores. They suggested that initial water saturation has a negligible impact on wettability alteration of chalk cores.

2.1.3. Crude Oil Composition

An appropriate evaluation of crude oil composition is a prerequisite for any successful LSW process. Polarity of crude oil is related to the presence of heteroatoms such as sulphur, nitrogen, and oxygen, which are found in the functional groups of acidic and basic organic molecules in the crude oil, such as asphaltene and resins, which can affect rock wettability [94]. Asphaltenes are the most dense, most aromatic, and most polar component of crude oil, having a great importance in predicting the chemistry and properties of crude oil [95,96]. Asphaltenes have a complex nature and their molecular structures are not well known [97], and associated clogging in reservoir rock and oil wells, reduces oil production and has significant negative effects on the economics of this industry [98]. The amount of carboxylic material in crude oil can be quantified by the acid number (AN), determined as the number of milligrams of potassium hydroxide required to neutralize 1 g of crude oil (mg KOH/g) [99]. The AN and the chemical properties of the carboxylic material in the crude oil can strongly affect the wetting properties of carbonate [100,101]. Austad and colleagues [61] suggested that low salinity effect is dependent on a mixed-wet state, and such influence increases as the AN of crude oil is increased. However, no direct correlation was established between the AN of a crude oil and improvement in oil recovery by LSW [55]. Standnes and Austad [102,103] showed that oil recovery and water wetness increases as the AN of the oil decreases (Figure 1). Even though the base number (BN) of crude oils are believed to have a minor impact on wettability alteration, Puntervold et al. [92] observed an improvement in water-wetness of chalk by increasing the BN of crude oil with a given AN. It was argued that the formation of an acid–base complex in the crude oil can make the acidic material less active toward the carbonate rock surface.

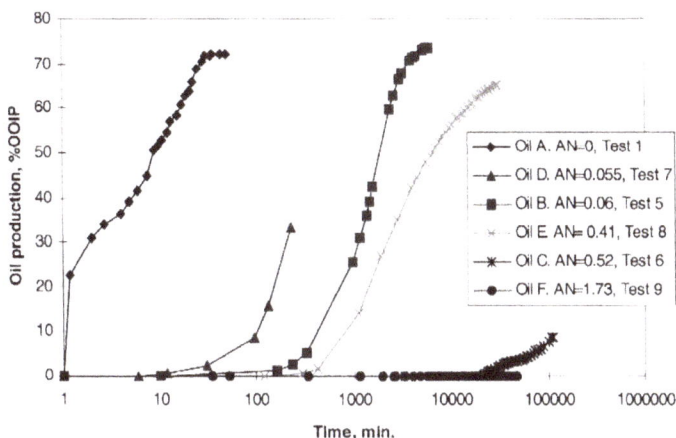

Figure 1. Oil production from a chalk core at 40 °C, when the core was saturated with crude oils with different AN from [102]. Reproduced with permission from Standnes and Austad, *J. Pet. Sci. Eng.*; published by Elsevier, 2000.

The effect of acidic material on wettability alteration of chalk was systematically examined by extracting water-soluble acids (naphthenic acids) from a crude oil a with high AN [104]. The results indicate that the core, saturated with oil depleted in water-soluble acids, was more water-wet, showing that water-soluble acids can influence the stability of the initial water film at oil–rock interfaces. In a different study, Yi and Sarma [80] observed 17% additional LSW oil recovery by using a heavy oil, in comparison with 9% for a thin field oil. They attributed such difference to the ability of large molecules in the viscous oil to reach equilibrium with the rock surface and re-establish the mixed or oil-wet condition, as opposed to a thin oil, where wettability restoration would take a longer time and the final wettability state would be less oil-wet. This suggests that LSW can be more efficient in the case with heavier oil compared to lighter oil, as the more oil-wet initial state can be altered to a more water-wet condition and LSW has a higher chance to make a significant wettability alteration.

The effects of various chemical compounds (carboxylic acids, alcohols, SO_4^{2-}, sulphonates, amines, amino acids, and carboxylated polymers) on surface adsorption, and consequently modification of the rock wettability of carbonate minerals (calcite, dolomite, and magnesite), were investigated by Thomas and co-workers [105]. They reported that organic compounds can readily adsorb on carbonate surfaces from both organic and aqueous solvents and fatty acid can be adsorbed from aqueous solutions irreversibly. Additionally, it was found that carboxylic acids adsorb most strongly to the carbonate surface and their adsorption is stabilized by long straight chains, due to interactions between closed-packed hydrophobic layer, or by polymeric structures by providing multiple attachments sites, altering the wettability of rock to strongly oil-wet. In a separate study [106], it was shown that long chain carboxylic acids (naphthenic acid) could alter the wettability of chalk to a more oil-wet condition, comparing to short chain fatty acids (acetic acid). Rezaei Gomari et al. [107] studied the effect of different fatty acids, such as short chain (heptanoic acid), long chain (stearic and oleic acids), and naphthenic acids (cyclohexane-pentanoic and decahydronaphthalene-pentanoic acids) on wettability alteration of calcite crystal surfaces. Their experimental results showed that long chain saturated stearic acid, as well as long chain unsaturated oleic acid, are by far the most effective among the studied acidic species to alter the calcite surfaces to a more oil-wet state. Therefore, higher AN of crude oil, as well as the presence of long chain carboxylic acids, are suggested to play role in wettability alteration of carbonate surfaces toward a more oil-wet state.

2.1.4. Aging Time, Temperature, and Pressure

Reservoir temperature is an essential factor that affects the activation energy required for the chemical reaction for wettability improvement to occur by LSW [59]. Carbonate rock reservoirs are suggested to be more water-wet at higher temperatures [108]. It is believed that reservoir temperature and the AN of crude oil are related to each other: as the reservoir temperature increases, the AN decreases, due to increased decarboxylation of the acidic molecules at high temperature [109]. The effect of aging time and temperature on the wettability alteration of carbonate surfaces were studied by Heidari et al. [110], where they reached an intermediate-wet state at 50 °C after 10 days aging of carbonate slides, as opposed to 63 days aging at 25 °C. They suggested that aging is an essential parameter in wettability modification in order to produce similar oil-wet conditions to a reservoir; however they concluded that wettability of rock is mostly influenced by temperature rather than the time for which the core is aged with oil. Core preparation involves, first solvent cleaning using Dean–Stark extraction to remove contamination before petrophysical measurements (dimensions and porosity) and wettability restoration. Then, the core is saturated with the formation water under vacuum and permeability is determined by the formation water flooding using different flooding rates. Afterwards, crude oil is injected to the core to achieve initial water saturation, and the core plug is aged using steel aging cell under experimental temperature and pressure of about 200 psi [80]. For limestone core plugs to restore the reservoir wettability, Yi and Sarma suggested that, according to properties of the rock core and fluids, a minimum aging time of three weeks is required [80]: they suggested that a deficiency in aging time may result in improved oil recovery in the secondary oil recovery mode, but limited wettability modification in the tertiary mode. In contrast, it was suggested that aging time and temperature have a negligible impact on oil recovery from chalks, as the oil–rock–brine system reaches equilibrium quickly [111,112]. Pressure may also affect wettability of carbonate rock reservoirs, due to changes in solubility of compounds in the crude oil. Solubility of asphaltenes decreases, as the pressure decreases toward the bubble point of the crude oil, which results in precipitation and adsorption of the polyaromatic compounds on the rock surface [113].

2.1.5. Rock Mineralogy

Different carbonate minerals, such as chalk, limestone (reservoir and outcrop), and dolomite possess different surface areas and available binding sites toward PDIs. Chalk and reservoir limestone cores are reactive toward the active ions present in seawater, resulting in wettability modification. However, reactivity of reservoir limestone cores toward PDIs (Ca^{2+}, Mg^{2+}, and SO_4^{2-}) can vary. For example, it was shown that wettability of oil-wet reservoir limestone cores can be improved by firstly cleaning with toluene and methanol, and then flooding with synthetic seawater at 130 °C, resulting in an improvement of the water-wet fraction by approximately 30% [114]. Furthermore, the results show that oil recovery from low permeable reservoir limestone was improved considerably, from 8% to 37% of OOIP, after switching the injected brine from formation water to seawater in both secondary and tertiary oil recovery modes. However, in a different investigation, Fathi and colleagues [115] showed that it was impossible to improve the water-wetness of an offshore oil-wet reservoir limestone, even though the reservoir crude oil had a very low AN. Different behavior of outcrop limestone from reservoir limestone in a wettability modification process was shown by Ravari et al. [116]. Two outcrop limestones collected from different regions were examined, and both of them showed that wettability can be modified in the presence of crude oil. However, no reactivity toward PDIs was observed on the outcrop limestone surfaces, while wettability modification and improved oil recovery could not be obtained using seawater at high temperature. This could be due to outcrop limestone acting oil-wet without being in contact with oil, or outcrop limestone acting water-wet, but showing no reactivity towards PDIs due to structural differences between outcrop limestone with that of reservoir limestone, or due to adsorption of polymeric hydrophilic materials on the rock surface. It was concluded that care must be taken to use outcrop limestone as a model for carbonate reservoir rock in systematic studies of wettability alteration and oil recovery. However, even different sources of limestone showed different surface reactivity towards variation in brine salinity, which could be due to their different grain structure and degree of crystallinity [25]. Chalk is made of pure biogenic $CaCO_3$ (fragmentary portions of calcite skeletones generated by plankton) and is known to have higher reactivity towards PDIs than limestone [59]. It is suggested that seawater has higher potential to act as a wettability modifier towards chalk surfaces compared to limestone [59]. NaCl depleted seawater was recognized as a better wettability modifier toward chalk surfaces than seawater [117]. Moreover, Austad et al. [62] observed no low salinity effect for chalk cores that do not contain anhydrite and suggested that much higher oil recovery (up to 20% of OOIP) in limestone cores can be obtained if the core contains a high amount of anhydrite. Similarly, even though Zahid et al. [69] observed considerable oil recovery improvement in a tertiary mode for reservoir carbonate (limestone) cores, by reducing concentration of seawater at 90 °C, they did not notice any oil recovery improvement in anhydrite free chalk cores. Additionally, different types of outcrop chalk cores are shown to have different degree of reactivity towards SO_4^{2-} ions [118]. Dolomite surfaces are suggested to behave similar to calcite surfaces in wettability alteration toward LSW [30,31,62,119]. However, it was suggested that, adsorption of stearic acid molecules on dolomite surfaces is stronger than on calcite surfaces, due to the positive surface charge of dolomite in diluted seawater [31]. Consequently, diluted seawater is less efficient in altering the surface charge of dolomite and releasing adsorbed carboxylic acid materials when compared to calcite [31]. Mahani and co-workers [25,26] explained that different carbonate rocks exhibited different surface reactivity and wettability alteration by variation in the brine salinity, pH, and temperature. It was shown that chalk surfaces had the highest surface reactivity towards alteration in brine concentration and pH, while dolomite surfaces showed the lowest [26]. Furthermore, a more noticeable improvement in wettability of dolomite surfaces at elevated temperature was observed compared to limestone surfaces [25]. Lastly, it is suggested that the optimal brine concentration for LSW that produces the highest oil recovery varies in different layers of the same reservoir and is dependant on the rock mineralogy and properties [22].

2.1.6. Rock Porosity and Permeability

Carbonate reservoir wettability is also affected by rock porosity and relative permeability (the permeability difference between the fractures and the matrix block). Porosity is directly linked to the amount of oil in place and permeability is linked to the rate at which the crude oil can be recovered [105]. Evaluation of porosity and permeability in carbonates is multifarious, due to complex pore systems and their chemical reactivity. If the natural fractures have a high degree of connectivity, the viscous displacement of oil from the matrix blocks, by the injected brine, decreases, due to its negative impact on capillary pressure, which prevents brine imbibing spontaneously from the fractures through the matrix blocks. In contrast, displacement of oil from the matrix block, with injected brine, results in an increase in the water saturation in matrix blocks of a reservoir and relative permeability. Therefore, if the difference between permeability of the matrix and fractures is moderate, then a large pore-volume of seawater flows into the matrix blocks during waterflooding [120]. Naturally fractured outcrop chalk cores are characterized by their high porosity and low permeability. However, limestone is less homogeneous than chalk, in terms of porosity and permeability, and has smaller surface area [121]. Korsnes and co-workers [122] studied the mechanical properties of high porosity outcrop chalk using synthetic seawater, modified seawater, and distilled water at 130 °C. Their experimental results showed an increase in compaction, by a factor of 2.7, when seawater, including SO_4^{2-} ions, was used, due to chemical reactions at the chalk surface, rather than pure chemical dissolution of chalk. Additionally, reductions in permeability, by a factor of 2.5, when brine contained SO_4^{2-} ions was observed compared to 1.7 for brine without SO_4^{2-}. They suggested that no considerable reduction in permeability, due to compaction of chalk with high porosity, was observed, by using different brines, and seawater does not alter permeability of the chalk even at high temperature. Even though porosity and permeability of carbonate rocks are important factors that influence oil recovery efficiency, additional previous studies on the effect of LSW on permeability and wettability alteration were not available in the open literature, hence, the discussion on their impact on oil recovery in carbonates could not be extended further.

2.2. Injected Brine Parameters

2.2.1. Ionic Composition and Temperature

The effects of Ca^{2+}, Mg^{2+}, and SO_4^{2-}, three primary PDIs toward carbonate surfaces, have been extensively studied in the literature. It was reported [123] that increasing the concentration of SO_4^{2-} ions at four times the concentration present in seawater can improve oil recovery from chalk cores, from 15% to 55% of OOIP after 30 days at 70 °C (Figure 2a). In a similar work, Zhang and co-workers [111] reported that increasing the concentration of Ca^{2+} ions at four times the concentration present in seawater whilst keeping the SO_4^{2-} concentration the same as in the seawater, could increase oil recovery of chalk cores from 28% to 60% of the OOIP, after 30 days at 70 °C (Figure 2b). Additionally, they suggested that increasing the concentration of Ca^{2+} and SO_4^{2-} ions in seawater and the initial brine altered the wettability of chalk surfaces, and improved oil recovery. However, high concentrations of Ca^{2+} ions did not enhance oil recovery when cores were aged with a crude oil of high AN. The effect of Ca^{2+} and SO_4^{2-} ions on wettability modification of chalk surfaces were investigated, and it was found that adsorption of SO_4^{2-}, and co-adsorption of Ca^{2+}, onto chalk surfaces with increasing temperature, could result in enhanced water-wetness conditions and improved oil recovery [58,124]. However, the surface adsorption decreases as the ratio of Ca^{2+}/SO_4^{2-} increases, at temperatures beyond 100 °C, due to precipitation of anhydrite ($CaSO_4$). Additionally, reduction of brine salinity and increasing concentration of SO_4^{2-} ions are shown to improve oil recovery in sandstone cores with a low clay content, containing high amount of dolomite and anhydrite and the interstitial dolomite crystals are argued to play a role in low salinity oil recovery mechanism [119].

Activity of SO_4^{2-}, as a PDI and wettability modifier of chalk surfaces, could increase as temperature increases from 70–130 °C [125]. Oil recovery from a chalk core at two different temperatures (100 °C and 130 °C) for a fixed oil by varying concentration of SO_4^{2-} ions in a synthetic seawater is shown in Figure 3 [123]. However, individual observations need to be carefully considered with respect to a system of interest as anticipated gains may not always be obtained (Figure 4) [111]. Systematic series of studies was carried out to investigate the effect of divalent cations (Mg^{2+} as well as Ca^{2+}) and an anion (SO_4^{2-}) on the wettability alteration of chalk surfaces [63,120,122,126]. They revealed that, at high temperatures (above 70 °C), Mg^{2+} can substitute Ca^{2+} on the chalk, and the quantity of carboxylate complexes and the degree of substitution increases as temperature increases. Similarly, it was shown that (Mg^{2+} ions can substitute Ca^{2+} on reservoir limestone cores at high temperature (130 °C), but the reactivity was less than for chalk [59]. It is possible that SO_4^{2-} cannot improve water-wetness and oil recovery alone in the absence of divalent cations (Mg^{2+} and Ca^{2+}). Additionally, Chandrasekhar and Mohanty [43], as well as Shehata et al. [127], suggested that Mg^{2+} and SO_4^{2-} ions are more efficient than Ca^{2+} ions in wettability alteration and improved oil recovery of reservoir limestone. Figure 5 shows that the equilibria established with specific seawater concentrations can be significantly affected by additional PDIs. Yi and Sarma [80] studied the effect of reducing ionic strength and increasing concentration of SO_4^{2-} ions on wettability alteration and oil recovery of reservoir limestone cores at 70–120 °C. They argued that, at 70 °C, lowering the concentration of brine is more efficient than increasing the concentration of SO_4^{2-} ions, whereas, at 120 °C, reducing water salinity and increasing SO_4^{2-} concentration resulted in much higher oil recovery. Additionally, at 90 °C, oil recovery can be improved by either lowering water salinity or increasing SO_4^{2-} concentration, while divalent cations had a negligible impact on wettability alteration [80]. However, it is shown that wettability of carbonate rock surfaces can be altered to a more water-wet state even at ambient temperature, depending on specific type of rock, and elevated temperature does not always lead to a more water-wet condition [25,128].

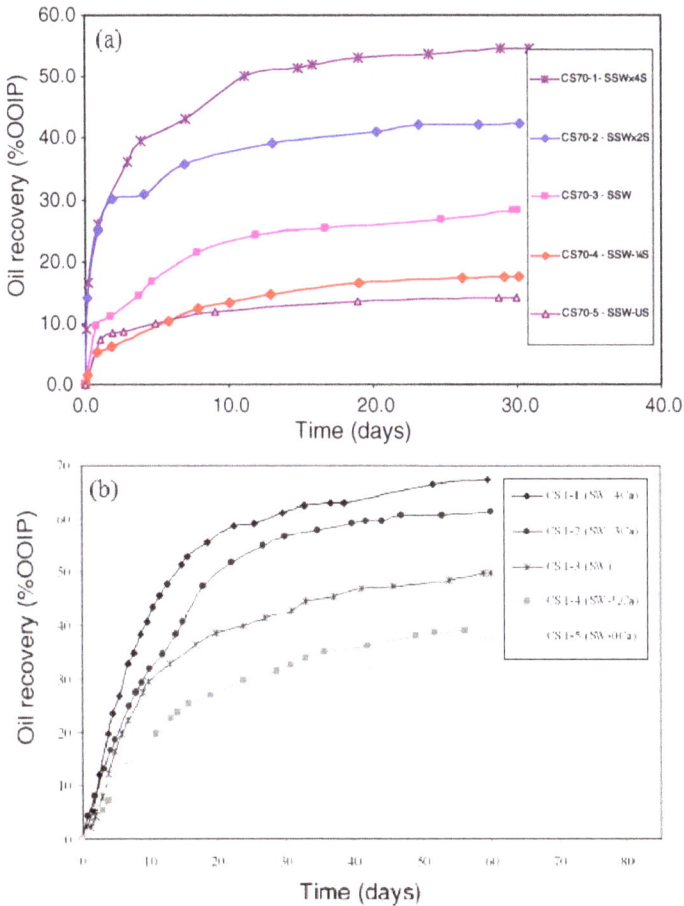

Figure 2. Oil recovery from a chalk core 70 °C when a synthetic seawater (SSW) with different concentrations of: (**a**) SO_4^{2-} (0–4 times the concentration present in synthetic seawater (SSW-US, SSW-1/2S, SSW, SSWx2S, and SSWx4S)) from [123] (reproduced with permission from Zhang et al., *Colloids Surf. A*; published by Elsevier, 2006); and (**b**) Ca^{2+} (0–4 times the concentration present in seawater (SW-0Ca, SW-1/2Ca, SW, SWx3Ca, and SWx4Ca)), but constant amount of SO_4^{2-} ions from [111] (reproduced with permission from Zhang et al., *Energy Fuels*; published by American Chemical Society, 2006), were used as imbibing fluids (AN = 0.55 mg KOH/g).

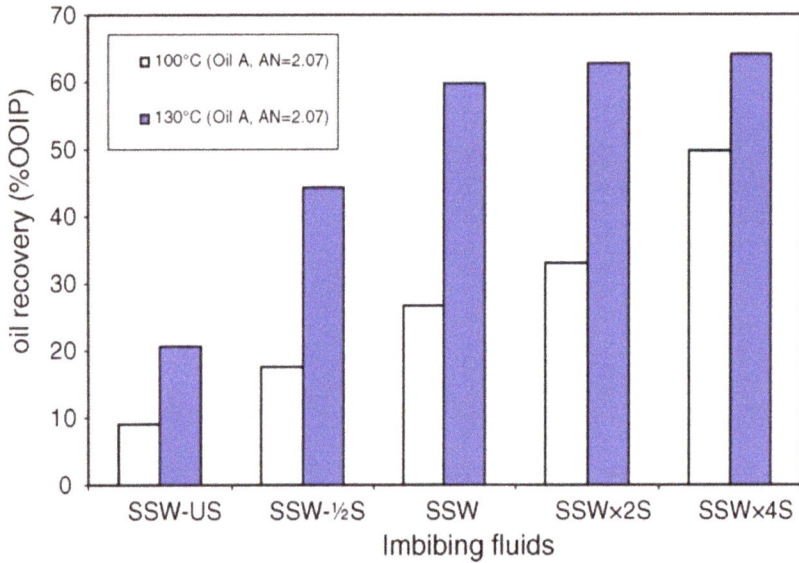

Figure 3. Effect of temperature (100 °C and 130 °C) and different concentrations of SO_4^{2-} ions (0–4 times the concentration present in synthetic seawater (SSW-US, SSW-1/2S, SSW, SSWx2S, and SSWx4S)) on oil recovery from a chalk core (AN = 2.07 mg KOH/g) from [123]. Reproduced with permission from Zhang et al., *Colloids Surf. A*; published by Elsevier, 2006.

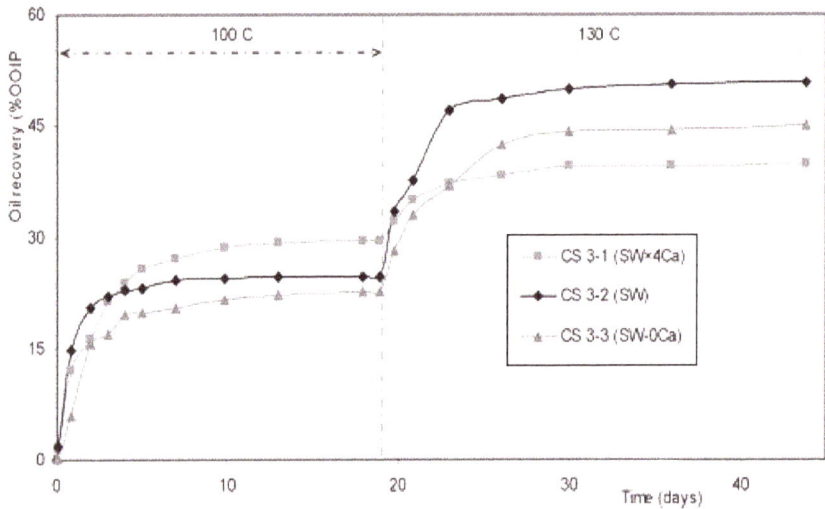

Figure 4. Effect of temperature (100 °C and 130 °C) and different concentrations of Ca^{2+} ions (0–4 times the concentration present in seawater (SW-0Ca, SW, and SWx4Ca), but constant amount of SO_4^{2-} ions) on oil recovery from a chalk core (AN = 2.07 mg KOH/g) from [111]. Reproduced with permission from Zhang et al., *Energy Fuels*; published by American Chemical Society, 2006.

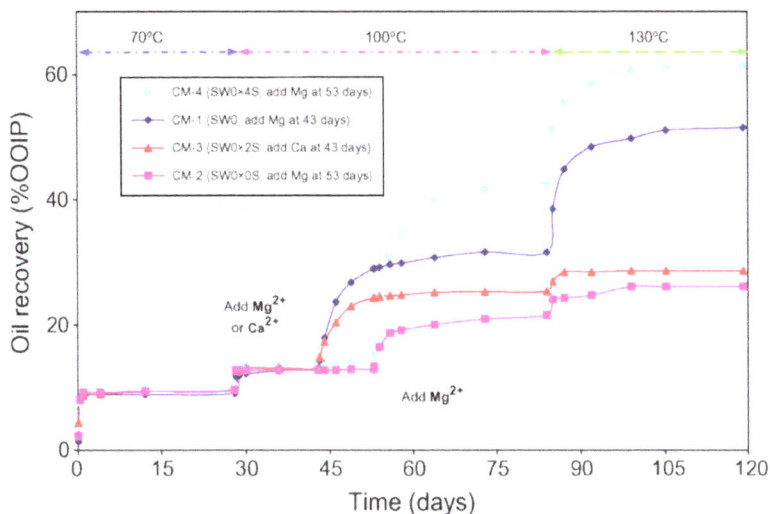

Figure 5. Effect of temperature (70 °C, 100 °C, and 130 °C) and PDIs (Ca^{2+}, Mg^{2+} and SO_4^{2-}) on oil recovery from a chalk core when modified seawater depleted in Ca^{2+}, Mg^{2+} ions and different concentrations of SO_4^{2-} ions was used as initial imbibing fluids (SW0x0S, SW0x2S, SW0, and SS0x4S) and Ca^{2+}, Mg^{2+} ions were added later as the same concentration present in seawater (AN = 2.07 mg KOH/g) from [63]. Reproduced with permission from Zhang et al., *Colloids Surf. A*; published by Elsevier, 2007.

As for the effect of non-active salt (NaCl), Fathi and co-workers suggested that the depletion of NaCl from seawater, as well as the presence of reactive ions (Ca^{2+}, Mg^{2+} and SO_4^{2-}), at elevated temperature (120 °C), are essential for wettability alteration of chalk surfaces and improved oil recovery [65,79,129]. They observed improvement in oil recovery by using NaCl depleted seawater and increasing SO_4^{2-} concentration up to four-times at 90 °C and a decrease in oil recovery, by increasing the concentration of NaCl in seawater up to four times at 100 °C (Figure 6). Similarly, improvement in oil recovery and water-wetness were observed by using NaCl depleted seawater containing four times the SO_4^{2-} concentration compared to seawater in a carbonate reservoir chalk cores [36]. However, it should be mentioned that, by virtue of removing NaCl from seawater, the total salinity is greatly reduced due to high concentration of this salt present in seawater. Therefore, it is expected that brine becomes more efficient and there is a possibility that NaCl is not a determinant salt in the brine.

Influence of polyatomic anions, e.g., borate (BO_3^{3-}) and phosphate (PO_4^{3-}), on the wettability alteration and improved oil recovery of limestone and dolomite cores free of anhydrite was examined, and it was found that they could facilitate 15–20% of OOIP additional oil recovery [27]. It is likely that considerable amounts of oil can be recovered when SO_4^{2-} ions are replaced by BO_3^{3-} and PO_4^{3-} ions in the injected brine when soft water (Ca^{2+} and Mg^{2+} depleted formation water) was used. Additionally, Meng and co-workers [130] showed that the presence of large amount of PO_4^{3-} ions (1000 mg/L) in seawater can alter wettability of limestone cores to a more water-wet condition at 90 °C. Similarly, it was suggested that BO_3^{3-} ions can substitute and replace SO_4^{2-} ions on chalk surfaces, but degree of substitution is limited at field scale due to formation of precipitate and aqueous speciation at reservoir pH [131]. It can be concluded, based on the previous studies, that the injection brine for LSW in carbonates should contain PDIs (Ca^{2+} and/or Mg^{2+}, SO_4^{2-}), depleted in NaCl, and operate at high temperatures (beyond 70 °C). However, there is a limit for increasing the concentration of SO_4^{2-} with increasing temperature, as high concentrations of SO_4^{2-} at high temperature results in precipitation of $CaSO_4$ and reduces oil recovery. As for limestone cores containing anhydrite, the SO_4^{2-}

ions generated due to dissolution of $CaSO_4$ can give rise to oil recovery, which is dependant on the brine salinity/content and temperature. The solubility of $CaSO_4$ increases as concentration of Ca^{2+} ions in the brine and temperature decreases, and concentration of NaCl increases [62]. In addition, concentration of SO_4^{2-} ions in the brine decreases as temperature increases, due to adsorption on the rock surface. On the other hand, the wettability modification process and oil recovery are improved by increasing temperature and decreasing concentration of NaCl. Therefore, the influence of non-active salt concentration and temperature contrast each other. Thus, for limestone, an optimum temperature for improved oil recovery in LSW is suggested to be in range of 90–110 °C [62].

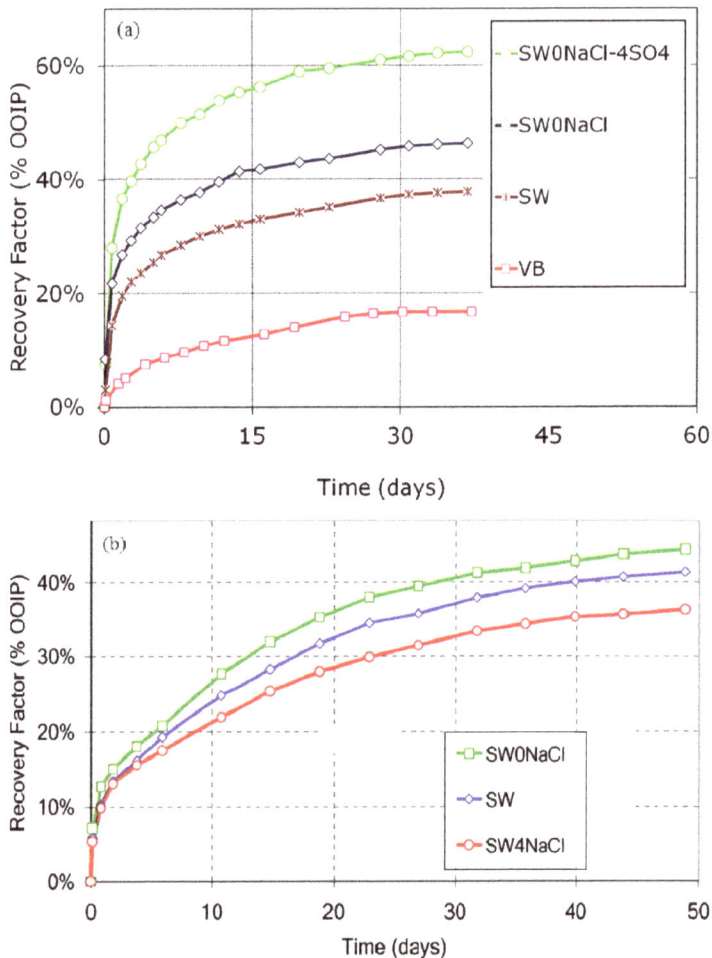

Figure 6. Oil recovery from a chalk core at (**a**) 90 °C (AN = 0.5 mg KOH/g) and (**b**) 100 °C (AN = 1.90 mg KOH/g) when formation water (VB), seawater (SW), NaCl depleted seawater (SW0NaCl), NaCl depleted seawater spiked with four times the concentration of SO_4^{2-} ions present in seawater (SW0NaCl-4SO4), and seawater spiked with four times the concentration of NaCl present in seawater (SW4NaCl), were used as imbibing fluids from [79] (reproduced with permission from Fathi et al., *Energy Fuels*; published by American Chemical Society, 2011) and [65] (reproduced with permission from Fathi et al., *Energy Fuels*; published by American Chemical Society, 2010).

2.2.2. Ionic Concentration

The effect of salinity on wettability alteration of carbonate rocks was studied using both diluted brine (formation water or seawater) and tuned brine. There are several experimental results reported in the literature that (up to 20-times) diluted seawater showed higher potential to improved oil recovery from carbonate reservoirs compared to the formation water and seawater [22,43,80]. For example, Yousef and co-workers [38,66,68] enhanced oil recovery from a carbonate reservoir (composite limestone cores) by using synthetic seawater (57,600 ppm) and up to 100-times diluted seawater. The highest levels of oil recovery were obtained using twice diluted seawater (28,800 ppm at 7–8.5% of OOIP) and 10 times diluted seawater (5760 ppm at 9–10% of OOIP). However, 20 times diluted seawater (2880 ppm) resulted in lower oil recovery improvement of 1–1.6% of OOIP and 100-times diluted seawater did not improve oil recovery. Therefore, the total additional oil recovery of approximately 18% was observed by using diluted seawater compared to the conventional waterflooding. Similarly, Sohal et al. [131] observed improvement in water-wetness of chalk by using up to 20-times diluted seawater, where higher dilution did not increase water-wet conditions. Effect of the formation water (160 kppm), seawater (32 kppm), and low salinity water (2 kppm) on oil recovery from limestone cores was studied by Tetteh et al. [81]. Their experimental results indicated 9.1% of OOIP incremental oil recovery by switching the injected brine from seawater to low salinity water. Moreover, Shehata and colleagues [127] observed a considerable improvement on oil recovery of limestone cores (50.4% of OOIP) by changing the concentration of injected brine from seawater in the secondary mode to deionized water in the tertiary mode, due to sudden changes in the injected brine ionic content between two recovery modes. However, it is worth noting that 20% diluted seawater did not improve oil recovery compared to seawater, while 50% diluted seawater showed a slight improvement. An improvement of 1–5% of OOIP with oil recovery in limestone cores that contain small amounts of anhydrite was reported in tertiary LSW, when diluting the high salinity formation brine (208,940 ppm) by 10 and 100 times [62]. Additionally, Zahid et al. [69] reported a significant improvement in oil recovery by using (up to 20-times) diluted synthetic seawater in reservoir carbonate (limestone) core plugs only at high temperature (90 °C). Furthermore, Alotaibi et al. [132] observed 8.6% of OOIP additional oil recovery using aquifer water (4 kppm) after the formation water (230 kppm) flooding. Similarly, it is shown that reducing the brine salinity (25-times diluted seawater) can improve rock water-wetness and increase oil recovery compared to seawater (44 kppm) and the formation water (180 kppm) flooding in reservoir limestone cores [133]. It was suggested that injection of seawater can also improve oil recovery in carbonates compared to conventional formation waterflooding, unlike sandstone reservoirs where only low salinity brine results in improved oil recovery.

The potential of enhanced oil recovery in carbonate cores at much higher concentrations than reported in literature was argued by Al-Harrasi et al. [134], based on 2, 5, 10 and 100 times diluted formation water, at 70 °C. They observed improved oil recovery even with high salinity twice diluted formation brine (9225 ppm), at 10% of OOIP, whilst the 100 times diluted formation brine (1944 ppm) resulted in maximum oil production and a faster oil production rate (up to 21% of OOIP). Additionally, Romanuka et al. [88] observed additional oil recovery from 4% to 20% of OOIP by lowering salinity of seawater (up to 10-times diluted) in limestone and dolomite core plugs. However, the highest oil recovery for chalk cores was found when using a higher concentration of SO_4^{2-} ions, rather than by reducing the ionic concentration. Furthermore, Al-Attar et al. [135] attempted to determine the optimum salinity of diluted seawater, where they observed improved oil recovery of 63–84.5% of OOIP from reservoir limestone, by varying the concentration of Ca^{2+} and SO_4^{2-} ions, which was attributed to intermediate wetting states. Their experimental results showed that increasing the concentration of SO_4^{2-} ions beyond 46.8 ppm has a negative effect on LSW processes. However, using 20-times diluted seawater (1600 ppm) resulted in a considerable decrease in oil recovery from chalk cores compared to seawater, from 60% to 15% of OOIP, due to a reduction in the amount of available PDIs [65]. It was suggested that wettability alteration is not a low salinity effect, and that seawater, up to a concentration of 33 kppm, can modify the wettability of rocks and improve oil recovery in LSW in carbonate [54,60,65].

Therefore, inconsistencies in the literature regarding the optimum concentration of the injected brine are reported. Even though low salinity brine has shown to improve oil recovery in most cases, and it is suggested that diluted brine without high quantities of PDIs can improve oil recovery from carbonates at elevated temperature [25]. It is also argued that reducing ionic concentration does not necessarily improve oil recovery and concentration of PDIs is claimed to play a more important role [60,65].

2.2.3. Surfactant

Wettability of carbonate surfaces can be altered toward a more water-wet state in the presence of surface active cationic surfactants dissolved in seawater, which help to enhance efficiency of LSW. Fourteen different types of surfactants were investigated to analyze the effect of surfactant on wettability alteration of oil-wet chalk cores [102]. The results suggested that cationic surfactants, of the type of alkyl trimethyl ammonium ($R–N^+(CH_3)_3$), were able to desorb and release carboxylic material from the chalk surface in an irreversible way, and consequently improve oil recovery by up to 70% of OOIP. It was also found that anionic surfactants were not able to desorb anionic organic carboxylic molecules or modify the wettability irreversibly. The mechanism for wettability modification, in this case, was proposed as ion-pair formation between the cationic surfactants and the negatively charged carboxylic layer adsorbed on the carbonate, which results in formation of cationic–anionic complexes, releasing carboxylates from the rock surface. In a separate study, the effects of a non-ionic surfactant (ethoxylated alcohol, EA) and a cationic surfactant (dodecyltrimethylammonium bromide, DTAB) on wettability alteration of oil-wet carbonate reservoir cores were examined [136]. The experimental results showed that DTAB was more efficient than EA in wettability alteration towards a more water-wet condition, by recovering 45% and 10% of OOIP, respectively. Hognesen et al. [125] suggested that the activity of SO_4^{2-} as a PDI and wettability modifier of chalk surfaces increases as temperature is increased, by addition of cationic surfactant (DTAB) to the studied brine. Furthermore, Jarrahian et al. [137] studied the effect of a cationic surfactant (DTAB), nonionic surfactant (Triton X-100), and anionic surfactant (SDS) on wettability alteration of dolomite surfaces using various analytical techniques. Their results indicated that DTAB is most efficient in wettability alteration of dolomite surfaces to a more water-wet condition. In addition, Triton X-100 and SDS changed the wettability to weak water-wet and neutral wet conditions, respectively. Lastly, it was shown that Mg^{2+} ions in the presence of a cationic surfactant (DTAB) are able to remove adsorbed carboxylic material from calcite surfaces and led to a more water-wet surface compared to the absence of this surfactant [49].

3. Investigation Techniques

3.1. Surface Wettability

In an oil reservoir, wettability describes to what degree oil or water have the tendency to spread over the rock surface. In other words, wettability is a measure of the affinity of a surface towards one contacting fluid over another immiscible fluid, which is also present in the bulk. Wettability is an essential factor in a multiphase flow and fluid distribution in reservoir rocks. It also affects the relative permeability of a rock–fluid system [36]. Electrostatic attraction between carbonate rock–brine and oil–brine interfaces facilitate the formation of a thin and unstable water film and brings oil in contact with the carbonate mineral surface, which consequently traps hydrocarbons in the reservoir rock pores and reduces the oil recovery efficiency [138]. In this section, we review different techniques to infer surface wettability of carbonate rocks, such as chromatographic wettability test, contact angle measurements, diffusion nuclear magnetic resonance (NMR), atomic force microscopy (AFM), zeta potential measurements, and surface complexation modelling, and present the findings from such measurements.

3.1.1. Chromatographic Wettability

The chromatographic wettability test is a technique developed by Strand et al. [124], based on the chromatographic separation of two water soluble compounds, such as non-adsorbing tracer (SCN^-), and a PDI (SO_4^{2-}), toward a carbonate surface. The aim is to validate the alteration in a water-wet fraction after exposing carbonate rock samples to various fluids (water or oil). The area between the relative concentration of SCN^- and SO_4^{2-}, as a function of the injected pore-volume for a chalk core at residual oil saturation, can then be related to the water-wet area of the core. However, the chromatographic separation is only suitable for coreflooding in SO_4^{2-} free carbonate cores at low temperature [131]. Core preparation and aging for the chromatographic test includes two different procedures, depending whether the initial water is present or not. The cores are placed in a container and aged with either crude oil (without the initial water) or brine (with the initial water saturation) and the porosity is calculated using the weight difference, oil density, and bulk volume. The cores are then prepared using the Hassler core holder and flooded with the crude oil with pressure of 25 bar and then aged with the crud oil in a closed container to establish initial reservoir conditions [124].

The chromatographic test was used to determine the influence of different PDIs and temperature on wettability alteration of chalk and limestone cores [36,58,59,65,124,129]. For example, Strand et al. [124] examined the effect of SO_4^{2-} ions on wettability alteration of chalk cores using crude oil samples with different AN and synthetic brines, based on the chromatographic wettability test. They compared the results with those obtained from Amott–Harvey spontaneous imbibition tests at room temperature, and concluded that the chromatographic wettability test could be an appropriate tool to be used in the neutral-wet state and at residual oil saturation without performing long-term imbibition tests. With the same methodology, they carried on to measure the adsorption of SO_4^{2-} ions on chalk cores as a function of Ca^{2+} concentrations and temperatures [58]. The results showed that adsorption of SO_4^{2-} ions increases with increasing temperature and concentration of Ca^{2+} ions, which is consistent with their imbibition and zeta potential measurement results. Fathi et al. [65] studied the effect of salinity and the presence of Ca^{2+}, Mg^{2+}, and SO_4^{2-} ions on wettability alteration of chalk surfaces and displacement of adsorbed carboxylic acid, at a range of temperatures (100–120 °C), where the order of improved water-wet fraction from diluted seawater, seawater and NaCl depleted seawater was confirmed by separate chromatographic wettability tests. Lastly, wettability of reservoir limestone cores and their reactivity toward PDIs (Ca^{2+}, Mg^{2+}, and SO_4^{2-}) in the temperature range of 20–130 °C was studied using the chromatographic test [59]. The experimental results indicated that the chromatographic wettability test is also applicable to limestone, where the PDIs interacted with the limestone surface and relative concentration of the PDIs affected the established chemical equilibrium. Additionally, Ca^{2+} and Mg^{2+} ions showed similar adsorption from a NaCl solution on the limestone surface at room temperature. However, by increasing temperature, Mg^{2+} ions adsorbed more strongly on the rock surface and at 130 °C, Ca^{2+} ions replaced Mg^{2+} ions on the rock surface. Moreover, interactions between Ca^{2+} and Mg^{2+} ions towards limestone surface in seawater was controlled by SO_4^{2-} ions and Ca^{2+} ions showed higher affinity to adsorb on the rock surface than Mg^{2+} ions, due to formation of ion-pair between Mg^{2+} and SO_4^{2-} ions and strong adsorption of SO_4^{2-} ions on the limestone surface.

3.1.2. Contact Angle

A contact angle measurement can quantitatively reflect the degree of wetting when a solid and two liquids are in contact, and is described as the angle between the solid and the liquid–liquid interface through the denser liquid phase [139]. Contact angle values show the equilibrium between the interfacial tensions of the two fluid phases and their adhesive attraction to the solid surface. The degree of liquid wettability increases when the contact angle of a liquid droplet on a solid surface decreases. In a system with reservoir rock, oil, and water, the rock could be: (i) water-wet when water is the spreading fluid (0° to 75°); (ii) intermediate/neutral-wet when the rock surface has no preference for either oil or water to spread on it (75° to 115°); and (iii) oil-wet when oil is the spreading fluid (115° to 180°). Additionally, weakly water-wet and weakly oil-wet are considered as 55° to 75°, and 115° to 135°, respectively [140]. However, contact angle measurement requires homogeneous surfaces without surface roughness. Therefore, a wide range of contact angle results are reported in the literature, which is due difference in surface roughness of each carbonate rock type. The aging procedure for the contact angle test involves first surface alteration of carbonate surfaces with or without an initial brine solution, and then with a crude oil to establish an oil-wet state. Following the treatment with oil, the carbonate surfaces may be aged with a brine to study its influence on the wettability. Contact angle can be measured by a drop of water or a drop of oil. However, it is mostly an oil drop in a bulk of water, due to optical transparency of the system. [29].

Abdallah et al. [29] examined the effect of ionic composition (Ca^{2+}, Mg^{2+}, and SO_4^{2-}) and concentration of modified seawater, on wettability alteration of cleaved calcite (limestone) crystals aged with a crude oil using static contact angle measurements. They measured the contact angle of deionized water on calcite crystals after surface aging in crude oil and different brines. Their results show that first SO_4^{2-} ions and then diluted seawater have notable impacts on wettability alteration of calcite surfaces to strongly water-wet. Weaker wettability alteration was observed when removing Mg^{2+} ions from the brine, indicating that Mg^{2+} ions must be present in the brine. The effect of different smart brines on carbonate (limestone) rock wettability was measured by Awolayo et al. [78] using different synthetic brine solutions (NaCl, Na_2SO_4, $NaHCO_3$, $MgCl_2$, and $CaCl_2$), and an oil from a carbonate reservoir, on polished rock plates cut from the same carbonate reservoir core plugs aged with a synthetic brine and then the reservoir oil to restore their wettability toward reservoir conditions at 95 °C. Contact angle measurement of oil droplets on the submerged aged plates in the studied brines, confirmed that the contact angle decreases with increasing exposure time to brine and increasing SO_4^{2-} concentration, suggesting that rock wettability was modified from a strongly oil-wet to a weakly oil-wet condition. They explained that by increasing the concentration of SO_4^{2-} ions, the water film on the rock surface becomes thicker and more stable; therefore, the contact between oil droplets and rock surfaces is prevented, due to the repulsive electrostatic forces at the oil–rock interfaces. Thereby, wettability of the rock is altered and carboxylic acid is desorbed from the rock surfaces, whereafter SO_4^{2-} ions occupy the free sites on the rock surface. Hamouda et al. [141] studied the effect of temperature (25–80 °C) on wettability alteration of calcite (limestone) crystals previously aged with stearic acid, using advancing contact angle measurements. Their experimental results indicated that, as temperature increases, contact angle decreases and the wettability of calcite surface improves from strongly oil-wet, at low temperature (160 ± 3° at 25 °C), to a more water-wet condition at higher temperature (90 ± 3° at 80 °C). The influence of brine chemistry and temperature on wettability alteration of dolomite surfaces aged with crude oil was evaluated by water advancing contact angle [142]. The results showed that, by using up to 15 times diluted seawater, wettability of the dolomite surfaces were altered from an oil-wet condition (158°) to a neutral wet condition (86°). Additionally, similar wettability alteration behaviour was observed by increasing SO_4^{2-} concentration up to two times and increasing the temperature from 26 to 121 °C. Yousef et al. [68] measured contact angles of oil droplets on carbonate (limestone) rock plates submerged in brines of various salinity at reservoir conditions, and observed a reduction in contact angle from about 90° to 60° for formation water and 100 times diluted seawater, respectively. Lastly, oil-droplet contact angle of reservoir limestone and dolomite surfaces as a function of brine concentration and temperature (25–100 °C) was measured [25]. The experimental results indicated that by increasing temperature the rock surfaces became more water-wet, although the degree of water-wetness was dependant on the type of rock. For example, limestone showed less response to increasing temperature by reducing the brine salinity, while wettability of dolomite surfaces improved considerably at elevated temperature when using low salinity brine, which consequently can result in additional oil recovery.

3.1.3. Nuclear Magnetic Resonance (NMR)

Nuclear magnetic resonance (NMR) has been used as a rapid, non-destructive, and non-invasive methodology for both laboratory and field applications. A novel bench-type NMR instrument is designed, where T2 relaxation rate of proton nuclear spins in fluid samples in a porous rock is measured and has been used to determine the petrophysical properties of reservoir rock, such as pore-size distribution, porosity, permeability, and rock wettability [68]. Yousef et al. [68] performed NMR T2 measurements on reservoir carbonate (limestone) rock samples used in their coreflooding study, before and after LSW treatment with a core water saturation (S_w) of 1 to determine the effect of brine salinity on pore geometry and reactivity of the formation water on carbonate rock surfaces. Their experimental results showed that injection of various diluted seawater in the carbonate core samples can have a significant impact on surface relaxation alteration of carbonate surfaces. Surface relaxivity is a geochemical property that explains the capacity of the grain surface to improve relaxation and is usually increased as the concentration of paramagnetic ions on the rock surface increases [68]. In addition, an improvement in connectivity among macro and micro pores was observed, which was attributed to the rock dissolution, due to alteration in the brine composition. To investigate the effect of post core cleaning (with toluene and methanol) on the measured T2, they performed another NMR measurements on new cores before and after the cleaning process. The results indicated that the cleaning process had a negligible effect on the NMR T2 distributions and confirmed that modification of the brine content is indeed the cause of the observed shift in the position of T2 distribution between NMR results taken before and after LSW. However, NMR T2 distribution, performed by Zahid et al. [69], using different diluted synthetic seawaters showed no significant changes in surface relaxation and shift in T2 distribution of both reservoir carbonate (limestone) and outcrop chalk cores.

3.1.4. Atomic Force Microscopy

The DLVO (Derjaguin, Landau, Verwey and Overbeek) theory is a classic mechanism explaining long-range interaction forces between the surfaces of two electrically charged particles in aqueous solution, based on London-dispersion (van der Waals) and electrostatic forces. Additionally, other contributions to the interfacial forces include hydrophobic, hydration, and steric forces, as well as hydrodynamic effects, which are known as non-DLVO (extended DLVO) forces [55,143,144]. All of these interfacial forces could contribute to the disjoining pressure of the water film which dictates its thickness. Generally speaking, attractive forces will have a negative contribution on the disjoining pressure and reduce the stability and thickness of the water film, whilst repulsive forces have a positive contributing to the disjoining pressure and produce more stable and thicker films, and a more water-wet rock surface [99]. The van der Waals forces between oil and carbonate surfaces are attractive, while the electrostatic force could be attractive or repulsive, based on the oil, brine and mineral compositions, as well as pH [99]. By reducing the ionic concentration, the electrostatic repulsive force increases and the EDLs at the oil–brine and rock–brine interfaces expand to become more diffuse (due to reduced ion density in the diffuse layer and increase of Debye screening length), and electrostatic screening becomes weaker, which causes the disjoining pressure to be more positive and prevents crude oil components from adhering to the rock surface and reduce adhesion force at rock–oil interfaces [25,26,42,145]. (However, this might not be always the case, as reducing brine salinity did not result in increased repulsion force for dolomite surfaces at lower temperature [25]). In contrast, by increasing the salinity, the magnitude of the electrostatic forces will be reduced and, at a certain concentration, the van der Waals attractive forces will overwhelm the repulsive electrostatic force and the water film will collapse, and consequently, the oil phase replaces the water phase and adheres to the rock surface [42,145]. In other words, a water film will remain stable if it can resist the pressure from the oil phase [85]. Therefore, rock surface wettability is dependent on the variation between the oil and water pressure (capillary pressure). Additionally, brines with lower ionic strength tend to reduce adhesion of the crude oil molecules on the rock surface by producing thicker water films and modifying rock wettability toward a more water-wet condition, thereby improving oil recovery.

Atomic force microscopy (AFM) can be used to directly measure adhesion forces, a combination of electrostatic and van der Waals forces, as well as forces, due to chemical bonds or acid–base interactions found in different fluids by generating force versus distance curves (force curves) [143]. Force curves provide valuable information about the solid properties, such as hardness, elasticity, surface energy and surface charge densities [143]. The adhesive force between two surfaces can be affected by the properties of the material (surface roughness and inhomogeneity), as well as contact geometry [143]. To generate surface topography or force curves, the surface of the sample is scanned by a tip mounted on a fine cantilever. Micro-fabricated silicon nitride (tips) are commonly used, as they provide high resolution of surface force measurements; however, they introduce a further complication when determining the surface geometry at the nanometre scale and in the presence of complex surface chemistry [146]. This challenge can be addressed by attachment of a smooth sphere, of micrometre size (2–10 µm), to the cantilever tip (colloidal probe), thereby considerably reducing the issues related to sample roughness, since the sample surface needs to be smooth only at a similar scale to the radius of the curvature at the end of the tip (5–50 nm) [146].

Most previous AFM force measurement studies [147–154] in aqueous electrolyte solutions have used mica or silica substrates, due to their fine surface roughness, as well as providing a good basis for comparison, due to several other investigations, undertaken on these samples using other techniques. Limited work has been done in examining the wettability alteration of carbonate rock surfaces (calcite and dolomite crystals), following the adsorption of polar components of crude oil, using AFM force measurements. Undertaking AFM force measurements on calcite or dolomite crystals can be challenging, again, due to the surface roughness of these substrates and their heterogeneous surfaces. Karoussi and co-workers [155] studied the surface characteristics of polished and cleaved calcite substrates coated with organic long chain or short chain carboxylic acids (stearic and heptanoic acids) that were exposed to solutions of different pH. Both surface topography and adhesion were acquired. Their investigation showed that cleaved calcite samples are more suitable for AFM measurement than polished calcite, due to the surface roughness of polished samples. Due to the coating, there was a thin layer of alkane present on the calcite surface, which prevented spontaneous recrystallization of the surface. Force measurements showed the adhesion forces of heptanoic acid treated samples was not responsive to the pH of the environment. However, cleaved calcite, treated with stearic acid, showed higher adhesion forces at pH 5 than pH 10. In a recent study [156], growth and dissolution of a stearic acid coated calcite crystal that was exposed to various aqueous salt solutions were investigated using amplitude-modulation atomic force microscope. It was believed that stearic acid particles tend to act as "pinning points" on the calcite crystal, interfering with the growth and dissolution process. When supersaturated brine was used, the growth of crystal around dense stearic acid patches was observed, including the patches into the crystal accompanied by ionic impurities present in the solution. When the brine was diluted, the freshly grown material was dissolved and the stearic acid patches, formerly incorporated into the crystal during the growth phase, were exposed. Sauerer et al. [48] measured the surface free energy of calcite and dolomite samples using AFM adhesion force measurements, in order to predict carbonate reservoir wettability. Adhesion forces, between the installed colloidal calcite and dolomite particles on the AFM tip (Figure 7) and the planar cleaved calcite and dolomite samples, both in air and in the presence of organic solvents (ethylene glycol, bromonaphthalene and heptane) were measured. Reduced adhesion forces were observed in the presence of polar and non-polar solvents compared to the same systems in air. This was attributed to the presence of adsorbed water layers at both the AFM tip and on the mineral substrates, due to humidity in the air environment. This experiment demonstrated the applicability of this approach for determination of surface total energies of mineral surfaces by AFM force measurement and, therefore, provided a valuable new tool for reservoir characterisation. Overall, AFM adhesion force measurements of carbonate surfaces has not been studied extensively, due to the challenges involved in determination of surface roughness, heterogeneity, tip characterisation, as well as approach/retract conditions. However, the methodology has a great potential as a high-throughput approach to quantitatively evaluate the surface energy of natural objects such as carbonate formation.

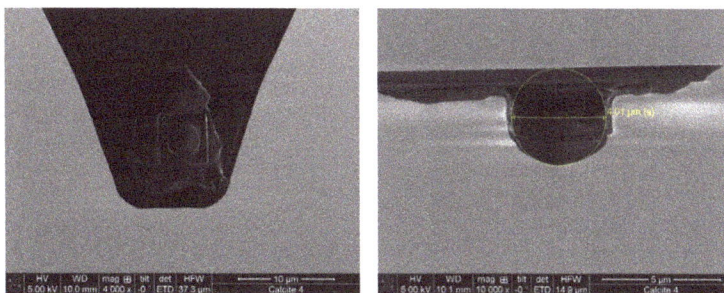

Figure 7. Scanning electron microscopy image of calcite particle fixed on a cantilever (calcite probe) used for force measurements from [48]. Reproduced with permission from Sauerer et al., *J. Colloid Interface Sci.*; published by Elsevier, 2016.

3.1.5. Zeta Potential

The surface charge of rock–brine or oil–brine interfaces is critical for better understanding the stability of water films and, therefore, the degree of rock wettability. Wettability of a rock is a function of the sign and magnitude of the electric surface charges at the two interfaces, which arises from the electrostatic interactions at the oil–rock–brine interfaces. Zeta potential is an essential parameter used to evaluate the electrokinetic behaviour of solid–liquid or liquid–liquid interfaces. Zeta potential indicates the electrokinetic properties of a particle, while the magnitude shows the stability of the colloidal system. Expansion of the EDL at the rock–brine and oil–brine interfaces will result in increased electrostatic repulsion between the two interfaces and consequently form a thick and stable film and a more water-wet rock surface. On the contrary, a thin and unstable water film could reduce the thickness of the EDL, resulting in the rock surface being more oil-wet. Additionally, the sign and magnitude of the zeta potential at rock–brine interfaces is dependent on the type and concentration of ions, pH, temperature, the presence of additives, such as surfactants, in the solution, as well as rock mineralogy [26]. However, zeta potential measurements require uniformly sized particles. Also, there are difficulties in measuring zeta potential of particles at high salinity, temperature, and pressure.

Previous zeta potential measurements have primarily focused on calcite surfaces as a representative of carbonate rock reservoirs [58,63,111,123,157–161], with considerable work using zeta potentials measured as a function of pH. The zeta potential measurements of carbonate surfaces are routinely inconsistent in both trend and magnitude, which could be attributed to the specific systems examined, such as sources of calcite. In addition, the complex dissolution mechanism of calcite in water makes it difficult to reach equilibrium in solution. The zeta potential of calcite is mostly measured at high pH (above 7) to prevent dissolution of the mineral. Carbonate surfaces are aged with carboxylic acid materials using two methods: dry and wet aging [137,157,162]. In the dry method, carbonate surfaces are immersed into dry carboxylic acid materials or carboxylic acid material dissolved in an oil phase (e.g., n-heptane, n-decane, toluene), while, in the wet method, the carbonate surfaces are aged with carboxylic material in the presence of water film, where carbonate surfaces are added to deionized water (in the presence or absence of electrolytes) and then carboxylic acid materials dissolved in an oil phase are added to the system. The suspensions in both methods are shaken for a period of time to ensure adsorption of fatty acid molecules on the rock surface and then the aged powder is dried.

Zeta potential of calcite particles was measured as a function of concentration of monovalent (Na^+) and multivalent (Ca^{2+}, Ba^{2+} and La^{3+}) ions, over a range of pH [163]. It was suggested that Na^+ act as indifferent ions towards calcite surfaces, whilst Ba^{2+} showed similar behaviour to that of Ca^{2+}, increasing the positive magnitude of calcite zeta potential with increasing concentration. However, the presence of Ba^{2+} ions resulted in a greater magnitude of zeta potential compared to Ca^{2+}, due to the higher hydration energy of Ca^{2+} ions. However, La^{3+} ions showed an opposite behaviour to that of the divalent cations, resulting in reduced zeta potential values with increasing ionic strength, due to the lacking of specific adsorption on the calcite surfaces. The effects of concentration of two types of divalent ions, namely Ca^{2+} and SO_4^{2-}, were investigated for their impact on surface charge of milled chalk in NaCl brine [58]. It was suggested that surface charge of chalk surfaces is controlled by the relative concentration of Ca^{2+} and SO_4^{2-} and the chemical mechanism for surface charge modification was proposed as adsorption of SO_4^{2-} ions, in parallel with the co-adsorption of Ca^{2+}, with increasing temperature on the conditioned chalk surface. Similarly, the effect of PDIs (Ca^{2+}, Mg^{2+}, and SO_4^{2-}) on surface charge and wetting properties of chalk was studied by Zhang et al. [63,111,123] using zeta potential measurements. It was suggested that Mg^{2+} ions can substitute Ca^{2+} ions from chalk surfaces at high temperature, and increase the positive surface charge density.

Mahani et al. [25,26,164] suggested that surface charge alteration is likely to be the main mechanism for LSW in carbonates, where they examined calcite, limestone, chalk and dolomite surfaces. They observed that, regardless of carbonate rock type, zeta potential is positive for the high salinity formation water (180 kppm) at pH 6.5–8.5, while diluted seawater (1750 ppm) and addition of SO_4^{2-} ions lead to more negative surface charges and more water-wet conditions. Changes in pH influenced the zeta potential at low concentrations of brine, due to the presence of fewer active ions compared to H^+ and OH^- ions. Additionally, when being exposed to diluted seawater, surface charge of carbonate (limestone) rock particles was found to become more negative, resulting in greater interactions with water molecules and rock wettability alteration [37,67].

Zeta potential measurements of rock–brine and oil–brine interfaces in the temperature range of 25–70 °C showed that, by increasing temperature, the observed trend as a function of salinity and pH remains the same for all rock types, although the magnitude of zeta potential changes [25]. As temperature increased, the magnitude of zeta potential of both rock–brine and oil–brine interfaces reduces towards the point of zero potential. This reduction in magnitude of zeta potential was more apparent for low salinity brine, which could be due to lower concentration of SO_4^{2-} ions, allowing Ca^{2+} and Mg^{2+} ions to adsorb on the rock surfaces [25]. Additionally, zeta potential measurements of calcite at elevated temperatures and pressures showed that the negative magnitude of zeta potential increases with increasing temperature and pressure, up to 45 psi, before decreasing at higher pressures, due to changes in solubility of minerals, which affects the EDL structure and consequently surface charges [165].

As for the effect of adsorbed carboxylic acid materials on surface charge modification of carbonates, Gomari and co-workers [157,158] observed that zeta potential surface charge of calcite was altered from positive to negative, in the presence of stearic and oleic acids (introduced onto the calcite particle using both dry and wet treatments). The ability for SO_4^{2-} and Mg^{2+} ions to displace the pre-adsorbed carboxylic acid materials from the calcite surface was suggested, based on zeta potential measurements. It was reasoned that ions causing positive carbonate surface charges are likely to change surface wettability of limestone and dolomite surfaces toward a more oil-wet condition, whereas negatively charged minerals increase oil recovery by releasing trapped crude oil molecules [159,166]. The effect of a mixture of ions (Ca^{2+}, Mg^{2+}, and SO_4^{2-}) on zeta potential of calcite and dolomite surfaces aged with stearic acid (using the dry method) was investigated by Kasha et al. [30]. The results confirmed that, as the concentration of divalent Ca^{2+} and Mg^{2+} ions increases, the zeta potential of the aged calcite was altered from negative to positive and resulted in higher positive charge in dolomite, while SO_4^{2-} ions caused the zeta potential of the carbonate surfaces to be more negative. Such a change in the measured zeta potential is highly likely, due to the affinity of Ca^{2+}, Mg^{2+}, and SO_4^{2-} ions being affected by the presence of other PDIs when using mixtures of ions. Additionally, Jackson et al. [145] measured zeta potentials of rock–brine and oil–brine interfaces using a reservoir limestone core sample and various oil components to investigate the effect of different salts (NaCl, $CaCl_2$, $MgCl_2$, and Na_2SO_4) on wettability alteration and oil recovery. It was suggested that in order to design an optimum brine composition for LSW in carbonates, zeta potential of both mineral-brine and oil–brine must be taken into account. The authors reported positive surface charge for the oil-brine system for the first time. It was concluded that LSW results in oil recovery improvement only if two mineral-brine and oil–brine interfaces have the same sign of zeta potential (both positive or both negative), which results in repulsive electrostatic forces and positive contribution to disjoining pressure and acts to stabilize the water film on the mineral surface (Figure 8). Therefore, the studies that reported failure in improving oil recovery by performing LSW, may have had a positively charged oil–brine interface, which had not been identified [145]. Lastly, zeta potential of calcite and dolomite surfaces aged with stearic acid was studied using synthetic diluted seawater [31]. The examined calcite surfaces maintained negative zeta potential in the presence of deionized water and in the all tested brines, excluding diluted seawater with higher concentration of cations (Ca^{2+} and Mg^{2+}), which resulted in a positive zeta potential. In contrast, dolomite surfaces showed positive zeta potential in deionized water and the diluted brines, except for diluted seawater containing two and four times higher concentrations of SO_4^{2-}. It was concluded that diluted seawater is less efficient in altering surface charge of dolomite, due to the tendency of stearic acid to adsorb more strongly on dolomite surfaces.

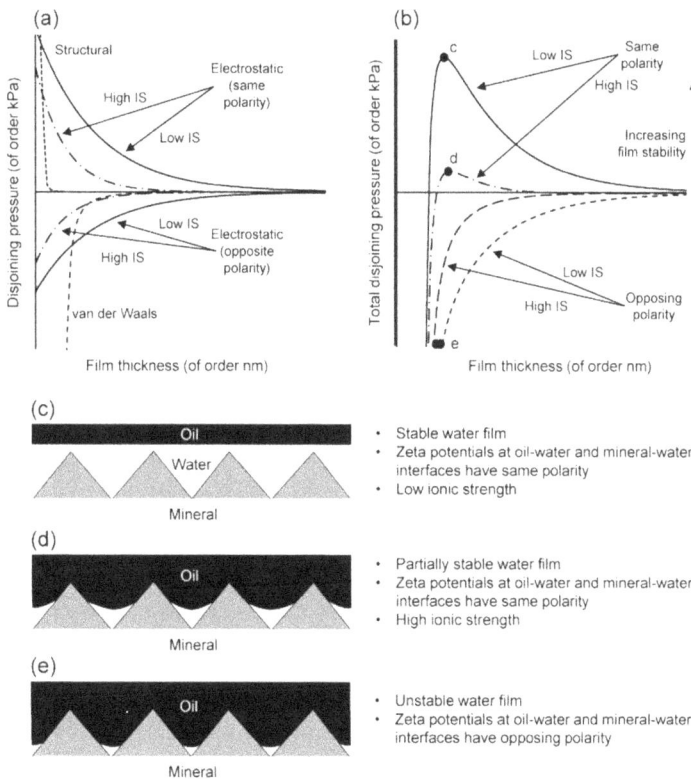

Figure 8. Effect of disjoining pressure on wettability alteration of carbonate surfaces: (**a**) schematic view of contribution of electrostatic and van der Waals forces on disjoining pressure; (**b**) total disjoining pressure at low and high ionic strength (IS) when rock–brine and oil–brine interfaces have the same/opposite sign of zeta potential; and (**c–e**) schematic models showing stability of water film between oil and mineral interfaces (water-wet, mixed-wet, and oil-wet conditions, respectively). Reproduced from reference [145].

3.1.6. Surface Complexation Modelling

A surface complexation model (SCM) simulates the chemical equilibrium, based on the chemical reactions occurring at the mineral-solution interfaces, which results in the formation of surface complexes. These surface complexes are responsible for the mineral surface charge and their equilibrium constants are assumed to be similar to the analogue reactions in the bulk solution [167]. The SCM is used to describe adsorption of inorganic ions on mineral surfaces using either inner or outer-sphere surface complexes and fit these models to adsorption data to distinguish between inner- and outer-sphere adsorption mechanisms [168]. The SCM has been used to understand surface charges of carbonate particles and calculate surface potential that can be validated by experimentally measured zeta potential values. Hiroth and co-workers [85,169,170] developed a geochemical model to investigate the effect of water chemistry on surface charge, considering mineral precipitation and dissolution of calcium carbonate rock, based on a calculation for formation water and seawater at 70–130 °C by determining the expected dissolved Ca^{2+} ions in the brine effluent after LSW against oil recovery from a outcrop core. It was suggested that the model could precisely predict the surface potential of calcite, and adsorption of SO_4^{2-} ions from pore water, and concluded that changes in surface potential cannot describe all aspects of oil recovery improvement by alteration of pore water chemistry and temperature. Instead, mineral dissolution could be the main mechanism.

A Basic–Stern surface complexation model, based on zeta potential results for calcite, considering outer-sphere complexes of ions other than protons and hydroxide, was developed by Heberling et al. [171]. It was suggested that the model can successfully predict zeta potential results with acceptable values for the inner Helmholtz capacitance, which is in agreement with the estimated Stern layer thickness, based on surface diffraction data. In another attempt, Mahani and colleagues [167] built a surface complexation model relating the surface reaction to their experimental zeta potential results for different types of carbonate rock, as a function of ionic concentration and pH. Their model confirmed that zeta potential increases with increasing pH, due to formation of surface species, which strongly affect the total surface charge. Additionally, at higher concentration, the surface charge of carbonates is less sensitive to changes in pH, resulting in a smaller variation in zeta potential. It was suggested that this particular model could capture the changes in the surface charge of carbonate rocks by altering salinity and pH. Lastly, Song et al. [172] used a double layer surface complexation to model experimental results obtained from synthetic calcite zeta potential measurements using mixed PDIs (Ca^{2+}, Mg^{2+}, SO_4^{2-}, and CO_3^{2-}) and different CO_2 partial pressures to systematically analyse the contribution of charged species, and better understand the role of electrostatic forces in wettability alteration of calcite. It is claimed that their model can precisely predict calcite zeta potential in brine containing PDIs and apply it to predict zeta potential at low and high pressures of CO_2.

3.2. Interfacial Tension

Distributions of fluid in the reservoir porous media are also dependant on the forces at liquid–liquid interfaces, which are quantified by interfacial tension (IFT) measurements. The effect of ionic content, as well as temperature and pressure on the IFT of crude oil–brine systems have been extensively studied. For example, the influence of NaCl concentration on the IFT of an oil phase (*n*-dodecane) and water was examined [173], where a critical salt concentration of NaCl (5 wt %) resulted in the lowest IFT. It was concluded that lowering the ionic concentration of brine will not always decrease the IFT. Moreover, Hamouda and Karoussi [155] measured IFT of stearic acid in an n-decane-water system including $MgCl_2$ and Na_2SO_4, where they observed lower IFT in the presence of Mg^{2+} ions compared to SO_4^{2-} or distilled water by increasing temperature from 0–80 °C. In a different work, Meng and co-workers [130] examined the influence of PO_4^{3-} ions on the IFT and wettability alteration of carbonate reservoirs by increasing temperature from 25–90 °C, based on IFT and contact angle measurements. The brine with a higher concentration of PO_4^{3-} ions showed higher wettability alteration, as well as a reduction in the IFT compared to brine with a lower concentration of PO_4^{3-}. This could be due to strong ability of these ions to desorb crude oil from pore throats and enhanced ionic exchange at both rock–fluid and fluid–fluid interfaces.

The effect of monovalent and divalent cations (NaCl, CaCl$_2$ and MgCl$_2$) in range of concentration of 0–45 kppm on crude oil–brine interactions using an Iranian crude oil, as well as extracted asphaltene and resin from the crude oil was investigated using the IFT measurements [174]. The experimental results indicated that NaCl has a slight influence on IFT of asphaltene and resin, whereas the effect of CaCl$_2$ and MgCl$_2$ on the IFT was more apparent. At low concentration of the divalent salts, asphaltene resulted in a higher reduction in IFT compared to resin, while at higher salts concentration, the reduction in IFT was more dominant in case of resin. Additionally, MgCl$_2$ resulted in higher reduction in IFT of resin than CaCl$_2$, due to the higher affinity of Mg^{2+} ions to bond with the oxygen molecules present in resin. In contrast, Ca^{2+} resulted in greater IFT reduction in asphaltene than Mg^{2+}. This could be due to a higher chance of forming of complexes between Ca^{2+} and asphaltene molecules because of their molecular sizes. It was concluded that oil–brine interactions are dependant on the presence of natural surface active compounds in crude oil (both asphaltene and resin), as well as type and concentration of ions in brine. Similarly, it was shown that IFT of crude oil–brine interfaces is dependant on asphaltene content of crude oil as a function of monovalent and divalent cation (NaCl and MgCl$_2$) concentrations [175]. In a different work, the effect of different ions (Ca^{2+}, Mg^{2+}, SO$_4^{2-}$), as well as diluted and softened seawater on dynamic IFT of crude oil–brine interfaces was studied [176]. The experimental results indicated that in all brines, IFT reduced over time (about 12 mN/m) until it reached a stable value depending on the type of brine. It was suggested that under dynamic conditions, active ions present in the brine can migrate to the surface until reaching equilibrium over time. Therefore, interfacial properties were claimed as an important contribution to the LSW mechanisms, which is affected by crude oil composition, type of ions, temperature, and contact time. However, brine composition was suggested to have a more significant effect on IFT than brine concentration.

The influence of brine salinity on the IFT of crude oil–brine was also studied by using diluted reservoir fluids [177], where a reduction in IFT (critical spreading tension) for up to 50% volume diluted reservoir brine was observed; however, further dilution of the brine increased the IFT. The authors suggested that a critical salinity of brine should be proposed to enhance oil recovery, based on IFT results. Similarly, Okasha et al. [178], investigated the effect of diluted brine on IFT of an oil–brine interface, where they observed a reduction in the IFT by decreasing the ionic concentration to twice and four-times diluted reservoir brine, using different crude oils under reservoir conditions, as well as reduction in IFT by increasing temperature and increase in IFT by increasing pressure. They suggested that, based on the IFT results, LSW may have a potential to improve oil recovery in carbonates. In contrast, Alameri et al. [179] observed an increase in the IFT of oil–brine by reducing the salinity of the brine from formation water to various diluted seawater and deionized water, which was attributed to reduction of pH of the brines. Additionally, the effect of diluted seawater on IFT of oil–brine interfaces was studied by Yousef et al. [38,68], when various diluted brines and crude oils at reservoir conditions were examined. The authors only observed reduction in IFT by diluting the fluid from formation water to seawater and no changes in IFT were observed by using different diluted seawater. It was concluded that LSW has no significant influence on the IFT and therefore fluid–fluid interactions. Lastly, dynamic IFT of crude oil–brine interfaces using a formation water, seawater and diluted seawater was studied [81]. The experimental results showed that the equilibrium IFT reduced significantly in the case of seawater, but not formation water and diluted seawater. Similar observations were reported in the literature where seawater resulted in the lower IFT compared to low salinity brine [38,68,174,180]. This was attributed to the presence of SO$_4^{2-}$ ions in seawater and its effect to reduce dynamic IFT of seawater-crude oil interfaces [81]. Therefore, it was suggested that reduction of IFT and improvement of the capillary number could be a mechanism for seawater flooding but not LSW.

3.3. Recovery Factor

Recovery factor is reported as a percentage of the extracted crude oil that is initially in place, and it is affected by the displacement mechanism. The main purpose of EOR processes is to maximize the recovery factor. In this section, different techniques that have measured the recovery factor by performing LSW, such as imbibition, coreflooding, and reservoir modelling are summarized.

3.3.1. Imbibition

Spontaneous imbibition is a process where a non-wetting phase is displaced by a wetting phase in a porous medium, due to capillary forces (combined effect of interfacial tension, wettability, and pore throat size). Therefore, it is a reliable laboratory technique that not only is used to determine the initial rock wettability conditions, but also the ability of different fluids to modify the wettability of rock surface towards a more water-wet state [22,181]. Wettability alteration of rock is quantitatively measured by the amount of brine being absorbed into or oil being displaced from the porous material. However, the spontaneous imbibition test is very time-consuming, as it takes months to collect results using this method. In addition, the chance of precipitation in the presence of reactive anions (e.g., SO_4^{2-}) is not taken into account in the spontaneous imbibition test [131].

A range of imbibition experiments on chalk [65,92,93,103,111,125] and reservoir limestone [59,125] were performed by Austad and co-workers to better understand the wettability alteration mechanism during LSW processes involving carbonates. Their findings for chalk cores were already discussed in previous sections. As for the limestone cores, the effect of SO_4^{2-} ions in the imbibing fluid was investigated, where synthetic seawater resulted in 15% OOIP additional oil recovery from a reservoir limestone core (aged with a crude oil (AN = 0.05 and BN = 0.44 mg KOH/g) for four weeks at 90 °C), compared to SO_4^{2-} depleted seawater at 120 °C [59]. Similarly, Ligthelm et al. [20] observed 5% improvement in oil recovery (total recovery of 17% OOIP) from a microcrystalline limestone core compared to the formation water by increasing concentration of SO_4^{2-} ions (up to 54 times) and reducing concentration of Ca^{2+} ions (about 22 times), to avoid critical solubility and precipitation limit of $CaSO_4$. However, Hognesen et al. [125] observed no improvement in oil recovery in spontaneous imbibition test on unfractured limestone cores (aged with a crude oil (AN = 1.7 mg KOH/g) for four weeks at 90 °C) by increasing SO_4^{2-} concentration to three times the concentration present in a synthetic seawater at 120 °C. Furthermore, Yi and Sarma [80] performed spontaneous imbibition on limestone core plugs from a carbonate reservoir (aged with a crude oil for three weeks) using diluted seawater, where they observed total oil recovery production of 23% of OOIP. Their results indicated that rock wettability is altered toward a more water-wet condition by LSW at 70 °C.

Imbibition measurements can be performed under various conditions. For example, Romanuka et al. [88] performed Amott spontaneous imbibition tests on samples collected from oil-bearing zones, namely outcrop chalks, limestone, and dolostone (aged with different crude oils (AN = 0.92 and BN = 0.16 mg KOH/g, AN = 0.42 and BN = 0.1 mg KOH/g, and AN = 0.07 and BN = 1.77 mg KOH/g) for four weeks at reservoir temperature), with increasing concentration of SO_4^{2-}. The tests were conducted under simulated reservoir conditions, by controlling key parameters, such as formation water and crude oil composition, as well as temperature. Their results showed increased oil recovery, from 4% to 20% of OOIP, in tertiary oil recovery mode, due to an improved water-wet condition. Kazankapov et al. [36] performed spontaneous imbibition tests on carbonate reservoir cores (aged with a crude oil (AN = 2.8 and BN = 0.74 mg KOH/g) for four weeks at 90 °C) acquired from the Caspian sea using NaCl depleted seawater, containing four times the SO_4^{2-} concentration at 90 °C, where they observed an improvement of 10% of OOIP in oil recovery compared to seawater. Lastly, Al Harrasi et al. [134] observed an increase in oil recovery, by 16–21% of OOIP, by performing spontaneous imbibition on carbonate reservoir core samples aged with a crude oil for 20–30 days, using synthetic diluted seawater at 70 °C. Even though spontaneous imbibition shows direct evidence of oil recovery and wettability alteration by LSW and indicates the efficiency of low salinity water as imbibing brines, it was argued that spontaneous imbibition results cannot be used directly to evaluate the potential recovery of LSW on the field scale, especially for non-fractured reservoirs, as the recovery factor during LSW is less than that of spontaneous imbibition test. For this reason, coreflooding tests are used to quantify the effect of brine salinity and composition on incremental oil recovery during LSW.

3.3.2. Coreflooding

Coreflooding experiments are used to better understand the wettability mechanism of improved oil recovery by performing LSW. Most previous coreflooding studies on carbonate cores [22,27,43,80,179], were performed at low flooding rate condition (about 1 ft/day equal to 0.3 cm^3/min), where up to 40% of OOIP additional oil recovery was reported, in comparison to conventional formation waterflooding. However, in this condition, capillary force is the dominant factor, with flow and oil production being strongly influenced by capillary end effects, which results in marginal overestimation of residual oil saturation and suppression of the water end point relative permeability [182]. Therefore, the reduction of capillary end effects, due to wettability alteration toward a more water-wet state in the LSW process, could have positive impacts on the improved oil recovery, which is not the case for real reservoir conditions, as there is no capillary end effect at field scale. This has to be taken into consideration when interpreting coreflooding results. For example, Gupta et al. [27] performed coreflooding experiments on limestone and dolomite cores (aged with a crude oil (AN = 0.11 mg KOH/g) for six weeks at reservoir temperature) with flooding rates of 1–2 ft/day (0.1 cm^3/min), where they observed 7–9% improvement in oil recovery in limestone cores by switching imbibing fluid from the formation water to seawater, 5–9% additional oil recovery in dolomite and non-fractured limestone cores by imbibing the core with seawater with added SO_4^{2-} ions, compared to the formation water, and 15% and 20% improvement in oil recovery by addition of BO_3^{3-} and PO_4^{3-} to synthetic seawater, respectively. Moreover, Nasralla and co-workers [22] performed quantitative and qualitative coreflood tests on reservoir limestone cores (aged with a crude oil for 28 days at reservoir temperature (100 °C)) intentionally at low flooding rates (0.25 cm^3/min or 0.05 cm^3/min) in order to examine capillary pressure. They observed up to 7% of OOIP improvement in oil recovery using 10-times diluted seawater and confirmed the higher potential of diluted seawater compared to seawater and the formation water to improve oil recovery in the secondary and tertiary modes.

To overcome capillary end effects, Al-Harrasi et al. [134] examined carbonate cores (aged with a crude oil for 20–30 days) at 70 °C with higher injection rates (0.4, 0.7 and 1.3 cm^3/min) during high salinity waterflooding, before injecting low salinity brine, where an improvement of oil recovery from 3% to 5% of OOIP by using diluted formation water was observed. They reported low salinity effects within first pore-volume of the injected brine for such a high salinity for the first time. Additionally, Shehata et al. [127] used outcrop limestone cores (aged with a crude oil (AN = 0.18 and BN < 0.01 mg KOH/g) for 20 days at reservoir temperature (90 °C)) in their coreflooding experiments, operated at secondary and tertiary modes at 90 °C, with injection rates of 0.25, 0.5, 1.0, and 2.0 cm^3/min to avoid capillary end effects. The experimental results showed that deionized water and 50% diluted seawater improved oil recovery up to 3% of OOIP in the tertiary mode, compared to seawater in the secondary mode. Laboratory coreflooding experiments were performed using composite cores from a Saudi Arabian carbonate (limestone) reservoir (aged with a crude oil (AN = 0.25 mg KOH/g) for six weeks), and various diluted synthetic seawaters with injection rates of 1, 2 and 4 cm^3/min, reflecting reservoir conditions, such as temperature, pressure, and initial formation water [38,68]. The authors observed additional oil recovery, of up to 10% of OOIP, showing substantial potential of LSW to improve oil recovery from carbonates. Moreover, reductions in the pressure drop across the examined carbonate core and, consequently capillary forces by reducing the brine salinity was observed, which was attributed to modification of oil–rock–brine interactions. Lastly, limestone outcrop cores (aged with a crude oil (AN = 0.17 and BN = 0.11 mg KOH/g) for 40 days) were flooded with a synthetic formation water, seawater and diluted seawater with flooding rate of 1–2 cm^3/min [81]. The experimental results showed an additional 9.1% of OOIP improvement in oil recovery by LSW (2 kppm, 49.1%), compared to seawater flooding (33 kppm, 40%) in the tertiary mode. The coreflooding experiment was also performed using a non-aged limestone core to minimize the effect of wettability alteration, where improvement in oil recovery by LSW was still observed (additional 3%), indicating that there might be another mechanism apart from wettability alteration involved in LSW. It was concluded that both wettability alteration and fluid–fluid interactions could be two potential mechanisms for improved oil recovery by LSW.

3.3.3. Reservoir Modelling

Reservoir modelling in carbonates includes consideration of the effect of LSW on wettability alteration by simulating capillary pressure and relative permeability, as a function of ionic concentration/composition or temperature. Such models have been developed using experimental spontaneous imbibition or coreflooding experimental results. Reservoir modelling is a very valuable tool, as it helps to verify and validate the experimental LSW results to predicts wettability process at reservoir conditions (high temperature and pressure), where experimental work may not be possible. Yu et al. [183] developed a numerical model to simulate dynamic wettability of oil-wet chalk cores, taking into account molecular diffusion, adsorption of a wettability modifier agent (SO_4^{2-} ions), gravity, and capillary pressure. The correlation between capillary pressure and relative permeability and its influence on wettability alteration was calculated by interpolation. Their modelling results indicated that the adsorption of SO_4^{2-} ions onto the rock surface induced wettability alteration, thus modifying the dynamic capillary pressure curve to higher values from oil-wet to water-wet state. Additionally, it was shown that the capillary pressure depends on both initial water saturation and the SO_4^{2-} concentration, as it was altered from negative (at initial water saturation) to positive towards to a more water-wet condition. The simulated capillary pressure and relative permeability results were claimed to be consistent with their spontaneous imbibition experimental results, leading to the conclusion that wettability alteration is dominated by diffusion, as shown by the initially limited recovery rate and wettability alteration, which is a key element that must be taken into account in simulation studies. Aladasani and colleagues [184] developed a model for LSW in carbonate reservoirs to calculate the capillary pressure and relative permeability, as a function of phase saturations and salt as a mass component in the water phase, based on the published coreflooding experiment results by Yousef et al. [38] and validated their simulation results with the coreflooding experimental results. Their simulation indicated that, in a neutral-wetting state, incremental oil recovery could be controlled by low capillary pressure, and an increase in relative permeability is the principal mechanism for LSW. Al-Shalabi et al. [185,186] simulated reservoir wettability to advocate that EDL expansion is a primary mechanism for LSW in carbonates, based on recently published coreflood experiments conducted by Yousef et al. [38]. Their simulation suggested that wettability modification has a minor impact on water relative permeability, which is consistent with an EDL expansion mechanism. Qiao et al. [187] modelled wettability alteration in carbonates by using a reaction network to capture the competitive surface reactions between carboxylic groups, cations, and SO_4^{2-} ions. Their model was developed, based on the published experimental results from spontaneous imbibition tests on Stevns Klint chalk [65], and then used to predict oil recovery for various conditions, where increased recovery, up to 30% of OOIP, was observed. Their simulation results indicated that increasing SO_4^{2-} anions concentration can improve oil recovery up to 40% of OOIP, whereas increasing the concentration Ca^{2+} cations reduced oil recovery by 5% of OOIP. Additionally, they predicted an optimum brine concentration of 0.096 mol/kg for SO_4^{2-} ions, a moderate concentration of cations, and a total ionic salinity of injected brine of 0.2 mol/kg for LSW of chalks.

3.4. Field Studies

There is only a limited amount of reports available in the open literature, concerning LSW treatment in carbonates at field scale. Performing pilot waterflooding on the chalk Ekofisk field located in the Norwegian sector of the North Sea, Thomas and co-workers [188,189] demonstrated the potential of LSW in carbonates, where a significant increase in oil recovery rate was observed (30% over a two-year waterflooding period). The Ekofisk field is a thick, heterogeneous, low permeability, and fractured chalk reservoir with a high temperature (130 °C) with a crude oil AN of about 0.1 mg KOH/g. Fathi et al. [65] showed, via laboratory investigation, that oil recovery can be improved by 10% of the OOIP by using NaCl depleted seawater. Additionally, because the reservoir is running at a high temperature, SO_4^{2-} ions should not be added to the injected brine, due to potential precipitation of anhydrite. Furthermore, Webb et al. [15] reported that seawater improved oil recovery from the Valhall field in the North Sea considerably (40%) compared to SO_4^{2-} depleted formation water. The Valhall field is an oil-wet fractured chalk field with a temperature of 90 °C with a crude oil AN of 0.35 mg KOH/g. Therefore, due to the low reservoir temperature, considerable oil recovery improvement can be achieved by removing NaCl and adding up to four times the inherent SO_4^{2-} ionic concentration to the injected seawater, which indicated oil recovery improvement up to 25% of the OOIP at the laboratory scale [65].

Additionally, significant oil recovery improvement was reported by injection of seawater from the Persian Gulf, transported by pipelines to the desert, into a Saudi Arabian non-fractured limestone reservoir [3]. Moreover, Yousef et al. [37] reported the first ever field trials of LSW in Saudi Arabian carbonate reservoirs to examine the effect of ionic salinity (up to 5 kppm) on oil recovery for secondary and tertiary recovery modes, where a single well chemical tracer was used to determine the residual oil saturation after LSW. Their results showed great potential of LSW in tertiary and secondary oil recovery modes, where reduction of residual oil, by seven units beyond conventional waterflooding, was observed. Wettability alteration is confirmed as a primary mechanism for improved oil recovery and argued that variation in oil recovery efficiency varies for carbonate reservoirs, due to a different reservoir temperature and the chemistry of the initial formation water, oil properties and reservoir heterogeneity.

4. Proposed Mechanisms

Based on the literature, there are several mechanisms elaborated on improving oil recovery when performing LSW, which are dependent on the type of crude oil, reservoir conditions, and the properties of the formation brine [76]. Wettability alteration of rock surfaces towards a more desirable state for oil to be recovered during LSW is widely accepted as a primary reason for improved oil recovery efficiencies during LSW. Even though most of the published work [36,38,58,63,65,66,88,123] links improved oil recovery to wettability alteration to a more water-wet condition by creating thicker and more stable water film, only a handful of reports [135,184,190,191] attributed wettability alteration to a mixed-wet condition. Almost all studies agree that the injection of low salinity water alters the wettability of rock to improve oil recovery. However, there are inconsistencies regarding the fundamental mechanisms of wettability alteration in carbonates, and the following section outlines the principles of each mechanism. Therefore, all mechanisms mentioned in this section can lead to wettability alteration. In other words, wettability alteration is the result of LSW, but not the cause of it.

4.1. Multicomponent Ionic Exchange

The mechanism of multicomponent ionic exchange (MIE) in chalk [20,59,61–63,121,192,193] and reservoir limestone [20,46,59] was first proposed by Austad and co-workers. They suggested that the presence of MIE between the injected brine and the rock surface results in a reduction of ionic bonding between oil molecules and rock surface, which is a dominant mechanism of improved oil recovery during LSW (Figure 9). The mechanism of MIE in carbonates is attributed to the exchange of anions, including adsorption of potential determining anions (SO_4^{2-}) and co-adsorption of divalent cations (Ca^{2+} and Mg^{2+}) onto the rock surface, resulting in desorption, and consequently release of negatively charged fatty acid components of crude oil from the rock surface. According to this mechanism, SO_4^{2-} ions act as a catalytic agent and adsorb onto the carbonate rock surface, decreasing the positive surface charge density. This minimizes electrostatic repulsive forces, and results in co-adsorption of Ca^{2+} and cations on the rock surface. The Ca^{2+} ions can then react with carboxylic acid groups that are bonded to the rock surface, thus breaking the attractive interactions between the oil and rock interface. Consequently, carboxylic acid components are released from the rock surface and the wettability of the rock surface is altered to a more water-wet condition. Additionally, it is suggested that, at high temperature (above 90 °C), Mg^{2+} ions can substitute Ca^{2+} ions on the carbonate surface and, therefore, displace surface bonded Ca^{2+} ions that are also bonded to carboxylic acid molecules, causing them to be released in the form of calcium-carboxylate complexes, thereby further improving oil recovery [65,129]. As a result, the presence of divalent cations (Ca^{2+} and Mg^{2+}) in brine can affect the ability of anions (SO_4^{2-}) to adsorb on carbonate surfaces [29,30]. Additionally, it was explained that the mechanism of MIE could also occur at high brine salinity given that the injected brine includes a different relative concentration of active ions compared to the formation water [60]. The mechanism of MIE for carbonate (limestone) cores containing anhydrite is suggested to be similar to chalk cores, the only difference is that SO_4^{2-} ions are generated from the rock matrix, due to anhydrite dissolution [62]. The MIE mechanism is suggested to be valid for dolomite as well [62]. However, the MIE was not observed by Ravari et al. [116] using outcrop limestone, where negligible increase in concentration of Ca^{2+} ions and decrease in concentration of Mg^{2+} and SO_4^{2-} ions were observed. Even higher temperature and increasing contact time between the rock surface and the injected brine did not result in substitution of Ca^{2+} ions with Mg^{2+} ions. Moreover, Zahid and co-workers [69] observed the MIE for outcrop chalk cores by LSW, but not for anhydrite free reservoir carbonate (limestone) cores. However, they claimed that MIE did not result in any additional oil recovery from the chalk cores. Therefore, the mechanism of MIE is dependent on the rock mineralogy and is not valid for all carbonate rock types.

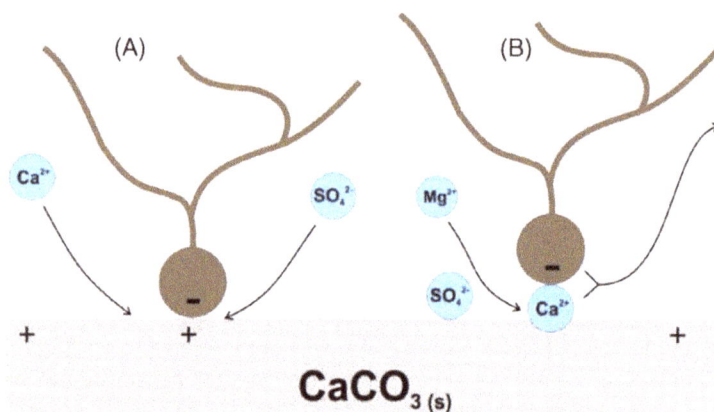

Figure 9. Schematic model for proposed wettability alteration mechanism by LSW when: (**A**) Ca^{2+} and SO_4^{2-} ions act as PDIs toward carbonate surface at low and high temperature; and (**B**) Mg^{2+} and SO_4^{2-} ions act as PDIs toward carbonate surface at high temperature from [63]. Reproduced with permission from Zhang et al., *Colloids Surf. A*; published by Elsevier, 2007.

4.2. Rock Dissolution

Dissolution of calcite was proposed as a mechanism for wettability alteration of chalks, by Hiorth et al. [85,169,170], using a geochemical thermodynamic model, based on experimental spontaneous imbibition tests conducted by Austad and co-workers [52,63,64,125,194]. Hiorth et al. suggested that changes in surface potential cannot describe the improvement in oil recovery with changes in pore water chemistry or temperature. Instead, they rationalized that lowering the concentration of Ca^{2+} ions could result in dissolution of calcium carbonates to restore the equilibrium with the brine. Such chemical dissolution of calcite will release the adsorbed polar components of crude oil from the rock surface, consequently improving water-wetness. Additionally, anhydrite and dolomite dissolutions are suggested to play a role in LSW improved oil recovery [62,195]. Although some researchers [27,30,31,38,43,66,67,69,80] have identified that rock dissolution may be one of the possible mechanisms for LSW in carbonates, Austad et al. [126] strongly disagreed with the mechanism of calcite dissolution proposed by Hiroth et al. [169], based on published experimental results, where they questioned the applicability of the geochemical model to calculate the chemical equilibrium between calcite and seawater during few days of spontaneous imbibition at high temperature (70–130 °C). It was argued that under these conditions a large portion of minerals will be supersaturated, resulting in the precipitation of minerals in the solution; therefore, a chemical equilibrium at rock–brine interfaces cannot be calculated precisely to define mineral dissolution by their geochemical model. Furthermore, it was argued that mineral dissolution cannot be a mechanism for improved oil recovery, based on spontaneous imbibition and corefooding tests where incremental oil recovery was observed for samples from the same geological formation without dissolution of salt mineral [88,133]. Mahani et al. [26,164] demonstrated that improved oil recovery happens in the absence of mineral dissolution, arguing that surface charge modification is probable to be the primary mechanism for LSW, which is good news for field application, as calcite dissolution will not contribute at a reservoir scale. Therefore, even though rock dissolution is suggested by some authors, based on the thermodynamic geochemical modelling, currently there is no experimental result that validates this mechanism.

4.3. Fines Migration

A mechanism of fines migration and stripping of oil bearing particles from rock surfaces in sandstones was proposed by Bernard et al. [196], and Tang and Morrow [12]. Both teams suggested that fines migration could help with improving the water-wetness condition, as released movable particles can potentially block some pore throats, in turn diverting the fluid flow and increasing the microscopic sweep efficiency, while reducing permeability. Therefore, the fines migration can lead to improvement in rock water-wetness condition. Additionally, an increase in pressure drop is experienced during LSW processes in carbonates, as reported in the literature [63,68–70]. However, Lager et al. [54] suggested that fines migration is not a mechanism of low salinity oil recovery improvement, but a phenomenon of MIE, based on coreflooding experiments under reservoir conditions, where improved oil recovery, without fines migration or permeability reduction, was observed. RezaeiDoust et al. [60] supported the argument and suggested that diversion of the original flow path would be a more important result than wettability alteration by fines migration. The mechanism of fines migration in LSW was initially proposed for sandstone reservoirs; however, later, this mechanism also has been suggested for carbonates by some researchers [69,70,80], who performed experimental coreflooding tests. Additionally, it is suggested that LSW can result in anhydrite and dolomite dissolution, and consequently fines migration [119,154,195]. Therefore, to acknowledge fines migration as a mechanism for carbonates, petrophysical analysis of rock samples before and after performing LSW is required [80].

4.4. Reduction of Interfacial Tension and Role of pH

Increased pH and reduced interfacial tension (IFT) can result in reduction of the residual oil saturation, and are a possible mechanism during LSW [16]. It is believed that reduction of IFT results in an increase in capillary number, and leads to lower residual oil saturation and improved oil recovery. The residual oil saturation is determined by the ratio of viscous force to capillary force (capillary number) at the end of the waterflooding process. In LSW treatment, reduced salinity and ionic composition have insignificant effect on viscosity; therefore, changes in capillary force will have the dominant impact on the residual oil saturation. Capillary force is a primary mechanism, causing the injected brine to be imbibed into the matrix micropores where oil is bound, and it becomes stronger as the wettability of the rock is altered toward a more water-wet condition. To reduce residual oil saturation, a substantial reduction in capillary force, which is determined by fluid–fluid and fluid–rock interactions, is required. However, whether performing LSW affects IFT (fluid–fluid interactions), contact angle (fluid–rock interactions), or both of them is not clearly established in the literature. Meng and co-workers [130] concluded that wettability alteration and reduction of IFT could be the two dominant mechanisms for LSW in carbonates. However, Yousef et al. [38,68] claimed that brine salinity has a negligible impact on the IFT of crude oil and seawater compared to contact angle measurements. Therefore, it is probable that LSW has a more significant impact on rock–brine interactions (contact angle) than oil–brine interactions (IFT). Furthermore, some experimental studies showed no correlation between LSW and an increase in pH [16,72]. Therefore, even though the mechanism of reduction of IFT in carbonates is suggested by some authors [130,173,178], others [18,25,38,68,131,134] argued that variation in pH value and reduction of IFT are not the primary causes of LSW oil recovery improvement.

4.5. Fluid–Fluid Interactions and Formation of Microemulsions

Interactions at oil–brine interfaces and formation of water micro-dispersions was proposed as a potential mechanism for LSW oil recovery improvement by Sohrabi and co-workers [197–199] using micromodels at reservoir conditions. It was suggested that when low salinity brine as opposed to high salinity formation water comes into contact with crude oil, water micro-dispersions are formed inside the oil phase, which results in oil recovery improvement due to depletion of natural surface active compound from the oil–brine interface, as well as swelling of high salinity formation water droplets. It was explained that depletion of surface active materials from oil–brine interface, results in changes in the balance between binding and repulsive forces at rock–brine and oil–brine interfaces and consequently alters wettability. In addition, formation of water micro-dispersions in case of low salinity brine and their coalescence at the high salinity formation water, leads to swelling of the formation water, redistribution of fluids, and consequently release of trapped crude oil. It should be noted that injection of low salinity brine only improved oil recovery when a mixed-wet system and a high salinity formation water was used. Additionally, Alvardo et al. [176,200] suggested that improved oil recovery during LSW is due to fluid–fluid interactions and an increase in oil–brine interfacial viscoelasticity, which suppress the trapped oil by snap-off. It was explained that by reducing concentration of the injected brine, regardless of type of ions present in the brine, viscoelasticity of the film at the oil–brine interface is increased and the effect of interfacial film breakage (snap-off) is decreased. Therefore, the oil phase becomes more continuous and larger oil droplets block the pore throat and improve flow of the low salinity brine in non-swept oil bearing pore throats, thus decrease the residual oil saturation and increase oil recovery. Similarly, increase in oil–brine interfacial viscoelasticity at low salt concentration, resulting in reduction of pressure fluctuations and decrease of the oil phase snap-off and, consequently enhanced oil recovery was proposed as a mechanism for LSW [175]. Furthermore, osmosis was proposed as a mechanism for LSW, where the oil phase inside a porous medium acts as a semipermeable membrane and only passes brine with low ionic strength due to sudden changes in osmosis gradient [201]. It was suggested that osmosis expansion of the formation water, which results in relocation of oil phase inside a porous medium could be a potential mechanism of LSW. Lastly, a mechanism of formation of water-in-oil microdispersions (with diameter of 50 μm) due to oil–brine interactions, which results in increased sweep efficiency was suggested for improved oil recovery by LSW [81]. It was argued that the presence of divalent cations (e.g., Ca^{2+}) in the injected brine results in stability of water-in-oil microdispersions in order to maintain the continuous oil phase and block the pore throats to improve sweep efficiency.

4.6. Expansion of Electric Double Layer

A mechanism of electric double layer (EDL) expansion was first suggested by Ligthelm et al. [20], who explained that brine with a high salinity, containing a high level of multivalent cations, can interact with negatively charged oil surfaces, thereby screening any repulsive forces and compressing the EDL. This results in a more oil-wet state and suppresses oil recovery. They suggested that the wettability alteration towards increasing water-wetness is responsible for the low salinity oil recovery, due to the increased thickness of the EDL surrounding the oil droplets and rock particles. Expansion of the EDL increases the electrostatic repulsive forces between oil–brine and rock–brine interfaces and, therefore, creates a thicker and more stable water film on the rock surface, thereby altering the wettability by increasing the water-wet condition. This was further confirmed by Fathi and co-workers who explained that, as the concentration of NaCl in seawater is much higher than for PDIs (Ca^{2+}, Mg^{2+}, and SO_4^{2-}), the EDL surrounding a charged rock surface contains a large amount of ions that are inactive in the wettability alteration process and do not form part of the inner Stern layer, which prevents access of active ions to the rock surface [79]. Therefore, NaCl depleted seawater results in EDL expansion and wettability alteration [179]. Additionally, increasing the pH above the carbonate point of zero charge density, changes the rock–brine interface from positive to negative, resulting in repulsive electrostatic forces between the rock–brine and the negatively charged oil–brine interfaces. This results in an increase in the disjoining pressure and expansion of the EDL [202]. Similarly, it was explained that when rock–brine and oil–brine interfaces have the same sign of electrostatic charge, the electrostatic repulsive force will increase, resulting in an increase in the disjoining pressure and creating a thicker and more water-wet film [99,145]. The mechanism of EDL expansion has been suggested, by some researchers [25,42,65,87,164,179,184,186,203], for wettability alteration and LSW oil recovery improvement in carbonates (Figure 10).

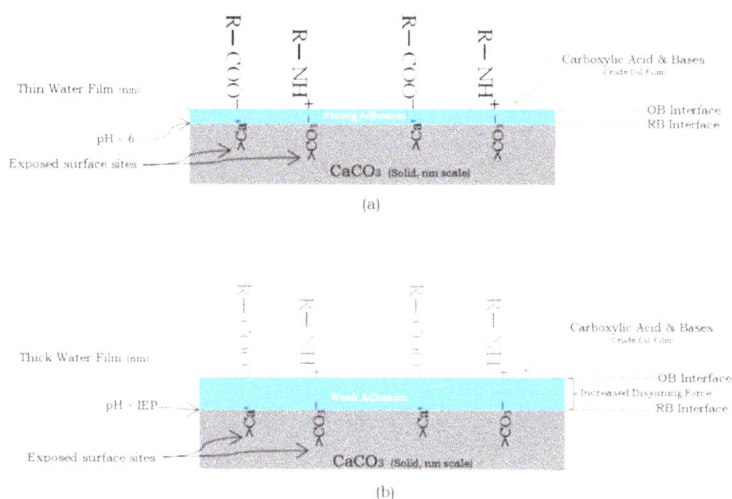

Figure 10. Schematic view of suggested mechanism for wettability alteration by expansion of EDL: (**a**) original wetting condition; and (**b**) low salinity brine condition from [42]. Reproduced with permission from Sohal et al., *Energy Fuels*; published by American Chemical Society, 2016.

5. Conclusions

This review collates evidence to show that LSW has a great potential to improve oil recovery in carbonates. However, LSW is a very complicated technique that is dependent on different variables, such as the specific reservoir conditions and oil–rock–brine interactions. Therefore, it is difficult to apply a brine formulation that suits all systems and conditions and, in order to design an optimized injection brine for a specific type of reservoir, it is essential to gain a mechanistic understanding of the wettability alteration at oil–rock–brine interfaces under controlled laboratory conditions by investigating the effect of individual ions and mixture of ions, as well as different oil components on surface charge modification and wettability alteration of different carbonate mineral surfaces, as the suggested primary mechanisms differ depending on the type of carbonate minerals.

A review of the available literature shows that mechanisms involved at both rock–brine and oil–brine interfaces, such as wettability alteration and fluid–fluid interactions, can play a role in

Colloids Interfaces **2018**, *2*, 20

enhanced oil recovery during LSW. Surface charge modification of carbonate mineral surfaces, due to MIE between PDIs, present in the injected brine, and carbonate rock surfaces, as well as expansion of EDL, due to modification of electrostatic forces between rock surface and carboxylic acid materials in line with the DLVO theory, results in an increase in the disjoining pressure and wettability alteration of rock surfaces by creating thicker and more stable water film on the rock surface, and, therefore, desorption and release of crude oil components from the rock surface and, consequently, improved oil recovery. Alternatively, dissolution of carbonate minerals resulting in increased concentration of Ca^{2+} ions in the brine leads to alteration of the brine pH and therefore expansion of EDL and wettability alteration to improve oil recovery. Additionally, alteration of the brine composition could modify oil–brine interactions and improve oil recovery as a result of fluid–fluid interactions and formation of microemulsions, by enhancing the sweep efficiency. Therefore, a single mechanism is incapable of explaining all oil production outcomes observed for LSW, especially in consideration of the large variation in key parameters between reservoirs, as there are many different variables involved in LSW in carbonates. There are also issues in the underpinning empirical data that have been obtained for such systems, as the nature of oil–rock–brine interactions is complex and there are inconsistencies in the experimental studies reported in the literature. Consequently, there might either be a number of mechanisms, or even an undiscovered mechanism, playing a role in LSW improved oil recovery processes and this work suggests greater rigour is required in experimental studies, which may also benefit from the use of advanced techniques, including atomic force microscopy, together with symbiotic modelling studies at realistic reservoir conditions.

Acknowledgments: MHD thanks the Energy Technology Partnership (ETP) for a PhD scholarship (ETP 120). Z.J.Z. thanks the support from EPSRC Impact Acceleration Account via the University of Birmingham.

Conflicts of Interest: The authors declare no conflict of interest.

References

1. Kokal, S.; Al-Kaabi, A. Enhanced oil recovery: Challenges & opportunities. *World Pet. Coun. Off. Publ.* **2010**, *64*.
2. Terry, R.E. *Encyclopedia of Physical Science and Technology*, 3rd ed.; Academic Press: San Diego, CA, USA, 2001; Volume 18.
3. Sheng, J. *Enhanced Oil Recovery Field Case Studies*; Elsevier Science: San Diego, CA, USA, 2013.
4. Lyons, W.C.; Plisga, G.J. *Standard Handbook of Petroleum and Natural Gas Engineering*; Gulf Professional Publishing: Houston, TX, USA, 2011.
5. Lake, L.W. *Enhanced Oil Recovery*; Prentice Hall Englewood Cliffs: Upper Saddle River, NJ, USA, 1989.
6. Reiter, P.K. A Water-Sensitive Sandstone Flood Using Low Salinity Water. Ph.D. Thesis, University of Oklahoma, Norman, OK, USA, 1961.
7. Jadhunandan, P.P. Effects of Brine Composition, Crude Oil, and Aging Conditions on Wettability and Oil Recovery. Ph.D. Thesis, Department of Petroleum Engineering, New Mexico Institute of Mining & Technology, Socorro, NM, USA, 1990.
8. Jadhunandan, P.; Morrow, N. Spontaneous imbibition of water by crude oil/brine/rock systems. *In Situ (USA)* **1991**, *15*, 4.
9. Jadhunandan, P.; Morrow, N.R. Effect of wettability on waterflood recovery for crude-oil/brine/rock systems. *Soc. Pet. Eng. Reserv. Eng.* **1995**, *10*, 40–46. [CrossRef]
10. Yildiz, H.O.; Morrow, N.R. Effect of brine composition on recovery of Moutray crude oil by waterflooding. *J. Pet. Sci. Eng.* **1996**, *14*, 159–168. [CrossRef]
11. Tang, G.; Morrow, N.R. Salinity, temperature, oil composition, and oil recovery by waterflooding. *Soc. Pet. Eng. Reserv. Eng.* **1997**, *12*, 269–276. [CrossRef]
12. Tang, G.Q.; Morrow, N.R. Influence of brine composition and fines migration on crude oil/brine/rock interactions and oil recovery. *J. Pet. Sci. Eng.* **1999**, *24*, 99–111. [CrossRef]
13. Webb, K.; Black, C.; Al-Ajeel, H. Low salinity oil recovery-log-inject-log. In *Proceedings of the Middle East Oil & Gas Show and Conference, 9–12 June 2003*; Society of Petroleum Engineers: Manama, Bahrain, 2003.

14. Webb, K.; Black, C.; Edmonds, I. Low salinity oil recovery–The role of reservoir condition corefloods. In Proceedings of the IOR 2005-13th European Symposium on Improved Oil Recovery, Budapest, Hungary, 25–27 April 2005.

15. Webb, K.J.; Black, C.J.J.; Tjetland, G. A laboratory study investigating methods for improving oil recovery in carbonates. In Proceedings of the International Petroleum Technology Conference, Doha, Qatar, 21–23 November 2005.

16. McGuire, P.; Chatham, J.; Paskvan, F.; Sommer, D.; Carini, F. Low salinity oil recovery: An exciting new EOR opportunity for Alaska's North Slope. In Proceedings of the Society of Petroleum Engineers Western Regional Meeting, Irvine, CA, USA, 30 March–1 April 2005.

17. Robertson, E.P. *Low-Salinity Waterflooding to Improve Oil Recovery-Historical Field Evidence*; Technical Report; Idaho National Laboratory (INL): Idaho Falls, ID, USA, 2007.

18. Lager, A.; Webb, K.J.; Collins, I.R.; Richmond, D.M. LoSal enhanced oil recovery: Evidence of enhanced oil recovery at the reservoir scale. In *Society of Petroleum Engineers Symposium on Improved Oil Recovery*; Society of Petroleum Engineers: Tulsa, OK, USA, 2008.

19. Seccombe, J.; Lager, A.; Jerauld, G.; Jhaveri, B.; Buikema, T.; Bassler, S.; Denis, J.; Webb, K.; Cockin, A.; Fueg, E. Demonstration of low-salinity EOR at interwell scale, Endicott field, Alaska. In *Society of Petroleum Engineers Improved Oil Recovery Symposium*; Society of Petroleum Engineers: Tulsa, OK, USA, 2010.

20. Ligthelm, D.J.; Gronsveld, J.; Hofman, J.; Brussee, N.; Marcelis, F.; van der Linde, H. Novel waterflooding strategy By manipulation Of injection brine composition. In Proceedings of the EUROPEC/EAGE Conference and Exhibition, Amsterdam, The Netherlands, 8–11 June 2009; Society of Petroleum Engineers: Tulsa, OK, USA, 2009.

21. Nasralla, R.A.; Sergienko, E.; Masalmeh, S.K.; van der Linde, H.A.; Brussee, N.J.; Mahani, H.; Suijkerbuijk, B.M.; Al-Qarshubi, I.S. Potential of low-salinity waterflood to improve oil recovery in carbonates: Demonstrating the effect by qualitative coreflood. *Soc. Pet. Eng. J.* **2016**, *21*, 1643–1654. [CrossRef]

22. Nasralla, R.A.; Mahani, H.; van der Linde, H.A.; Marcelis, F.H.; Masalmeh, S.K.; Sergienko, E.; Brussee, N.J.; Pieterse, S.G.; Basu, S. Low Salinity waterflooding for a carbonate reservoir: Experimental evaluation and numerical interpretation. *J. Pet. Sci. Eng.* **2018**, *164*, 640–654. [CrossRef]

23. Vledder, P.; Gonzalez, I.E.; Carrera Fonseca, J.C.; Wells, T.; Ligthelm, D.J. Low salinity water flooding: Proof of wettability alteration on a field wide scale. In Proceedings of the Society of Petroleum Engineers Improved Oil Recovery Symposium, Tulsa, OK, USA, 24-28 April 2010.

24. Mahani, H.; Sorop, T.; Ligthelm, D.J.; Brooks, D.; Vledder, P.; Mozahem, F.; Ali, Y. Analysis of field responses to low-salinity waterflooding in secondary and tertiary mode in Syria. In Proceedings of the Society of Petroleum Engineers EUROPEC/EAGE Annual Conference and Exhibition, Vienna, Austria, 23–26 May 2011.

25. Mahani, H.; Menezes, R.; Berg, S.; Fadili, A.; Nasralla, R.; Voskov, D.; Joekar-Niasar, V. Insights into the impact of temperature on the wettability alteration by low salinity in carbonate rocks. *Energy Fuels* **2017**, *31*, 7839–7853. [CrossRef]

26. Mahani, H.; Keya, A.L.; Berg, S.; Nasralla, R. Electrokinetics of carbonate/brine interface in low-salinity waterflooding: Effect of brine salinity, composition, rock type, and pH on ζ-potential and a surface-complexation model. *Soc. Pet. Eng. J.* **2017**, *22*, 53–68.

27. Gupta, R.; Smith, G.G.; Hu, L.; Willingham, T.; Lo Cascio, M.; Shyeh, J.J.; Harris, C.R. Enhanced waterflood for carbonate reservoirs-impact of injection water composition. In Proceedings of the Society of Petroleum Engineers Middle East Oil and Gas Show and Conference, Manama, Bahrain, 25–28 September 2011.

28. Gupta, R.; Lu, P.; Glotzbach, R.; Hehmeyer, O. A Novel, field-representative enhanced-oil-recovery coreflood method. *Soc. Pet. Eng. J.* **2015**, *20*, 442–452. [CrossRef]

29. Abdallah, W.; Gmira, A. Wettability assessment and surface compositional analysis of aged calcite treated with dynamic water. *Energy Fuels* **2013**, *28*, 1652–1663. [CrossRef]

30. Kasha, A.; Al-Hashim, H.; Abdallah, W.; Taherian, R.; Sauerer, B. Effect of Ca^{2+}, Mg^{2+} and SO_4^{2-} ions on the zeta potential of calcite and dolomite particles aged with stearic acid. *Colloids Surf. A Physicochem. Eng. Asp.* **2015**, *482*, 290–299. [CrossRef]

31. Al-Hashim, H.; Kasha, A.A.; Abdallah, W.; Sauerer, B. Impact of modified seawater on zeta potential and morphology of calcite and dolomite aged with stearic acid. *Energy Fuels* **2018**, *32*, 1644–1656. [CrossRef]

32.	Soraya, B.; Malick, C.; Philippe, C.; Bertin, H.J.; Hamon, G. Oil recovery by low-salinity brine injection: Laboratory results on outcrop and reservoir cores. In Proceedings of the Society of Petroleum Engineers Annual Technical Conference and Exhibition, New Orleans, LA, USA, 4–7 October 2009.

33.	Cissokho, M.; Bertin, H.; Boussour, S.; Cordier, P.; Hamon, G. Low salinity oil recovery on clayey sandstone: Experimental study. *Petrophysics* **2010**, *51*.

34.	Ashraf, A.; Hadia, N.; Torsaeter, O.; Tweheyo, M.T. Laboratory investigation of low salinity waterflooding as secondary recovery process: Effect of wettability. In Proceedings of the Society of Petroleum Engineers Oil and Gas India Conference and Exhibition, Mumbai, India, 20–22 January 2010.

35.	Skrettingland, K.; Holt, T.; Tweheyo, M.T.; Skjevrak, I. Snorre low-salinity-water injection–coreflooding experiments and single-well field pilot. *Soc. Pet. Eng. Reserv. Eval. Eng.* **2011**, *14*, 182–192. [CrossRef]

36.	Kazankapov, N. Enhanced oil recovery in Caspian carbonates with "smart water". In Proceedings of the Society of Petroleum Engineers Russian Oil and Gas Exploration & Production Technical Conference and Exhibition, Moscow, Russia, 14–16 October 2014.

37.	Yousef, A.A.; Al-Saleh, S.; Al-Jawfi, M.S. Improved/enhanced oil recovery from carbonate reservoirs by tuning injection water salinity and ionic content. In Proceedings of the Society of Petroleum Engineers Improved Oil Recovery Symposium, Tulsa, OK, USA, 14–18 April 2012.

38.	Yousef, A.A.; Al-Saleh, S.; Al-Kaabi, A.U.; Al-Jawfi, M.S. Laboratory investigation of novel oil recovery method for carbonate reservoirs. In Proceedings of the Canadian Unconventional Resources and International Petroleum Conference, Calgary, AB, Canada, 19–21 October 2010; Society of Petroleum Engineers: Tulsa, OK, USA, 2010.

39.	Shaker Shiran, B.; Skauge, A. Enhanced oil recovery (EOR) by combined low salinity water/polymer flooding. *Energy Fuels* **2013**, *27*, 1223–1235. [CrossRef]

40.	Matthiesen, J.; Bovet, N.; Hilner, E.; Andersson, M.P.; Schmidt, D.; Webb, K.; Dalby, K.N.; Hassenkam, T.; Crouch, J.; Collins, I. How naturally adsorbed material on minerals affects low salinity enhanced oil recovery. *Energy Fuels* **2014**, *28*, 4849–4858. [CrossRef]

41.	Jerauld, G.R.; Webb, K.J.; Lin, C.Y.; Seccombe, J. Modeling low-salinity waterflooding. In Proceedings of the Society of Petroleum Engineers Annual Technical Conference and Exhibition, San Antonio, TX, USA, 24–27 September 2006.

42.	Sohal, M.A.; Thyne, G.; Søgaard, E.G. Review of recovery mechanisms of ionically modified waterflood in carbonate reservoirs. *Energy Fuels* **2016**, *30*, 1904–1914. [CrossRef]

43.	Chandrasekhar, S.; Sharma, H.; Mohanty, K.K. Wettability alteration with brine composition in high temperature carbonate rocks. In Proceedings of the Society of Petroleum Engineers Annual Technical Conference and Exhibition, Dubai, UAE, 26–28 September 2016.

44.	Akbar, M.; Vissapragada, B.; Alghamdi, A.H.; Allen, D.; Herron, M.; Carnegie, A.; Dutta, D.; Olesen, J.R.; Chourasiya, R.; Logan, D. A snapshot of carbonate reservoir evaluation. *Oilfield Rev.* **2000**, *12*, 20–21.

45.	Marathe, R.; Turner, M.L.; Fogden, A. Pore-scale distribution of crude oil wettability in carbonate rocks. *Energy Fuels* **2012**, *26*, 6268–6281. [CrossRef]

46.	Myint, P.C.; Firoozabadi, A. Thin liquid films in improved oil recovery from low-salinity brine. *Curr. Opin. Colloid Interface Sci.* **2015**, *20*, 105–114. [CrossRef]

47.	Gomari, K.R.; Hamouda, A. Effect of fatty acids, water composition and pH on the wettability alteration of calcite surface. *J. Pet. Sci. Eng.* **2006**, *50*, 140–150. [CrossRef]

48.	Sauerer, B.; Stukan, M.; Abdallah, W.; Derkani, M.H.; Fedorov, M.; Buiting, J.; Zhang, Z.J. Quantifying mineral surface energy by scanning force microscopy. *J. Colloid Interface Sci.* **2016**, *472*, 237–246. [CrossRef] [PubMed]

49.	Karimi, M.; Al-Maamari, R.S.; Ayatollahi, S.; Mehranbod, N. Mechanistic study of wettability alteration of oil-wet calcite: The effect of magnesium ions in the presence and absence of cationic surfactant. *Colloids Surf. A Physicochem. Eng. Asp.* **2015**, *482*, 403–415. [CrossRef]

50.	Gomari, K.R.; Denoyel, R.; Hamouda, A. Wettability of calcite and mica modified by different long-chain fatty acids (C18 acids). *J. Colloid Interface Sci.* **2006**, *297*, 470–479. [CrossRef] [PubMed]

51.	Badri, M.; El-Banbi, A.; Montaron, B. The Schlumberger commitment to carbonate reservoirs. *Middle East Asia Reserv. Rev.* **2009**, 4–15.

52.	Puntervold, T.; Strand, S.; Austad, T. Coinjection of seawater and produced water to improve oil recovery from fractured North Sea chalk oil reservoirs. *Energy Fuels* **2009**, *23*, 2527–2536. [CrossRef]

53. Montaron, B. Defining the challenges of carbonate reservoirs. *Middle East Asia Reserv. Rev.* **2009**, *9*, 16–23.
54. Lager, A.; Webb, K.; Black, C. Impact of brine chemistry on oil recovery. In Proceedings of the IOR 2007-14th European Symposium on Improved Oil Recovery, Cairo, Egypt, 22 April 2007.
55. Lager, A.; Webb, K.J.; Black, C.; Singleton, M.; Sorbie, K.S. Low salinity oil recovery-an experimental investigation1. *Petrophysics* **2008**, *49*.
56. Seccombe, J.C.; Lager, A.; Webb, K.J.; Jerauld, G.; Fueg, E. Improving wateflood recovery: LoSalTM EOR field evaluation. In Proceedings of the Society of Petroleum Engineers Symposium on Improved Oil Recovery, Tulsa, Oklahoma, USA, 20–23 April 2008.
57. Zhang, Y.; Morrow, N.R. Comparison of secondary and tertiary recovery with change in injection brine composition for crude-oil/sandstone combinations. In Proceedings of the Society of Petroleum Engineers/DOE Symposium on Improved Oil Recovery, Tulsa, OK, USA, 22–26 April 2006.
58. Strand, S.; Høgnesen, E.J.; Austad, T. Wettability alteration of carbonates—Effects of potential determining ions (Ca^{2+} and SO_4^{2-}) and temperature. *Colloids Surf. A Physicochem. Eng. Asp.* **2006**, *275*, 1–10. [CrossRef]
59. Strand, S.; Austad, T.; Puntervold, T.; Høgnesen, E.J.; Olsen, M.; Barstad, S.M.F. "Smart water" for oil recovery from fractured limestone: A preliminary study. *Energy Fuels* **2008**, *22*, 3126–3133. [CrossRef]
60. RezaeiDoust, A.; Puntervold, T.; Strand, S.; Austad, T. Smart water as wettability modifier in carbonate and sandstone: A discussion of similarities/differences in the chemical mechanisms. *Energy Fuels* **2009**, *23*, 4479–4485. [CrossRef]
61. Austad, T.; Strand, S.; Høgnesen, E.; Zhang, P. Seawater as IOR fluid in fractured chalk. In Proceedings of the Society of Petroleum Engineers International Symposium on Oilfield Chemistry, The Woodlands, TX, USA, 2–4 February 2005.
62. Austad, T.; Shariatpanahi, S.; Strand, S.; Black, C.; Webb, K. Conditions for a low-salinity enhanced oil recovery (EOR) effect in carbonate oil reservoirs. *Energy Fuels* **2011**, *26*, 569–575. [CrossRef]
63. Zhang, P.; Tweheyo, M.T.; Austad, T. Wettability alteration and improved oil recovery by spontaneous imbibition of seawater into chalk: Impact of the potential determining ions Ca^{2+}, Mg^{2+}, and SO_4^{2-}. *Colloids Surf. A Physicochem. Eng. Asp.* **2007**, *301*, 199–208. [CrossRef]
64. Strand, S.; Puntervold, T.; Austad, T. Effect of temperature on enhanced oil recovery from mixed-wet chalk cores by spontaneous imbibition and forced displacement using seawater. *Energy Fuels* **2008**, *22*, 3222–3225. [CrossRef]
65. Fathi, S.J.; Austad, T.; Strand, S. "Smart water" as a wettability modifier in chalk: The effect of salinity and ionic composition. *Energy Fuels* **2010**, *24*, 2514–2519. [CrossRef]
66. Yousef, A.A.; Al-Salehsalah, S.H.; Al-Jawfi, M.S. New recovery method for carbonate reservoirs through tuning the injection water salinity: Smart waterflooding. In Proceedings of the Society of Petroleum Engineers EUROPEC/EAGE Annual Conference and Exhibition, Vienna, Austria, 23–26 May 2011.
67. Yousef, A.A.; Al-Saleh, S.; Al-Jawfi, M.S. The impact of the injection water chemistry on oil recovery from carbonate reservoirs. In Proceedings of the Society of Petroleum Engineers EOR Conference at Oil and Gas West Asia, Muscat, Oman, 16–18 April 2012.
68. Yousef, A.A.; Al-Saleh, S.H.; Al-Kaabi, A.; Al-Jawfi, M.S. Laboratory investigation of the impact of injection-water salinity and ionic content on oil recovery from carbonate reservoirs. *Soc. Pet. Eng. Reserv. Eval. Eng.* **2011**, *14*, 578–593. [CrossRef]
69. Zahid, A.; Shapiro, A.A.; Skauge, A. Experimental studies of low salinity water flooding carbonate: A new promising approach. In Proceedings of the Society of Petroleum Engineers Enhanced Oil Recovery Conference at Oil and Gas West Asia, Muscat, Oman, 16–18 April 2012.
70. Winoto, W.; Loahardjo, N.; Xie, S.X.; Yin, P.; Morrow, N.R. Secondary and tertiary recovery of crude oil from outcrop and reservoir rocks by low salinity waterflooding. In Proceedings of the Society of Petroleum Engineers Improved Oil Recovery Symposium, Tulsa, OK, USA, 14–18 April 2012.
71. Agbalaka, C.C.; Dandekar, A.Y.; Patil, S.L.; Khataniar, S.; Hemsath, J.R. Coreflooding studies to evaluate the impact of salinity and wettability on oil recovery efficiency. *Transp. Porous Media* **2009**, *76*, 77. [CrossRef]
72. Zhang, Y.; Xie, X.; Morrow, N.R. Waterflood performance by injection of brine with different salinity for reservoir cores. In Proceedings of the Society of Petroleum Engineers Annual Technical Conference and Exhibition, Anaheim, CA, USA, 11–14 November 2007.
73. Nasralla, R.A.; Nasr-El-Din, H.A. Double-layer expansion: Is it a primary mechanism of improved oil recovery by low-salinity waterflooding? *Soc. Pet. Eng. Reserv. Eval. Eng.* **2014**, *17*, 49–59. [CrossRef]

74. Rivet, S.; Lake, L.W.; Pope, G.A. A coreflood investigation of low-salinity enhanced oil recovery. In Proceedings of the Society of Petroleum Engineers Annual Technical Conference and Exhibition, Florence, Italy, 19–22 September 2010.

75. Nasralla, R.A.; Nasr-El-Din, H.A. Impact of electrical surface charges and cation exchange on oil recovery by low salinity water. In Proceedings of the Society of Petroleum Engineers Asia Pacific Oil and Gas Conference and Exhibition, Jakarta, Indonesia, 20–22 September 2011.

76. Austad, T.; RezaeiDoust, A.; Puntervold, T. Chemical mechanism of low salinity water flooding in sandstone reservoirs. In Proceedings of the Society of Petroleum Engineers improved oil recovery symposium, Tulsa, OK, USA, 24–28 April 2010.

77. Morrow, N.; Buckley, J. Improved oil recovery by low-salinity waterflooding. *J. Pet. Technol.* **2011**, *63*, 106–112. [CrossRef]

78. Awolayo, A.; Sarma, H.; AlSumaiti, A.M. A laboratory study of ionic effect of smart water for enhancing oil recovery in carbonate reservoirs. In Proceedings of the Society of Petroleum Engineers EOR Conference at Oil and Gas West Asia, Muscat, Oman, 31 March–2 April 2014.

79. Fathi, S.J.; Austad, T.; Strand, S. Water-based enhanced oil recovery (EOR) by "smart water": Optimal ionic composition for EOR in carbonates. *Energy Fuels* **2011**, *25*, 5173–5179. [CrossRef]

80. Yi, Z.; Sarma, H.K. Improving waterflood recovery efficiency in carbonate reservoirs through salinity variations and ionic exchanges: A promising low-cost "smart-waterflood" approach. In Proceedings of the Abu Dhabi International Petroleum Conference and Exhibition, Abu Dhabi, UAE, 11–14 November 2012.

81. Tetteh, J.T.; Rankey, E.; Barati, R. Low Salinity Waterflooding Effect: Crude Oil/Brine Interactions as a Recovery Mechanism in Carbonate Rocks. In *OTC Brasil*; Offshore Technology Conference: Rio de Janeiro, Brazil, 2017.

82. Tian, H.; Wang, M. Electrokinetic mechanism of wettability alternation at oil-water-rock interface. *Surf. Sci. Rep.* **2018**, *72*, 369–391. [CrossRef]

83. Afekare, D.A.; Radonjic, M. From mineral surfaces and coreflood experiments to reservoir implementations: Comprehensive review of low-salinity water flooding (LSWF). *Energy Fuels* **2017**, *31*, 13043–13062. [CrossRef]

84. Purswani, P.; Tawfik, M.S.; Karpyn, Z.T. Factors and mechanisms governing wettability alteration by chemically tuned waterflooding: A review. *Energy Fuels* **2017**, *31*, 7734–7745. [CrossRef]

85. Hiorth, A.; Cathles, L.; Madland, M. The impact of pore water chemistry on carbonate surface charge and oil wettability. *Transp. Porous Media* **2010**, *85*, 1–21. [CrossRef]

86. Crabtree, M.; Eslinger, D.; Fletcher, P.; Miller, M.; Johnson, A.; King, G. Fighting scale—Removal and prevention. *Oilfield Rev.* **1999**, *11*, 30–45.

87. Shariatpanahi, S.F.; Strand, S.; Austad, T. Initial wetting properties of carbonate oil reservoirs: Effect of the temperature and presence of sulfate in formation water. *Energy Fuels* **2011**, *25*, 3021–3028. [CrossRef]

88. Romanuka, J.; Hofman, J.; Ligthelm, D.J.; Suijkerbuijk, B.; Marcelis, F.; Oedai, S.; Brussee, N.; van der Linde, H.; Aksulu, H.; Austad, T. Low salinity EOR in carbonates. In Proceedings of the Society of Petroleum Engineers Improved Oil Recovery Symposium, Tulsa, OK, USA, 14–18 April 2012.

89. Khatib, Z.; Salanitro, J. Reservoir souring: Analysis of surveys and experience in sour waterfloods. In Proceedings of the Society of Petroleum Engineers Annual Technical Conference and Exhibition, San Antonio, TX, USA, 5–8 October 1997.

90. Bódi, T. Direct and indirect connate water saturation determination methods in the practice of Riaes Tibor Bódi. In Proceedings of the Conference & Exhibition on Earth Sciences and Environmental Protection, Miskolc-Egyetemváros, Hungary, 27–29 September 2012.

91. Morrow, N.R.; Xie, X. Oil recovery by spontaneous imbibition from weakly water-wet rocks. *Petrophysics* **2001**, *42*.

92. Puntervold, T.; Strand, S.; Austad, T. Water flooding of carbonate reservoirs: Effects of a model base and natural crude oil bases on chalk wettability. *Energy Fuels* **2007**, *21*, 1606–1616. [CrossRef]

93. Puntervold, T.; Strand, S.; Austad, T. New method to prepare outcrop chalk cores for wettability and oil recovery studies at low initial water saturation. *Energy Fuels* **2007**, *21*, 3425–3430. [CrossRef]

94. Denekas, M.; Mattax, C.; Davis, G. Effects of crude oil components on rock wettability. *Soc. Pet. Eng.* **1959**, *216*, 330–333.

95. Mullins, O.C.; Sabbah, H.; Eyssautier, J.; Pomerantz, A.E.; Barré, L.; Andrews, A.B.; Ruiz-Morales, Y.; Mostowfi, F.; McFarlane, R.; Goual, L. Advances in asphaltene science and the Yen–Mullins model. *Energy Fuels* **2012**, *26*, 3986–4003. [CrossRef]

96. Breure, B.; Subramanian, D.; Leys, J.; Peters, C.J.; Anisimov, M.A. Modeling asphaltene aggregation with a single compound. *Energy Fuels* **2012**, *27*, 172–176. [CrossRef]

97. Sabbaghi, S.; Jahanmiri, A.; Ayatollahi, S.; Shariaty Niassar, M.; Mansoori, G.A. Characterization of asphaltene using potential energy and nanocalculation. *Iran. J. Chem. Chem. Eng. (IJCCE)* **2008**, *27*, 47–58.

98. Akbarzadeh, K.; Hammami, A.; Kharrat, A.; Zhang, D.; Allenson, S.; Creek, J.; Kabir, S.; Jamaluddin, A.; Marshall, A.G.; Rodgers, R.P. Asphaltenes—Problematic but rich in potential. *Oilfield Rev.* **2007**, *19*, 22–43.

99. Dubey, S.; Doe, P. Base number and wetting properties of crude oils. *Soc. Pet. Eng. Reserv. Eng.* **1993**, *8*, 195–200. [CrossRef]

100. Zhang, P.; Austad, T. The relative effects of acid number and temperature on chalk wettability. In Proceedings of the Society of Petroleum Engineers International Symposium on Oilfield Chemistry, The Woodlands, TX, USA, 2–4 February 2005.

101. Fathi, S.J.; Austad, T.; Strand, S. Effect of water-extractable carboxylic acids in crude oil on wettability in carbonates. *Energy Fuels* **2011**, *25*, 2587–2592. [CrossRef]

102. Standnes, D.C.; Austad, T. Wettability alteration in chalk: 1. Preparation of core material and oil properties. *J. Pet. Sci. Eng.* **2000**, *28*, 111–121. [CrossRef]

103. Standnes, D.C.; Austad, T. Wettability alteration in chalk: 2. Mechanism for wettability alteration from oil-wet to water-wet using surfactants. *J. Pet. Sci. Eng.* **2000**, *28*, 123–143. [CrossRef]

104. Fathi, S.J.; Austad, T.; Strand, S.; Puntervold, T. Wettability alteration in carbonates: The effect of water-soluble carboxylic acids in crude oil. *Energy Fuels* **2010**, *24*, 2974–2979. [CrossRef]

105. Thomas, M.M.; Clouse, J.A.; Longo, J.M. Adsorption of organic compounds on carbonate minerals: 1. Model compounds and their influence on mineral wettability. *Chem. Geol.* **1993**, *109*, 201–213. [CrossRef]

106. Mwangi, P.; Thyne, G.; Rao, D. Extensive experimental wettability study in sandstone and carbonate-oil-brine systems: Part 1—Screening tool development. In Proceedings of the International Symposium of the Society of Core Analysts, Napa Valley, CA, USA, 16–19 September 2013; pp. 16–19.

107. Gomari, K.R.; Hamouda, A.; Davidian, T.; Fargland, D. Study of the effect of acidic species on wettability alteration of calcite surfaces by measuring partitioning coefficients, IFT and contact angles. *Contact Angle Wettabil. Adhes.* **2006**, *4*, 351–367.

108. Rao, D.N. Wettability effects in thermal recovery operations. In Proceedings of the Society of Petroleum Engineers/DOE Improved Oil Recovery Symposium, Tulsa, OK, 21–24 April 1996.

109. Shimoyama, A.; Johns, W.D. Formation of alkanes from fatty acids in the presence of $CaCO_3$. *Geochim. Cosmochim. Acta* **1972**, *36*, 87–91. [CrossRef]

110. Heidari, M.A.; Habibi, A.; Ayatollahi, S.; Masihi, M.; Ashoorian, S. Effect of time and temperature on crude oil aging to do a right surfactant flooding with a new approach. In Proceedings of the Offshore Technology Conference-Asia, Kuala Lumpur, Malaysia, 25–28 March 2014.

111. Zhang, P.; Tweheyo, M.T.; Austad, T. Wettability alteration and improved oil recovery in chalk: The effect of calcium in the presence of sulfate. *Energy Fuels* **2006**, *20*, 2056–2062. [CrossRef]

112. Sohal, M.A.; Kucheryavskiy, S.; Thyne, G.; Søgaard, E.G. Study of ionically modified water performance in the carbonate reservoir system by multivariate data analysis. *Energy Fuels* **2017**, *31*, 2414–2429. [CrossRef]

113. Buckley, J. Asphaltene precipitation and crude oil wetting. *Soc. Pet. Eng. Adv. Technol. Ser.* **1995**, *3*, 53–59. [CrossRef]

114. Austad, T.; Strand, S.; Puntervold, T.; Ravari, R.R. New method to clean carbonate reservoir cores by seawater. Presented at the SCA2008-15 International Symposium of the Society of Core Analysts, Abu Dhabi, UAE, 29 October–2 November 2008; Volume 29.

115. Fathi, S.J.; Austad, T.; Strand, S.; Frank, S.; Mogensen, K. Evaluation of EOR potentials in an offshore limestone reservoir: A case study. In Proceedings of the Eleventh International Symposium on Reservoir Wettability, Calgary, AB, Canada, 6–8 September 2010; pp. 7–9.

116. Ravari, R.R. Water-Based EOR in Limestone by Smart Water: A Study of Surface Chemistry. Ph.D. Thesis, University of Stavanger, Stavanger, Norway, 2011.

117. Shariatpanahi, S.; Strand, S.; Austad, T.; Aksulu, H. Wettability restoration of limestone cores using core material from the aqueous zone. *Pet. Sci. Technol.* **2012**, *30*, 1082–1090. [CrossRef]

118. Fernø, M.; Grønsdal, R.; Åsheim, J.; Nyheim, A.; Berge, M.; Graue, A. Use of sulfate for water based enhanced oil recovery during spontaneous imbibition in chalk. *Energy Fuels* **2011**, *25*, 1697–1706. [CrossRef]

119. Pu, H.; Xie, X.; Yin, P.; Morrow, N.R. Application of coalbed methane water to oil recovery from Tensleep sandstone by low salinity waterflooding. In Proceedings of the Society of Petroleum Engineers Symposium on Improved Oil Recovery, Tulsa, OK, USA, 20–23 April 2008.

120. Korsnes, R.; Strand, S.; Hoff, Ø.; Pedersen, T.; Madland, M.; Austad, T. Does the chemical interaction between seawater and chalk affect the mechanical properties of chalk. In *Multiphysics Coupling and Long Term Behaviour in Rock Mechanics*; CRC Press: Boca Raton, FL, USA, 2006; pp. 427–434.

121. Puntervold, T. Waterflooding of Carbonate Reservoirs: EOR by Wettability Alteration. Ph.D. Thesis, University of Stavanger, Stavanger, Norway, 2008.

122. Korsnes, R.; Madland, M.; Austad, T. Impact of brine composition on the mechanical strength of chalk at high temperature. In *Eurock, Proceedings of the International Symposium of the International Society for Rock Mechanics, Liège, Belgium, 9–12 May 2006*; Taylor & Francis/Balkema: Ghent, Belgium, 2006; pp. 133–140.

123. Zhang, P.; Austad, T. Wettability and oil recovery from carbonates: Effects of temperature and potential determining ions. *Colloids Surf. A Physicochem. Eng. Asp.* **2006**, *279*, 179–187. [CrossRef]

124. Strand, S.; Standnes, D.; Austad, T. New wettability test for chalk based on chromatographic separation of SCN^- and SO_4^{2-}. *J. Pet. Sci. Eng.* **2006**, *52*, 187–197. [CrossRef]

125. Hognesen, E.J.; Strand, S.; Austad, T. Waterflooding of preferential oil-wet carbonates: Oil recovery related to reservoir temperature and brine composition. In Proceedings of the Society of Petroleum Engineers Europec/EAGE Annual Conference, Madrid, Spain, 13–16 June 2005.

126. Austad, T.; Strand, S.; Puntervold, T. Is wettability alteration of carbonates by seawater caused by rock dissolution? In Proceedings of the SCA International Symposium, Noordwijk, The Netherlands, 27–30 September 2009; p. 2009-43.

127. Shehata, A.M.; Alotaibi, M.B.; Nasr-El-Din, H.A. Waterflooding in carbonate reservoirs: Does the salinity matter? *Soc. Pet. Eng. Reserv. Eval. Eng.* **2014**, *17*, 304–313. [CrossRef]

128. Brady, P.V.; Thyne, G. Functional wettability in carbonate reservoirs. *Energy Fuels* **2016**, *30*, 9217–9225. [CrossRef]

129. Fathi, S.J.; Austad, T.; Strand, S. Water-based enhanced oil recovery (EOR) by "smart water" in carbonate reservoirs. In Proceedings of the Society of Petroleum Engineers Enhanced Oil Recovery Conference at Oil and Gas West Asia, Muscat, Oman, 16–18 April 2012.

130. Meng, W.; Haroun, M.; Sarma, H.; Adeoye, J.; Aras, P.; Punjabi, S.; Rahman, M.; Al Kobaisi, M. A novel approach of using phosphate-spiked smart brines to alter wettability in mixed oil-wet carbonate reservoirs. In Proceedings of the Abu Dhabi International Petroleum Exhibition and Conference, Abu Dhabi, UAE, 9–12 November 2015.

131. Sohal, M.A.; Thyne, G.; Søgaard, E.G. Novel application of the flotation technique to measure the wettability changes by ionically modified water for improved oil recovery in carbonates. *Energy Fuels* **2016**, *30*, 6306–6320. [CrossRef]

132. Alotaibi, M.B.; Azmy, R.; Nasr-El-Din, H.A. Wettability challenges in carbonate reservoirs. In Proceedings of the Society of Petroleum Engineers Improved Oil Recovery Symposium, Tulsa, Oklahoma, USA, 24–28 April 2010.

133. Nasralla, R.A.; Sergienko, E.; Masalmeh, S.K.; van der Linde, H.A.; Brussee, N.J.; Mahani, H.; Suijkerbuijk, B.; Alqarshubi, I. Demonstrating the potential of low-salinity waterflood to improve oil recovery in carbonate reservoirs by qualitative coreflood. In Proceedings of the Abu Dhabi International Petroleum Exhibition and Conference, Abu Dhabi, UAE, 10–13 November 2014.

134. Al Harrasi, A.; Al-maamari, R.S.; Masalmeh, S.K. Laboratory investigation of low salinity waterflooding for carbonate reservoirs. In Proceedings of the Abu Dhabi International Petroleum Conference and Exhibition, Abu Dhabi, UAE, 11–14 November 2012.

135. Al-Attar, H.H.; Mahmoud, M.Y.; Zekri, A.Y.; Almehaideb, R.; Ghannam, M. Low-salinity flooding in a selected carbonate reservoir: Experimental approach. *J. Pet. Explor. Prod. Technol.* **2013**, *3*, 139–149. [CrossRef]

136. Standnes, D.C.; Nogaret, L.A.; Chen, H.L.; Austad, T. An evaluation of spontaneous imbibition of water into oil-wet carbonate reservoir cores using a nonionic and a cationic surfactant. *Energy Fuels* **2002**, *16*, 1557–1564. [CrossRef]

137. Jarrahian, K.; Seiedi, O.; Sheykhan, M.; Sefti, M.V.; Ayatollahi, S. Wettability alteration of carbonate rocks by surfactants: A mechanistic study. *Colloids Surf. A Physicochem. Eng. Asp.* **2012**, *410*, 1–10. [CrossRef]

138. Hirasaki, G.; Zhang, D.L. Surface chemistry of oil recovery from fractured, oil-wet, carbonate formations. *Soc. Pet. Eng. J.* **2004**, *9*, 151–162. [CrossRef]

139. Yuan, Y.; Lee, T.R. Contact angle and wetting properties. In *Surface Science Techniques*; Springer: New York, NY, USA, 2013; pp. 3–34.

140. Anderson, W. Wettability Literature survey-part 1: Rock/oil/brine interactions and the effects of core handling on wettability. *J. Pet. Technol.* **1986**, 1125–1144. [CrossRef]

141. Hamouda, A.A.; Rezaei Gomari, K.A. Influence of temperature on wettability alteration of carbonate reservoirs. In Proceedings of the Society of Petroleum Engineers/DOE Symposium on Improved Oil Recovery, Tulsa, OK, USA, 22–26 April 2006.

142. Kafili Kasmaei, A.; Rao, D. Is wettability alteration the main cause for enhanced recovery in low-salinity waterflooding? In Proceedings of the Society of Petroleum Engineers Improved Oil Recovery Symposium, Tulsa, Oklahoma, USA, 12–16 April 2014.

143. Butt, H.J.; Cappella, B.; Kappl, M. Force measurements with the atomic force microscope: Technique, interpretation and applications. *Surf. Sci. Rep.* **2005**, *59*, 1–152. [CrossRef]

144. Liang, Y.; Hilal, N.; Langston, P.; Starov, V. Interaction forces between colloidal particles in liquid: Theory and experiment. *Adv. Colloid Interface Sci.* **2007**, *134*, 151–166. [CrossRef] [PubMed]

145. Jackson, M.D.; Al-Mahrouqi, D.; Vinogradov, J. Zeta potential in oil-water-carbonate systems and its impact on oil recovery during controlled salinity water-flooding. *Sci. Rep.* **2016**, *6*, 37363. [CrossRef] [PubMed]

146. Butt, H.J.; Jaschke, M.; Ducker, W. Measuring surface forces in aqueous electrolyte solution with the atomic force microscope. *Bioelectrochem. Bioenerg.* **1995**, *38*, 191–201. [CrossRef]

147. Seiedi, O.; Rahbar, M.; Nabipour, M.; Emadi, M.A.; Ghatee, M.H.; Ayatollahi, S. Atomic force microscopy (AFM) investigation on the surfactant wettability alteration mechanism of aged mica mineral surfaces. *Energy Fuels* **2010**, *25*, 183–188. [CrossRef]

148. Zargari, S.; Ostvar, S.; Niazi, A.; Ayatollahi, S. Atomic force microscopy and wettability study of the alteration of mica and sandstone by a biosurfactant-producing bacterium Bacillus thermodenitrificans. *J. Adv. Microsc. Res.* **2010**, *5*, 143–148. [CrossRef]

149. Ducker, W.A.; Senden, T.J.; Pashley, R.M. Direct measurement of colloidal forces using an atomic force microscope. *Nature* **1991**, *353*, 239. [CrossRef]

150. Vigil, G.; Xu, Z.; Steinberg, S.; Israelachvili, J. Interactions of silica surfaces. *J. Colloid Interface Sci.* **1994**, *165*, 367–385. [CrossRef]

151. Hartley, P.G.; Larson, I.; Scales, P.J. Electrokinetic and direct force measurements between silica and mica surfaces in dilute electrolyte solutions. *Langmuir* **1997**, *13*, 2207–2214. [CrossRef]

152. Toikka, G.; Hayes, R.A. Direct measurement of colloidal forces between mica and silica in aqueous electrolyte. *J. Colloid Interface Sci.* **1997**, *191*, 102–109. [CrossRef] [PubMed]

153. Liu, J.; Xu, Z.; Masliyah, J. Studies on bitumen-silica interaction in aqueous solutions by atomic force microscopy. *Langmuir* **2003**, *19*, 3911–3920. [CrossRef]

154. Lebedeva, E.; Senden, T.; Knackstedt, M.; Morrow, N. Improved Oil Recovery from Tensleep Sandstone–Studies of Brine-Rock Interactions by Micro-CT and AFM. In Proceedings of the IOR 2009-15th European Symposium on Improved Oil Recovery, Paris, France, 27–29 April 2009.

155. Karoussi, O.; Skovbjerg, L.L.; Hassenkam, T.; Stipp, S.S.; Hamouda, A.A. AFM study of calcite surface exposed to stearic and heptanoic acids. *Colloids Surf. A Physicochem. Eng. Asp.* **2008**, *325*, 107–114. [CrossRef]

156. Ricci, M.; Segura, J.J.; Erickson, B.W.; Fantner, G.; Stellacci, F.; Voïtchovsky, K. Growth and dissolution of calcite in the presence of adsorbed stearic acid. *Langmuir* **2015**, *31*, 7563–7571. [CrossRef] [PubMed]

157. Gomari, K.R.; Hamouda, A.; Denoyel, R. Influence of sulfate ions on the interaction between fatty acids and calcite surface. *Colloids Surf. A Physicochem. Eng. Asp.* **2006**, *287*, 29–35. [CrossRef]

158. Rezaei Gomari, K.A.; Karoussi, O.; Hamouda, A.A. Mechanistic study of interaction between water and carbonate rocks for enhancing oil recovery. In Proceedings of the Society of Petroleum Engineers Europec/EAGE Annual Conference and Exhibition, Vienna, Austria, 12–15 June 2006.

159. Alroudhan, A.; Vinogradov, J.; Jackson, M. Zeta potential of intact natural limestone: Impact of potential-determining ions Ca, Mg and SO$_4$. *Colloids Surf. A Physicochem. Eng. Asp.* **2016**, *493*, 83–98. [CrossRef]

160. Monfared, A.D.; Ghazanfari, M.; Jamialahmadi, M.; Helalizadeh, A. Adsorption of silica nanoparticles onto calcite: Equilibrium, kinetic, thermodynamic and DLVO analysis. *Chem. Eng. J.* **2015**, *281*, 334–344. [CrossRef]

161. Al Mahrouqi, D.; Vinogradov, J.; Jackson, M.D. Zeta potential of artificial and natural calcite in aqueous solution. *Adv. Colloid Interface Sci.* **2017**, *240*, 60–76. [CrossRef] [PubMed]

162. Mihajlović, S.R.; Vučinić, D.R.; Sekulić, Ž.T.; Milićević, S.Z.; Kolonja, B.M. Mechanism of stearic acid adsorption to calcite. *Powder Technol.* **2013**, *245*, 208–216. [CrossRef]

163. Nyström, R.; Lindén, M.; Rosenholm, J.B. The influence of Na$^+$, Ca^{2+}, Ba^{2+}, and La^{3+} on the ζ potential and the yield stress of calcite dispersions. *J. Colloid Interface Sci.* **2001**, *242*, 259–263. [CrossRef]

164. Mahani, H.; Keya, A.L.; Berg, S.; Bartels, W.B.; Nasralla, R.; Rossen, W.R. Insights into the mechanism of wettability alteration by low-salinity flooding (LSF) in carbonates. *Energy Fuels* **2015**, *29*, 1352–1367. [CrossRef]

165. Rodríguez, K.; Araujo, M. Temperature and pressure effects on zeta potential values of reservoir minerals. *J. Colloid Interface Sci.* **2006**, *300*, 788–794. [CrossRef] [PubMed]

166. Alotaibi, M.B.; Nasr-El-Din, H.A.; Fletcher, J.J. Electrokinetics of limestone and dolomite rock particles. *Soc. Pet. Eng. Reserv. Eval. Eng.* **2011**, *14*, 594–603. [CrossRef]

167. Mahani, H.; Keya, A.L.; Berg, S.; Nasralla, R. The effect of salinity, rock type and ph on the electrokinetics of carbonate-brine interface and surface complexation modeling. In Proceedings of the Society of Petroleum Engineers Reservoir Characterisation and Simulation Conference and Exhibition, Abu Dhabi, UAE, 14–16 September 2015.

168. Goldberg, S. Sensitivity of surface complexation modeling to the surface site density parameter. *J. Colloid Interface Sci.* **1991**, *145*, 1–9. [CrossRef]

169. Hiorth, A.; Cathles, L.; Kolnes, J.; Vikane, O.; Lohne, A.; Madland, M. A chemical model for the seawater-CO$_2$-carbonate system–aqueous and surface chemistry. In Proceedings of the Wettability Conference, Abu Dhabi, UAE, 29 October–2 November 2008; pp. 27–28.

170. Hiorth, A.; Cathles, L.; Kolnes, J.; Vikane, O.; Lohne, A.; Madland, M. Chemical modelling of wettability change in carbonate rocks. Presented at the 10th Wettability Conference, Abu Dhabi, UAE, 7–28 October 2008; pp. 1–9.

171. Heberling, F.; Trainor, T.P.; Lützenkirchen, J.; Eng, P.; Denecke, M.A.; Bosbach, D. Structure and reactivity of the calcite–water interface. *J. Colloid Interface Sci.* **2011**, *354*, 843–857. [CrossRef] [PubMed]

172. Song, J.; Zeng, Y.; Wang, L.; Duan, X.; Puerto, M.; Chapman, W.G.; Biswal, S.L.; Hirasaki, G.J. Surface complexation modeling of calcite zeta potential measurements in brines with mixed potential determining ions (Ca^{2+}, CO$_3^{2-}$, Mg^{2+}, SO$_4^{2-}$) for characterizing carbonate wettability. *J. Colloid Interface Sci.* **2017**, *506*, 169–179. [CrossRef] [PubMed]

173. Alotaibi, M.; Nasr-El-Din, H. Effect of brine salinity on reservoir fluids interfacial tension. EUROPEC/EAGE Annual Conference and Exhibition, Amsterdam, The Netherlands, 8–11 June 2009. *Pap. Soc. Pet. Eng.* **2009**, *121569*, 1–14.

174. Lashkarbolooki, M.; Ayatollahi, S.; Riazi, M. Effect of salinity, resin, and asphaltene on the surface properties of acidic crude oil/smart water/rock system. *Energy Fuels* **2014**, *28*, 6820–6829. [CrossRef]

175. Chávez-Miyauchi, T.E.; Firoozabadi, A.; Fuller, G.G. Nonmonotonic elasticity of the crude oil–brine interface in relation to improved oil recovery. *Langmuir* **2016**, *32*, 2192–2198. [CrossRef] [PubMed]

176. Alvarado, V.; Garcia-Olvera, G.; Manrique, E. Considerations of adjusted brine chemistry for waterflooding in offshore environments. In Proceedings of the OTC Brasil, Offshore Technology Conference, Rio de Janeiro, Brazil, 27–29 October 2015.

177. Vijapurapu, C.S.; Rao, D.N. Effect of brine dilution and surfactant concentration on spreading and wettability. In Proceedings of the International Symposium on Oilfield Chemistry, Houston, TX, USA, 5–7 February 2003.

178. Okasha, T.M.; Alshiwaish, A. Effect of brine salinity on interfacial tension in Arab-D carbonate reservoir, Saudi Arabia. In Proceedings of the Middle East Oil and Gas Show and Conference, Manama, Bahrain, 15–18 March 2009.

179. Alameri, W.; Teklu, T.W.; Graves, R.M.; Kazemi, H.; AlSumaiti, A.M. Wettability alteration during low-salinity waterflooding in carbonate reservoir cores. In Proceedings of the Society of Petroleum Engineers Asia Pacific Oil & Gas Conference and Exhibition, Adelaide, Australia, 14–16 October 2014.

180. Xu, W. Experimental Investigation of Dynamic Interfacial Interactions at Reservoir Conditions. Master's Thesis, Louisiana State University, Baton Rouge, LA, USA, 2005.

181. Ma, S.; Zhang, X.; Morrow, N.; Zhou, X. Characterization of wettability from spontaneous imbibition measurements. *J. Can. Pet. Technol.* **1999**, *38*. [CrossRef]

182. Masalmeh, S. Impact of capillary forces on residual oil saturation and flooding experiments for mixed to oil-wet carbonate reservoirs. In Proceedings of the International Symposium of the Society of Core Analysts held in Aberdeen (SCA2012-11), Scotland, UK, 27–30 August 2012.

183. Yu, L.; Evje, S.; Kleppe, H.; Karstad, T.; Fjelde, I.; Skjaeveland, S.M. Analysis of the Wettability alteration process during seawater imbibition into preferentially oil-wet chalk cores. In Proceedings of the Society of Petroleum Engineers Symposium on Improved Oil Recovery, Tulsa, OK, USA, 20–23 April 2008.

184. Al-adasani, A.; Bai, B.; Wu, Y.S. Investigating low salinity waterflooding recovery mechanisms in carbonate reservoirs. In Proceedings of the Society of Petroleum Engineers Enhanced Oil Recovery Conference at Oil and Gas West Asia, Muscat, Oman, 16–18 April 2012.

185. Al Shalabi, E.W.; Sepehrnoori, K.; Delshad, M. Mechanisms behind low salinity water injection in carbonate reservoirs. *Fuel* **2014**, *121*, 11–19. [CrossRef]

186. Al Shalabi, E.W.; Sepehrnoori, K.; Delshad, M. Does the double layer expansion mechanism contribute to the LSWI effect on hydrocarbon recovery from carbonate rocks? In Proceedings of the Society of Petroleum Engineers Reservoir Characterization and Simulation Conference and Exhibition, Abu Dhabi, UAE, 16–18 September 2013.

187. Qiao, C.; Li, L.; Johns, R.T.; Xu, J. A mechanistic model for wettability alteration by chemically tuned water flooding in carbonate reservoirs. In Proceedings of the Society of Petroleum Engineers Annual Technical Conference and Exhibition, Amsterdam, The Netherlands, 27–29 October 2014.

188. Sylte, J.; Hallenbeck, L.; Thomas, L. Ekofisk formation pilot waterflood. In Proceedings of the Society of Petroleum Engineers Annual Technical Conference and Exhibition, Houston, TX, USA, 2–5 October 1988.

189. Hallenbeck, L.; Sylte, J.; Ebbs, D.; Thomas, L. Implementation of the Ekofisk field waterflood. *Soc. Pet. Eng. Form. Eval.* **1991**, *6*, 284–290. [CrossRef]

190. Sharma, M.; Filoco, P. Effect of brine salinity and crude-oil properties on oil recovery and residual saturations. *Soc. Pet. Eng. J.* **2000**, *5*, 293–300. [CrossRef]

191. Agbalaka, C.C.; Dandekar, A.Y.; Patil, S.L.; Khataniar, S.; Hemsath, J. The effect of wettability on oil recovery: A review. In Proceedings of the Society of Petroleum Engineers Asia Pacific Oil and Gas Conference and Exhibition, Perth, Australia, 20–22 October 2008.

192. Zhang, P.; Austad, T. Waterflooding in chalk–relationship between oil recovery, new wettability index, brine composition and cationic wettability modifier (Society of Petroleum Engineers94209). In Proceedings of the 67th EAGE Conference & Exhibition, Madrid, Spain, 13–16 June 2005.

193. Austad, T.; Strand, S.; Madland, M.V.; Puntervold, T.; Korsnes, R.I. Seawater in chalk: An EOR and compaction fluid. In Proceedings of the International Petroleum Technology Conference, Dubai, UAE, 4–6 December 2007.

194. Standnes, D.C. Enhanced Oil Recovery from Oil-Wet Carbonate Rock by Spontaneous Imbibition of Aqueous Surfactant Solutions. Ph.D. Thesis, Dibrugarh University, Assam, India, 2001.

195. Pu, H.; Xie, X.; Yin, P.; Morrow, N.R. Low-salinity waterflooding and mineral dissolution. In Proceedings of the Society of Petroleum Engineers Annual Technical Conference and Exhibition, Florence, Italy, 19–22 September 2010.

196. Bernard, G.G. Effect of floodwater salinity on recovery of oil from cores containing clays. In Proceedings of the Society of Petroleum Engineers California Regional Meeting, Los Angeles, CA, USA, 26–27 October 1967.

197. Emadi, A.; Sohrabi, M. Visual investigation of oil recovery by lowsalinity water injection: Formation of water micro-dispersions and wettabilityalteration. In Proceedings of the Society of Petroleum Engineers Annual Technical Conference and Exhibition, New Orleans, LA, USA, 30 September–2 October 2013.

198. Mahzari, P.; Sohrabi, M. Crude oil/brine interactions and spontaneous formation of micro-dispersions in low salinity water injection. In Proceedings of the Society of Petroleum Engineers Improved Oil Recovery Symposium, Tulsa, OK, USA, 12–16 April 2014.

199. Sohrabi, M.; Mahzari, P.; Farzaneh, S.A.; Mills, J.R.; Tsolis, P.; Ireland, S. Novel insights into mechanisms of oil recovery by use of low-salinity-water injection. *Soc. Pet. Eng. J.* **2017**, *22*, 407–416. [CrossRef]

200. Alvarado, V.; Moradi Bidhendi, M.; Garcia-Olvera, G.; Morin, B.; Oakey, J.S. Interfacial visco-elasticity of crude oil-brine: An alternative EOR mechanism in smart waterflooding. In Proceedings of the Society of Petroleum Engineers Improved Oil Recovery Symposium, Tulsa, OK, USA, 12–16 April 2014.

201. Sandengen, K.; Kristoffersen, A.; Melhuus, K.; Jøsang, L.O. Osmosis as mechanism for low-salinity enhanced oil recovery. *Soc. Pet. Eng. J.* **2016**, *21*, 1227–1235. [CrossRef]

202. Takamura, K.; Chow, R.S. The electric properties of the bitumen/water interface Part II. Application of the ionizable surface-group model. *Colloids Surf.* **1985**, *15*, 35–48. [CrossRef]

203. Al-Shalabi, E.; Sepehrnoori, K.; Pope, G.; Mohanty, K. A Fundamental model for predicting oil recovery due to low salinity water injection in carbonate rocks. In Proceedings of the Society of Petroleum Engineers Energy Resources Conference, Port of Spain, Spain, 9–11 June 2014.

colloids
and interfaces

MDPI

Article

Metal Ion Interactions with Crude Oil Components: Specificity of Ca²⁺ Binding to Naphthenic Acid at an Oil/Water Interface

Spencer E. Taylor * and Hiu Tung Chu

Centre for Petroleum and Surface Chemistry, Department of Chemistry, University of Surrey, Guildford, Surrey GU2 5XH, UK; tungchi0906@gmail.com
* Correspondence: s.taylor@surrey.ac.uk; Tel.: +44-1483-681999

Received: 17 August 2018; Accepted: 14 September 2018; Published: 18 September 2018

Abstract: On the basis of dynamic interfacial tension measurements, Ca^{2+} has been shown specifically to interact with naphthenic acid (NA) at the n-heptane/water interface, consistent with NA adsorption followed by interfacial complexation and formation of a more ordered interfacial film. Optimum concentrations of Ca^{2+} and NA have been found to yield lower, time-dependent interfacial tensions, not evident for Mg^{2+} and Sr^{2+} or for several alkali metal ions studied. The results reflect the specific hydration and coordination chemistry of Ca^{2+} seen in biology. Owing to the ubiquitous presence of Ca^{2+} in oilfield waters, this finding has potential relevance to the surface chemistry underlying crude oil recovery. For example, "locking" acidic components at water/oil interfaces may be important for crude oil emulsion stability, or in bonding bulk oil to mineral surfaces through an aqueous phase, potentially relevant for carbonate reservoirs. The relevance of the present results to low salinity waterflooding as an enhanced crude oil recovery technique is also discussed.

Keywords: dynamic interfacial tension; interfacial complexation; low salinity waterflooding; metal ion interactions; naphthenic acid; oil recovery

1. Introduction

Carboxylic acids are fundamental to the chemistry of all crude oils [1]. These are oil-soluble compounds with the general formula $C_nH_{2n+z}O_2$ containing acyclic ($z = 0$) and alicyclic ($z > 0$) hydrocarbyl groups; in crude oils, they are collectively referred to as 'naphthenic acid' (NA), since the earliest examples identified contained saturated cyclopentyl and cyclohexyl groups [2,3].

As is the case for other polar species present in petroleum, organic acids are often implicated in various upstream and downstream operational problems. Thus, during crude oil production, the surface activity of NA can lead to alterations in reservoir wettability [4,5] and increased stability of water-in-oil (w/o) emulsions [6]. Additionally, under high pH conditions, ionized acid groups stabilize oil-in-water (o/w) emulsions [7] when dissociation of oil-soluble acids produces anionic surface-active species which strongly affect properties, such as interfacial tension and interfacial rheology [8].

During crude oil refining, the acidity of NA (pK_a ~5 [9]) is responsible for corrosion in distillation units under the high temperature process conditions. The total acidity of the crude oil depends on its source, and oil-specific measures are often required to reduce potential damage when processing the highest acidity crudes (typically total acid number [TAN] values in excess of approximately 0.8 mg KOH/g) [10]. Naphthenic acids also exhibit specific interactions with calcium in certain oilfields, where Ca^{2+} coordination leads to precipitation and deposition of naphthenate-rich scale [11]. In this case, the naphthenates are tetrameric species, the original C80-tetra-acid member being identified as $C_{80}H_{142}O_8$, molecular weight 1230 g/mol [12]. Following this discovery, other similar members were identified as being responsible for naphthenate deposits in diverse locations [13,14]. Regarding the formation of these species, two points are noteworthy. First, if the tetra-acids are uncharged, it is likely that interactions must occur at an aqueous/oil interface. Second, from the composition of the deposits, there should be a specific affinity between the organic acid and aqueous Ca^{2+} ions, given the mixed cation composition of oilfield waters. This is also confirmed using a range of techniques and a synthetic tetra-acid, for which the strongest affinity was found for Ca^{2+} [15,16].

There is evidence in the literature to show that acidic species contribute to the structure and properties of crude oil–water interfaces [17]. Dissociation of carboxylic acids, for example, explains the pH-dependent behavior of crude oil/water interfaces [18,19] leading to increased o/w emulsion stability at high pH. It is also evident that crude oil/water interfacial properties can be significantly influenced by the presence of divalent metal ions. Thus, Alvarado et al. [20] reported the increased stabilization of water-in-crude oil emulsions in the presence of Ca^{2+} ions compared with Na^+ ions of the same ionic strength, suggesting that stronger interactions occur with Ca^{2+}, although the authors suggested that further investigation is necessary [21].

Carboxylic acid groups at crude oil interfaces have also been implicated in improved recovery mechanisms during low salinity waterflooding (involving the so-called "low salinity effect", LSE). One popular suggested LSE mechanism starts with the premise that oil is naturally bound to specific reservoir mineral surfaces via divalent metal ion bridging involving oil-based carboxylate groups [22–30]. It is considered that reducing the water salinity disrupts the bridges through ion exchange, resulting in reduced oil adhesion, thereby improving oil displacement and recovery.

Other studies into the LSE have suggested that interfacial properties should be sensitive to complexation across oil/water interfaces [31,32]. Indeed, the specificity of interfacial complexation between metal ions and oil-soluble components is important elsewhere in liquid–liquid extraction processes [33]. Early work [34–36] identified that interfacial reactions involving metal ions and spread carboxylic acid monolayers at the air-water interface are relatively slow, with half-lives of several minutes. More recently, de Ruiter et al. studied the evolution of the interfacial tension upon placing aqueous solutions of metal ions in contact with a decane solution of stearic acid (HSt) [31]. When the solution pH > pK_a, the acidity constant of stearic acid, distinct stages in the formation of metal stearate layers were noted, and interpreted as [31]: (i) deprotonation of adsorbed stearic acid (HSt), forming MSt_2 complexes at the interface; (ii) nucleation of the MSt_2 complexes, allowing further acid adsorption and complexation; (iii) partitioning of the three-dimensional MSt_2 structures into the oil phase.

The present study is concerned with the influence of alkali and alkaline earth metal ions (as the respective chloride salts) on oil/water interfaces containing a commercial (technical-grade) sample of NA. Previous work by Brandal et al. investigated the effects of divalent metal ions (Mg^{2+}, Ca^{2+}, Sr^{2+} and Ba^{2+}) on the interfacial tension behavior of different carboxylic acids under pH conditions where the acids were completely ionized [32]. However, we were keen to explore interactions between aqueous metal ions and the oil-soluble NA, initially in its unionized form as would be present in a native crude oil.

2. Experimental

2.1. Materials

Organic solvents (ethanol and *n*-heptane, both >98% purity) and the reagents naphthenic acid (technical grade), calcium chloride dihydrate ($CaCl_2 \cdot 2H_2O$), strontium chloride hexahydrate ($SrCl_2 \cdot 6H_2O$), magnesium chloride hexahydrate ($MgCl_2 \cdot 6H_2O$), sodium chloride (NaCl), potassium chloride (KCl), lithium chloride (LiCl), cesium chloride (CsCl), potassium hydroxide (KOH), calcium acetate ($Ca(OAc)_2 \cdot xH_2O$) were purchased from Sigma-Aldrich Co. Ltd. (Gillingham, Dorset, UK) and apart from *n*-heptane were used without further purification. The latter was passed through an alumina column to remove any surface-active impurities before use.

According to the supplier, the commercial naphthenic acid is a technical mixture of alkylated cyclopentane carboxylic acids, with a total acid number (TAN) value of 230.5 mg KOH/g, which equates to an average molecular mass of 56/0.2305 = 243 g/mol with the reasonable assumption that the mixture only contains monoprotic acids. However, consistent with our own analysis, Hindle et al. reported that this product contains a high proportion of saturated fatty acids, with relatively low levels of true naphthenic components [37]. Notwithstanding the relatively low content of authentic naphthenic structures, its composition does provide a representative range of carbon chain lengths to investigate interactions which mainly involve the carboxylic acid function of crude oils.

Deionized water (resistivity = 18.2 MΩ·cm) was from a Millipore Direct-Q system, with a pH of ~5.6 at room temperature (21 °C).

2.2. Methods

2.2.1. Synthesis of Calcium Naphthenate

Calcium naphthenate was synthesized based on a method used by Pereira et al. to prepare calcium salts of long-chain carboxylates [38]. Thus, KOH (0.752 g, 13.0 mmol) was dissolved in ethanol (20 mL) and heated gently on a hot-plate until boiling. NA (3.270 g, ~3.3 mmol based on the above TAN value) was then added and boiling maintained. Separately, Ca(OAc)$_2$·xH$_2$O (1.062 g, ~6.5 mmol) was dissolved in the minimum quantity of water with sonication, after which this solution was carefully added dropwise to the boiling ethanolic KOH/NA. The resultant mixture was then allowed to cool and evaporate, whereupon an amber-colored gel separated, together with residual water (~2 mL). The water was decanted, and the gel product was washed with an approximately equal volume of deionized water, heated to 80 °C, and then allowed to cool. Again, the residual water was decanted. This washing process was repeated one further time. The drained product was then stirred with petroleum ether (40–60 °C boiling range) until dissolved (~40 mL required). This solution was filtered (Whatman #1 filter paper), which also served to remove residual water from the resultant clear amber filtrate. Solvent evaporation yielded viscous product (3.079 g, 87%). The stoichiometry of the product was determined by thermogravimetric analysis (TA Instruments TGA Q500 thermogravimetric analyzer, Elstree, Hertfordshire, UK). The sample was heated from ambient temperature to 900 °C at 10 °C/min in air to yield a residue (calcium oxide) corresponding to 7.14% Ca. Using the average NA molecular mass of 243 calculated above, the product has a composition consistent with the expected 1:2 Ca:NA stoichiometry. Figure 1 shows a comparison of the infrared spectra (Bruker Alpha spectrometer with Diamond ATR (Bruker UK Ltd., Coventry, UK), each sample applied in neat form) for the original naphthenic acid and its calcium salt, from which the shift of the carbonyl band from 1700 to 1543 cm^{-1} is characteristic of neutralized carboxylic acids [39,40]. The small band at 1709 cm^{-1} in the calcium naphthenate spectrum indicates a trace amount of unreacted naphthenic acid remains in the product. Loss of bands at 1265 and 943 cm^{-1} are also consistent with calcium salt formation [40].

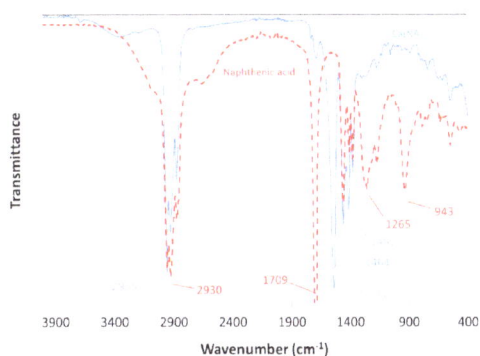

Figure 1. Attenuated Total Reflection (ATR)-infrared spectra of naphthenic acid (red, dashed) and its calcium salt (blue, continuous).

2.2.2. Dynamic Interfacial Tension Measurements

The dynamic tension of water/n-heptane interfaces was determined using an in-house dynamic drop volume (DDV) method. The general basis of this approach has been detailed by van Hunsel and Joos [41] and is similar to the method used by Deshiikan et al. [42]. Briefly, aqueous drops are formed at the tip of a capillary using different (known) flow rates, f, and the corresponding drop formation times, t_d, measured. By knowing f and t_d allows corresponding drop volumes to be determined under dynamic conditions. From a comparison with the Ward-Tordai equation, the adsorption time is $\sim^3/_7 t_d$ [43]. Strictly, drop volume (or weight) methods are used to determine equilibrium surface or interfacial tension for drops formed infinitely slowly, because at finite rates of drop formation hydrodynamic effects cause excess liquid to be incorporated into the detaching drop [41]. However, under dynamic conditions, the drop volume, $V(t_d)$, is related to the drop formation time by the empirical relationship attributed to McGee (see ref. [41]):

$$V(t_d) = V_0 + St_d^{-0.75} \tag{1}$$

in which V_0 is the drop volume at an infinitely slow formation time and S is the so-called McGee slope [41]. The latter is determined by measuring drop volumes at different flow rates for a series of pure liquids, under conditions of constant interfacial tension. Surface or interfacial tensions are then obtained using the established static pendant drop volume relationship:

$$\gamma = \frac{V_0 \Delta \rho g}{2 \pi r F} \tag{2}$$

in which r is the external tip radius, $\Delta \rho$ the density difference between the two phases, and F is a correction factor accounting for drop shape which was determined and tabulated as a function of $r/V_0^{1/3}$ by Harkins and Brown [44] and extended by Wilkinson [45]. The more recent polynomial developed by Lee et al. [46] shown in Equation (3), which correlates F with $r/V_0^{1/3}$ within the range $0 \geq r/V_0^{1/3} \geq 1.2$, was used in the present analysis.

$$F = 1 - 0.9121(r/V_0^{1/3}) - 2.109(r/V_0^{1/3})^2 + 13.38(r/V_0^{1/3})^3 - 27.29(r/V_0^{1/3})^4 + \\ 27.53(r/V_0^{1/3})^5 - 13.58(r/V_0^{1/3})^6 + 2.593(r/V_0^{1/3})^7 \tag{3}$$

By using pure low viscosity liquids of constant surface tension over the range of drop formation times, hydrodynamic contributions to drop formation are effectively eliminated, which allows the relationship between the McGee slope and V_0 to be defined. For our system, we found that the McGee slope is described well by

$$S = 0.22 V_0 \tag{4}$$

such that for a given drop time, Equation (2) becomes:

$$\gamma(t_d) = \frac{V(t_d) \Delta \rho g}{2 \pi r F \left(1 + 0.22 t_d^{-0.75}\right)} \tag{5}$$

In the present method, water or aqueous salt solution drops were formed below the surface of the n-heptane phase from the tip of a stainless steel syringe needle, the end of which was cleaned and roughened with a fine grade sand paper before each measurement to ensure effective contact of the aqueous phase is made with the outer diameter (3.413 mm). Drop volumes were in the range $\sim$30–120 mm^3. The pH of the unadjusted aqueous phases was in the range 5.4–5.7 consistent with Robertson et al. [47]. Drops are then produced at up to ten known rates using a pre-calibrated syringe pump (B. Braun, Melsungen, Germany) at 21 $\pm$ 1 °C, and the drop times calculated by averaging the time taken for several drops to be produced. Drop counting was performed visually with an estimated mechanical (stopwatch) response time of ±0.02 s for a single drop; this limits the realistic rate of formation typically to ~1 drop per second.

The procedure also requires the density difference between the liquid phases to be known under the same temperature conditions, and these were determined by pycnometry from the mass of a known volume of the respective liquids. Thus, by measuring t_d at different flow rates allows the determination of $\gamma(t_d)$ as a function of drop time using Equation (5). The combination of interfacial tension, drop volume, needle diameter and density difference translates for most of the measurements into values of the newly proposed Worthington number (Wo = $\Delta \rho g V_d / 2\pi r \gamma$) in the range ~0.7–0.9, for which a value of unity would be the ideal situation [48]. Additionally, replicate measurements using pure liquids with time-invariant interfacial tension, indicate a standard error of <3% of the quoted values, which is often better than ± 1 mN/m.

3. Results and Discussion

In advance of discussing the results of this study, it is pertinent to consider the ionization behavior of carboxylic acids at oil/water interfaces. Thus, on electrostatic grounds it would be expected that ionization equilibria will be affected by the presence of an interface, which would then be reflected in the relevant dissociation constants. To illustrate this, if the acid dissociation constant of a carboxylic acid in bulk (aqueous) solution is $pK_a(\infty)$, then the electrostatic image potential Ψ experienced by the acid group at a distance x from the interface is given by [49]:

$$\Psi(x) = -\frac{Ce}{2D_w x} \tag{6}$$

where $C = 1/4\pi\varepsilon_0$, e is the electronic charge (1.602×10^{-19} C), ε_0 is the permittivity of a vacuum (8.854×10^{-12} F/m), and D_w is the dielectric constant of water (~80).

By further considering the acid-base equilibrium as a function of distance from the interface, Dill and Bromberg [49] showed that the apparent pK_a at a distance x from the interface is related to the bulk value by

$$pK_a(x) = pK_a(\infty) - \frac{0.4343e\Psi(x)}{k_B T} = pK_a(\infty) + \frac{0.4343Ce^2}{2xD_w k_B T} \tag{7}$$

in which k_B is Boltzmann's constant (1.381×10^{-23} J/K) and T is the absolute temperature. This simple model suggests that adsorbed acid groups within, say, 0.2–0.5 nm of an oil/water interface, would experience an increase in pK_a of 0.3–0.8 units, purely on electrostatic grounds, without introducing specific ionic refinements.

However, by using more sophisticated computational approaches, Andersson et al. found that the pK_a of isolated acid molecules at an oil/water interface increased by an average of ~1 unit compared with the aqueous bulk solution value [50]. This reflects the greater affinity of the neutral molecule for the interface. The consequences of molecular confinement, at an interface, for example, were also probed using the same computational methods. In this situation, the possibility of deprotonated carboxylate groups stabilizing neighboring undissociated (neutral) carboxylic acid molecules arises through hydrogen bonding. Thus, the same workers also showed how the pK_a shift produced in the first molecule, as described above, is almost exactly compensated by the stabilizing effect induced on the second molecule [50].

Several experimental studies have identified increased pK_a values for acid groups present at surfaces and interfaces, including self-assembled monolayers [51], fatty acid micelles [52,53] and, more recently, at the air/water interface for fatty acid monolayers [54], the latter study also identifying the effect of chain length on the magnitude of the pK_a change. In the context of the present paper, this will mean that adsorbed NA molecules will remain largely undissociated as the effective (interfacial) pK_a is higher than the pH of the aqueous solutions.

3.1. NA Adsorption at the n-Heptane/Water Interface

Dynamic drop volume measurements between *n*-heptane containing NA and deionized water produce constant interfacial tensions over drop formation times in the approximate range ~0.3–500 s as shown in Figure 2a. The time-invariant values seen for different bulk NA concentrations (C_{NA}) are due to the changes in the composition of the *n*-heptane phase, reducing slightly from the value for the *n*-heptane/water interface. The results also indicate that adsorption is rapid compared with the experimental timescale, being complete in <0.3 s. Using a microfluidic technique, adsorption at the hydrocarbon/water interface has recently been shown [55] to occur on the millisecond timescale, more rapidly than previously thought.

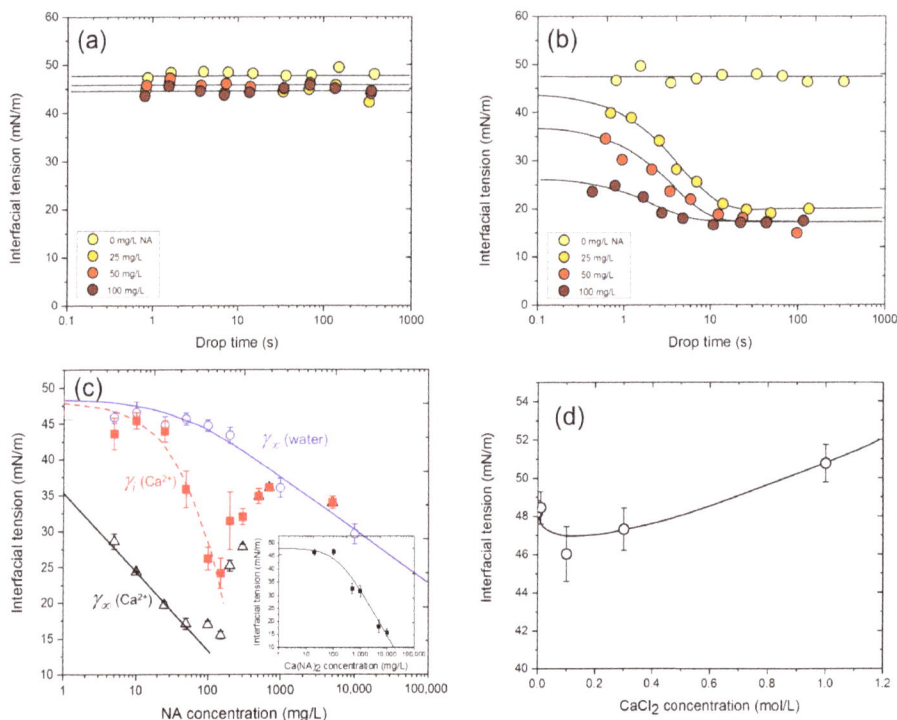

Figure 2. (**a**) Examples of the dynamic interfacial tension of naphthenic acid solutions in *n*-heptane against deionized water. (**b**) Corresponding results against water containing 0.3 mol/L Ca^{2+}. (**c**) Interfacial tension-log (naphthenic acid concentration) profiles for deionized water and the individual γ_i and γ_∞ contributions for 0.3 mol/L Ca^{2+} (see Equation (8)). The inset shows the corresponding plot for *n*-heptane solutions of Ca(NA)$_2$ against deionized water. All curves are fits to the Szyszkowski equation (Equation (9)). (**d**) The effect of Ca^{2+} concentration on the interfacial tension of the *n*-heptane/water interface. The line is based on the empirical equation $\gamma = \gamma_0 - A\gamma^{0.5} + B\gamma$ as described in the text.

Compared with the time-invariant interfacial tension behavior seen at the deionized water interface, the presence of 0.3 mol/L Ca^{2+} is accompanied by the appearance of time-dependent behavior, as shown in Figure 2b for equivalent naphthenic acid concentrations.

These latter profiles suggest that the approach to equilibrium of a growing drop in the presence of Ca^{2+} occurs in (at least) two stages. The first stage involves a rapid γ decrease, as described above, as the interface re-equilibrates from γ_0 ~48 mN/m to an "instantaneous" value (within the timescale of the present technique), γ_i, which, for 5 mg/L $\leq C_{NA} \leq$ 200 mg/L and 0.3 mol/L Ca^{2+}, is lower than the corresponding values obtained for the deionized water interface.

However, it is also evident in Figure 2b that within approximately the same C_{NA} range given above, the interfacial tension behavior exhibits a slower second process which finally results in an equilibrium value, γ_∞, i.e., $\gamma_0 \rightarrow \gamma_i$, followed by $\gamma_i \rightarrow \gamma_\infty$. The interfacial tension curves for the slower process, as shown in Figure 2b, are satisfactorily represented by first-order kinetic curves constructed according to Equation (8), each of which being characterized by a first-order rate constant, k.

$$\gamma = \gamma_\infty + (\gamma_i - \gamma_\infty) \exp(-kt) \tag{8}$$

Since γ_0 for each Ca^{2+} solution in the presence of NA is inaccessible using the present technique, we felt that it is reasonable to assume that it will be the same as the time-invariant value for the corresponding *n*-heptane/0.3 mol/L Ca^{2+} interface in the absence of NA.

Thus, in Figure 2c are shown plots of γ_i and γ_∞ for a Ca^{2+} concentration, C_{Ca} of 0.3 mol/L as a function of C_{NA}, together with the corresponding γ_∞ behavior at the deionized water interface, the latter being in good agreement with data published by Havre et al. [9] for a similar commercial NA sample. It is immediately apparent that the interfacial tension decreases in the presence of 0.3 mol/L Ca^{2+} in the aqueous phase for ~5 mg/L $\leq C_{NA} \leq$ 200 mg/L, and increases once again above this concentration, towards the deionized water isotherm.

In Figure 2c (inset), we also show interfacial tension data for the synthesized calcium salt, Ca(NA)$_2$. In this case, only very small time-dependent effects were observed, the error bars reflecting the spread of results over the timescale of the measurements. However, it is evident that the calcium salt follows the same general behavior as NA, inasmuch as the interfacial tension is largely unaffected up to ~100 mg/L, but thereafter the data diverge.

Within the NA concentration range considered, we find that the interfacial tension-log profiles in Figure 2c are satisfactorily represented by the Szyszkowski equation [56]:

$$\gamma = \gamma_0 - RT\Gamma_m \ln\left(\frac{C_{NA}}{a} + 1\right) \tag{9}$$

where C_{NA} is the bulk NA concentration in the organic phase, Γ_m is the interfacial concentration of NA at monolayer adsorption (mol/m^2) and a is a molecule-dependent constant (expressed as mol/L) related to the Gibbs energy of adsorption (ΔG^0) at infinite dilution by [57]:

$$a = 55.3 \exp\left(\frac{\Delta G^0}{RT}\right) \tag{10}$$

From Table 1 it is apparent that the "instantaneous" interfacial concentration of NA in the presence of Ca^{2+} is some ten times greater than in its equilibrium concentration. This translates into a much smaller apparent molecular area. The latter dimension is unreasonable, however, and most probably relates to a disordered interfacial accumulation of NA molecules in the initial stages, which would be consistent with the initial rapid response of the interfacial tension. The situation then transforms into a more structured interfacial film characterized by a more realistic molecular area, which is almost identical to the value at the deionized water interface. Moreover, the Gibbs energy of adsorption of NA is ~20 kJ/mol more favorable (i.e., lower) in the presence of Ca^{2+} ions, which presents an intriguing insight into the role played by Ca^{2+} ions in facilitating NA adsorption. Interestingly, this value is very close to the Gibbs energy for Ca^{2+} binding to hydrated carboxylate side chains of the protein β-lactoglobulin (-18 ± 1 kJ/mol) [58]. Whilst the interfacial packing of NA molecules appears to remain the same, the adsorption process itself is significantly more favorable, which suggests some degree of interaction occurs between NA and Ca^{2+} ions across the interface.

Table 1. Summary of the adsorption parameters for NA in the presence and absence of Ca^{2+} ions based on the Szyszkowski equation.

Interface Species	Γ_m (mol/m^2)	Molecular Area (nm^2)	ΔG^0 (kJ/mol)
NA	1.58×10^{-6}	1.05	−27.9
NA + Ca^{2+} (*i*)	1.66×10^{-5}	0.10	−27.5
NA + Ca^{2+} (∞)	1.60×10^{-6}	1.04	−48.7
Ca(NA)$_2$	3.34×10^{-6}	0.50	−29.1

Finally, as a relevant aside, in Figure 2d are shown the effects of C_{Ca} on the *n*-heptane/water interfacial tension in the absence of NA. It is well-known that high electrolyte concentrations generally show an increase in surface and interfacial tension as a consequence of "negative adsorption". However, relatively low electrolyte concentrations cause a small decrease in tension, a phenomenon known as the Jones-Ray effect, after the first researchers to report it, in 1937 [59]. While the original work focused on electrolyte effects on surface tension, some recent studies have considered the behavior at hydrocarbon/water interfaces [60], including the crude oil/water interface [61], because of their relevance to some of the applications mentioned earlier in the present paper.

In each of these earlier studies, interfacial tension minima have been seen at low C_{Ca}, which is also evident in the present results. Thus, Figure 2d shows that the interfacial tension reduction for <0.1 mol/L Ca^{2+} is ~1.4 mN/m, similar to values obtained elsewhere [60]. Suggested reasons for this effect have been debated, and have included cation [60], anion [62–64] or impurity [65] adsorption, or electrostatic effects influencing water structure at the interface [66,67]. The curve fitted through the experimental data in Figure 2d is of the form $\gamma = \gamma_0 - A\gamma^{0.5} + B\gamma$, where A and B are empirical constants (0.158 (mN/m)$^{0.5}$ and 0.217, respectively) and γ_0 is the interfacial tension of the *n*-heptane/water interface (48.4 mN/m).

3.2. NA Adsorption at the n-Heptane/Water Interface in the Presence of Metal Ions

With the exception of Ca^{2+}, none of the other metal ions studied has been found to exhibit time-dependent interfacial tensions using the present experimental method. Thus, in Figure 3a, NA adsorption at the *n*-heptane/metal chloride solution (1.0 mol/L) interface is indistinguishable from the deionized water curve shown in Figure 2c. These data include the two other *divalent* metal ions studied, Mg^{2+} and Sr^{2+}, which do not show the same interfacial tension behavior as a function of C_{NA} as seen for Ca^{2+}, i.e., interfacial tension minima that are dependent on C_{Ca} as highlighted in Figure 3b. As observed above, for NA concentrations beyond the minimum interfacial tension, the behavior returns to the curve defined by the other metal ion solutions and deionized water. This suggests a reduced influence of Ca^{2+} ions in lowering the Gibbs energy of adsorption of NA molecules (Table 1) as the interface becomes more saturated with NA. Crucially, however, it also suggests that the presence of specific calcium naphthenate species with higher surface activity may not be directly responsible for the increased interfacial activity of NA, but rather the Ca^{2+} ions promote NA adsorption, through specific interactions as will be discussed subsequently.

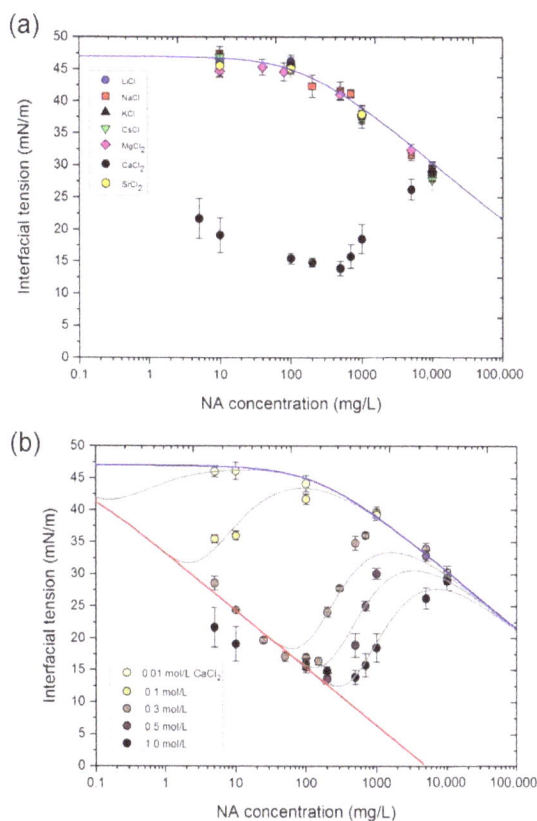

Figure 3. (a) Interfacial tension-log (naphthenic acid concentration) profiles at the *n*-heptane/water interface in the presence of the indicated cations (all chloride salts at 1 mol/L). (b) Corresponding profiles for different aqueous Ca^{2+} concentrations. The blue and red curves represent the respective limiting conditions for deionized water and a "Ca^{2+}-saturated" interface, and the gray curves are derived from Equation (13).

3.3. Dependence of Interfacial Tension on C_{NA} and C_{Ca}

The theoretical curves drawn through the data in Figure 3b were derived on the assumption that the interface can be described as a combination of two limiting situations, one based on pure water and the other containing a certain interfacial excess of Ca^{2+} (possibly representing a "saturated" interfacial film). The former condition is simply represented by deionized water data in Figure 2c, whilst the latter has to be obtained from the limiting interfacial tension data determined in the presence of different Ca^{2+} concentrations shown in Figure 3b (red line). As discussed above, both limiting conditions have been shown to be adequately described by the Szyszkowski equation (see Figure 2c). In the subsequent analysis, therefore, we have assumed that the overall adsorption conditions will be a combination of these two limiting cases, the extent to which each contributes being dependent on the (bulk) concentration ratio between NA and Ca^{2+}. Thus, the interfacial tensions produced by NA adsorption at a water interface and at a saturated Ca^{2+}-containing interface are given by Equations (9) and (11), respectively; it is unnecessary to define the bulk Ca^{2+} concentration responsible for the latter condition at this stage of the analysis. In Equation (11), Γ_m' and a' are the respective surface concentration and molecular constant of NA in the presence of the saturation concentration of Ca^{2+}.

$$\gamma = \gamma_0 - RT\Gamma_m\prime \ln\left(\frac{C_{NA}}{a\prime} + 1\right) \tag{11}$$

In the case of aqueous phases containing Ca^{2+} ions, and n-heptane phases containing NA, we assume that the extent of interaction at the water/n-heptane interface is a function of the concentration ratio C_{Ca}/C_{NA}, such that the contributions arising from pure water and aqueous Ca^{2+} solution interfaces are $[1 - \exp(\phi C_{Ca}/C_{NA})]$ and $\exp(\phi C_{Ca}/C_{NA})$, respectively, where the factor φ takes into account differences between concentration units as well as the relative effectiveness of Ca^{2+} and NA interactions at the interface. Thus, the general interfacial tension expression for different Ca^{2+} and NA concentrations is given by combining Equations (9) and (11), to give

$$\gamma = \gamma_0 - RT\left(\Gamma_m\left(1 - \exp\left(\frac{\phi C_{Ca}}{C_{NA}}\right)\right)\ln\left(\frac{C_{NA}}{a} + 1\right) + \Gamma_m\prime\exp\left(\frac{\phi C_{Ca}}{C_{NA}}\right)\ln\left(\frac{C_{NA}}{b} + 1\right)\right) \tag{12}$$

Initial attempts at fitting the data shown in Figure 3b to Equation (12) indicated that, perhaps unsurprisingly, $\Gamma_m \sim \Gamma_m\prime$, allowing the simplified Equation (13) to be used to generate the fitted curves shown in Figure 3b. For the purpose of this demonstration, the following values were used for all Ca^{2+} concentrations: γ_0 (47.0 mN/m); Γ_m (1.60×10^{-6} mol/m²), a (0.03 mol/L) and b (145 mol/L), leaving the interaction term φ (mg/mol) as the only adjustable parameter.

$$\gamma = \gamma_0 - RT\Gamma_m\left(1 - \exp\left(\frac{\phi C_{Ca}}{C_{NA}}\right)\right)\ln\left(\frac{C_{NA}}{a} + 1\right) + \exp\left(\frac{\phi C_{Ca}}{C_{NA}}\right)\ln\left(\frac{C_{NA}}{b} + 1\right) \tag{13}$$

It is evident from Figure 3b that this simplistic analysis produces reasonable fits to the experimental data and produces a sigmoidal variation of φ with the bulk Ca^{2+} concentration as shown in Figure 4. The behavior seen in Figure 3b reflects the increasing concentration of Ca^{2+}-NA interactions in the interface as the bulk Ca^{2+} concentration is raised. The appearance of a plateau region in Figure 4 at C_{NA} ~0.4 mol/L suggests that this bulk concentration corresponds to the interfacial "saturation" condition for Ca^{2+} ions, as alluded to earlier, and for which the maximum influence on NA adsorption is seen.

3.4. NA-Ca²⁺ Interfacial Tension Kinetics

The additional decrease in interfacial tension of NA at the n-heptane/water interface in the presence of Ca^{2+} ions shown in Figure 2b occurs at rates that are measurable using the present technique. Kinetic data associated with these changes are given in Figure 5 in the form of two plots. The first shows the effects of C_{NA} on the first-order rate constant k at constant C_{Ca} (0.3 mol/L; Figure 5a), and the second shows the variation in the second-order rate constant (=k/C_{NA}; for C_{NA} = 5 and 10 mg/L; Figure 5b) as a function of C_{Ca}.

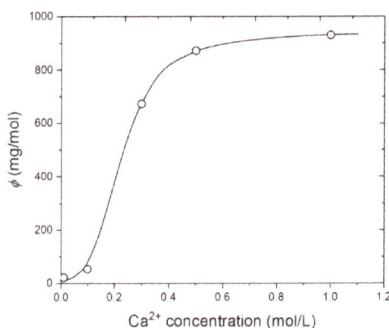

Figure 4. The effect of Ca^{2+} concentration on the interaction factor φ used in the construction of the fitted curves in Figure 3b.

Figure 5a illustrates the NA concentration range over which dynamic interfacial tension effects are observed for a Ca^{2+} concentration of 0.3 mol/L. This indicates that there is a C_{NA} "window", spanning almost three orders of magnitude, in which time-dependent interfacial processes are occurring. Moreover, for $C_{NA} < 50$ mg/L at this particular Ca^{2+} concentration, k is proportional to the NA concentration, the slope of the plot shown in the inset to Figure 5a representing the second-order rate constant for the process. As C_{NA} increases, however, not only is a maximum rate constant apparent at ~100 mg/L, but the uncertainties in the data are also seen to increase, as a result of the changes in the interfacial tension becoming smaller in this region. A comparison with the data in Figure 3b shows that the maximum k corresponds to the minimum interfacial tension of ~15 mN/m.

Also for the low C_{NA} region, the behavior of the second-order rate constant is shown as a function of C_{Ca} in Figure 5b. Here, it is seen that a minimum rate region is apparent at C_{Ca} ~0.5–1.0 mol/L, consistent with the onset of the plateau region in Figure 4.

We speculate that the cause of the slower reduction of interfacial tension shown in Figure 2b is due to complex formation involving Ca^{2+} and NA, and reorientation at the interface. This is evident only in the case of Ca^{2+}, but not for any of the other studied cations. With reference to the adsorption parameters in Table 1, it appears probable that the hydrophobic side of the interfacial region rapidly becomes saturated with randomly-distributed NA molecules upon contacting water containing Ca^{2+} ions. The increased local NA concentration then reconfigures into a more ordered adsorbed layer with the same adsorbed area/molecule as for NA at the *n*-heptane/deionized water interface. With respect to the adsorption data for pre-formed $Ca(NA)_2$ in Table 1, this finding suggests that complete complexation between NA and Ca^{2+} is unlikely to have taken place. The question then remains as to the role of Ca^{2+} ions in the adsorption process. To try to answer this, consideration needs to be given to the behavior of metal ions in aqueous solution, and especially with regard to interactions with carboxylates.

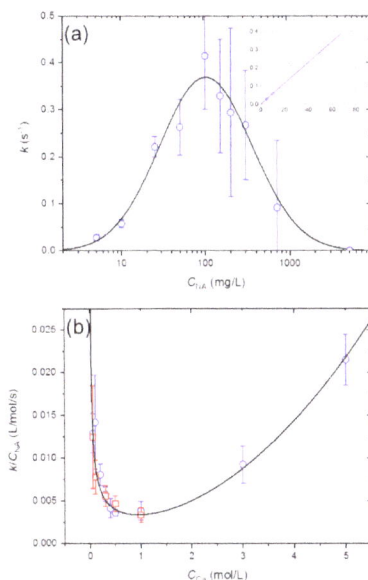

Figure 5. (**a**) The effect of NA concentration on the first-order kinetic rate constant for equilibration of the *n*-heptane/water interfacial tension in the presence of 0.3 mol/L Ca^{2+} ions. The inset shows the second-order plot for low NA concentrations. (**b**) The effect of Ca^{2+} concentration on the second-order rate constant in the low NA concentration range (red and blue symbols are for $C_{NA} = 5$ and 10 mg/L, respectively).

3.5. Interfacial Behavior of Ions and Metal Ion–Carboxylate Bonding

In order to provide a plausible explanation for the observed behavior described above, three particular factors need to be considered: (i) the behavior of hydrated cations at interfaces; (ii) exchange of ligated water in the inner coordination sphere; and (iii) specific interactions between cations and carboxylate groups.

It has long been known that aqueous ions influence interfacial behavior [68]. Typically, aqueous salts are considered to "negatively adsorb" at air/water or oil/water interfaces, producing an increase in the respective surface or interfacial tensions in accord with the Gibbs adsorption theory, and to an extent that is specific to each ion [69]. Ion theories have been developed over the years, with increasingly sophisticated models proposed to predict their behavior at interfaces [70,71].

Herein, we have considered several different monovalent and divalent cations and found that, of these, only Ca^{2+} affects the *n*-heptane/water interfacial tension in the presence of NA. All cations have a tendency to be hydrated, owing to their ionic charge [72], and as a consequence will be excluded from the interfacial region [73,74]. Therefore, we have to examine why Ca^{2+} behaves differently from the rest of the studied cations.

Ikeda et al. compared Ca^{2+} and Mg^{2+} ions in solution using constrained molecular dynamics, and found that the hydrated structure of Ca^{2+} is highly variable compared with that of Mg^{2+} [75]. Using density functional theory, Pavlov et al. identified that Ca^{2+} has a lower water binding energy compared with Mg^{2+} (as well as Be^{2+} and Zn^{2+}), notwithstanding the different water structures around the respective ions [76]. This implies that Ca^{2+} will show potentially greater lability of its inner-shell coordinated water molecules, which is of importance for adsorption as well as coordination to ligands.

Specific ion effects were originally classified based on the different abilities of salts to precipitate proteins from aqueous solution, a phenomenon which is accredited to Hofmeister in 1888 [77,78]. The "Hofmeister effect", as it came to be known, now refers more generally to the relative effects of either cations or anions on colloidal, surface or biological processes. The commonly-held view that has been developed, discussed and reviewed by many authors over the years is that ion hydration is fundamental to the Hofmeister effect [73,74,79]; this includes the behavior of ions at liquid interfaces [70]. The size of the ion and its nuclear charge control its hydration properties in aqueous solution and serve to define kosmotropic (strongly hydrated) and chaotropic (weakly hydrated) ions. Thus, kosmotropes are likely to possess high charge densities and their consequent strong hydration contributes to repulsion from an interface [73,74]. Adsorption of these ions would be accompanied by liberation of the strongly-bound hydration water, requiring large positive enthalpic contributions to the Gibbs energy which would not necessarily be outweighed by the corresponding entropy changes, i.e., $\Delta H_{ads} - T\Delta S_{ads} > 0$. On the other hand, water molecules bound to chaotropic ions are more easily lost on approaching an interface [73,74], making adsorption more favorable, i.e., $\Delta H_{ads} - T\Delta S_{ads} < 0$.

Considering the second factor mentioned above, that of the lability of the coordinated inner-shell water molecules, consistent with the theoretical investigations [75,76], it was found that water exchange rates from the innermost hydration shells are almost four orders of magnitude faster for Ca^{2+} than for Mg^{2+} ions [80]. This trend reflects the relative sizes of the anhydrous cations, and the greater charge density of Mg^{2+}. Using the same arguments, this would suggest that Sr^{2+} should behave in a similar fashion to Ca^{2+}, but we have not found this to be the case. We therefore turn to the third factor mentioned above, that of specific effects involving the coordination chemistry of Ca^{2+} and the carboxylate group.

In a recent example, Kherb et al. studied the strength of association of monovalent and divalent cations with carboxylate sites on a polypeptide biopolymer, and identified the divalent ion sequence $Zn^{2+} > Ca^{2+} > Ba^{2+} > Sr^{2+} > Mg^{2+}$, whilst monovalent cations showed weaker binding [81]. The inference made from this was that the most hydrated cations bind more strongly to carboxylate [81]. The structures of metal carboxylate complexes have been extensively studied over the years and have allowed the detailed examination of X-ray crystallographic data by Einspahr and Bugg [82] and Carrell et al. [83], which, together with other investigations [47,84–86] has culminated in the principal binding modes shown in Figure 6 being identified. Thus, specifically for Ca^{2+}, monodentate coordination with one carboxylate O atom and hydrogen bonded water, bidentate coordination, with both carboxylate O atoms, or bridging (see Figure 6a–c) are notable [82]. On the other hand, a major difference in bonding mode with carboxylic acids in the case of magnesium is the formation of uncoordinated ionic species of the form shown in Figure 6d [87]. The crystal structure of the magnesium complex of *p*-anisic acid (4-methoxybenzoic acid) contains a layer structure of alternating $[Mg(H_2O)_6]^{2+}$ cations and *p*-anisate anions with the lattice water functioning as a link between the layers. On the other hand, the monohydrated Ca(II) salt is a two-dimensional coordination polymer [87].

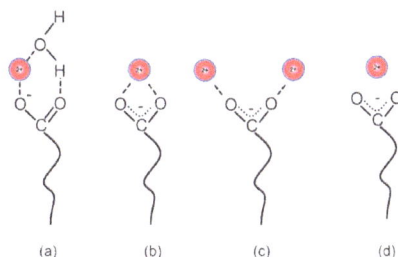

(a)   (b)   (c)   (d)

Figure 6. Bonding modes identified for divalent ion interactions with carboxylates [47,84–86]. 1:1 coordination is shown in binding modes (**a**) and (**b**), with bridging exhibited in (**c**). (**d**) shows ionic bonding.

The comparative behavior between Ca^{2+} and Mg^{2+} interactions at carboxylate monolayers has also been studied by Tang et al. using vibrational sum frequency generation spectroscopy [88,89]. Thus, binding of these ions to palmitic acid monolayers indicated different concentration-dependent metal ion binding modes [88], concluding that as the Ca^{2+} concentration increases (to 0.3 mol/L), complexation favors the 2:1 bridging configuration shown in Figure 6c, whereas Mg^{2+} adopts the ionic complex in Figure 6d consistent with solution and crystallographic results. Moreover, the stronger Ca^{2+} binding was shown to have a much greater impact on the underlying hydrogen bonding below the carboxylic acid monolayer [89], which is arguably consistent with the differences seen between Ca^{2+} and Mg^{2+} ions in the present study at the oil/water interface.

We have also looked to biological systems to provide an understanding of the differences between divalent ion coordination modes, especially with respect to carboxylate ligands found in a wide range of protein environments [90–94]. Williams provides a clear indication that the most significant factors responsible for ion binding in biological systems are, *inter alia*, ion charge, ion size, ligand donor atom, and preferred coordination geometry [95]. Thus, in connection with biological systems, binding of ions such as those considered in the present study is restricted to oxygen donor ligands, and specificity of Ca^{2+} binding (e.g., relative to Mg^{2+}) is due to its size and less-restricted steric preferences compared with other ions. Ca^{2+} shows a greater tendency for bidentate binding to carboxylate, compared with Mg^{2+} [91], and a lower affinity for water [92,94].

Ca^{2+} binding to carboxylate is also considered to be advantageous based on Collins' "matching water affinity" approach, according to which specific pairing interactions between ions, resulting in the formation of ion pairs, is most favorable for ions of comparable hydration energy [96]. The latter is quantified [97] by the apparent dynamic hydration number (ADHN) which, for Ca^{2+} and carboxylate are 2.1 and 2.0, respectively, which therefore match up surprisingly well (cf. 5.7 for Mg^{2+}). Kiriukhin and Collins consider that interaction between these two species will result in dissipation of the hydration shell, thereby benefitting coordination [97], which we suggest may also feature in specific Ca^{2+}-naphthenic acid interactions across an interface. Such interactions would be likely to require water to be displaced from the first coordination shell.

In summary, there is compelling evidence for specific Ca^{2+} interactions with carboxylic acids from biology. Therefore, a combination of evidence from biological and coordination chemistry appears to offer an interpretation of the present surface chemical findings, which may be of relevance in other practical and technological areas, such as our current interest in crude oil surface chemistry.

3.6. Implications for Crude Oil Surface Chemistry

Over the past two decades in particular, the importance of injection water composition on crude oil reservoir chemistry has been established through various laboratory coreflood investigations and field evaluations. Studies have shown that low salinity conditions contribute to improvements in oil recovery, but although various mechanisms have been proposed, a full understanding of the low salinity effect is still lacking. However, a common factor throughout some proposed mechanisms is the role of the ionic composition of the injection and formation waters [98] and especially their effects on reservoir wettability and the corresponding affinity of the oil for the rock surface [99]. Based on our present findings the specific role of Ca^{2+} is considered to be significant. From the data given in Figure 3b, for example, it is evident that the minimum interfacial tension of the present NA system is dependent on the Ca^{2+} concentration, as shown in Figure 7a. This representation of the data assumes that there is an excess of NA in the oil phase and that the bulk Ca^{2+} concentration governs the interfacial tension.

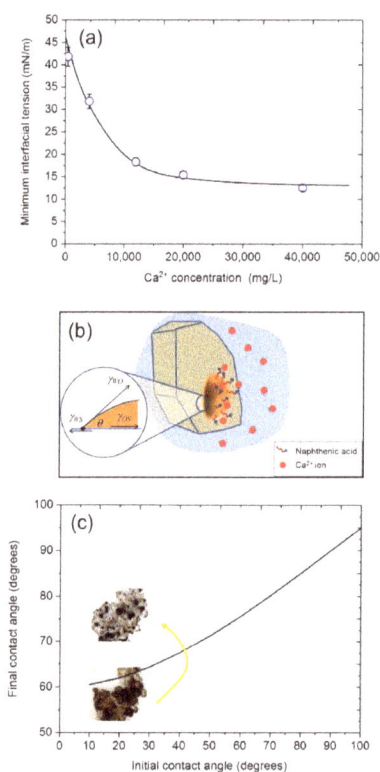

Figure 7. (a) The effect of Ca^{2+} ion concentration (expressed here in mg/L) on the minimum interfacial tension of the *n*-heptane/water interface when NA is in excess. (b) Schematic of an oil drop adhering to a rock surface in water showing the three-phase contact point defining the contact angle. (c) Demonstrating the effect of a decrease in Ca^{2+} concentration from 20,000 mg/L to 5000 mg/L on the final contact angle for different initial contact angles, and a simulated visual result on a sand surface to illustrate.

With reference to Figure 7b, showing a schematic of oil coating a solid (rock) surface, two equilibrium conditions can be envisaged during oil recovery. The first is the original condition within the reservoir, in which the crude oil and rock are in equilibrium with the formation brine, and the second is during waterflooding conditions where sea water, softened water or low salinity water will change the aqueous phase composition [100].

Young's equation is derived from the balance of forces acting on the contact line of a drop on a surface (Figure 7b), i.e.,

$$\gamma_{WS} = \gamma_{OS} + \gamma_{WO} \cos \theta \tag{14}$$

where the γ terms are the respective Gibbs energies of the water/solid (WS), oil/solid (OS) and oil/water (OW) interfaces and θ is the contact angle made by oil on the solid surface. We can therefore consider two hypothetical situations by way of illustrating the potential role of Ca^{2+} ions in oil recovery, based on an *initial* oil/rock system in equilibrium with formation water, and a *final* waterflooded system with reduced salinity, and a concomitant lower Ca^{2+} concentration.

We assume that Equation (14) can be applied to each situation, such that the change in Ca^{2+} concentration causes a resultant change in the water/oil interfacial tension, γ_{WO}. For this elementary analysis, we also assume that γ_{WS} and γ_{OS} remain relatively unaffected by the change in Ca^{2+} concentration. However, although we recognize that, in practice it is likely that wetting films comprising, for example, hydrophobic species [101] or hydration layers [102], can modify water/solid and oil/solid interfaces and their respective interfacial energies, predicting the effects on the contact angle, for example, is not easy. Therefore, we consider, as an example, the simple analysis involving the initial and final conditions as introduced above, for which it is apparent that

$$\gamma_{OW}^i \cos \theta_i = \gamma_{OW}^f \cos \theta_f \tag{15}$$

where i and f refer to initial and final states, respectively. As shown in Table 2, formation waters are generally more saline than sea water, the ionic composition being source-dependent but also dependent on the geologic period [103]. Thus, it is possible that during geologic time the oil/rock system will have been "conditioned" with a relatively highly saline brine containing a high Ca^{2+} concentration, say ~20,000 mg/L as an illustrative value (Table 2). Therefore, with reference to Figure 7a, dilution of the formation brine to ~5000 mg/L Ca^{2+} would approximately double the interfacial tension, such that

$$\cos \theta_f = (\gamma_{OW}^i / \gamma_{OW}^f) \cos \theta_i \approx 0.5 \cos \theta_i \tag{16}$$

Table 2. Na^+ and Ca^{2+} compositions of formation waters from different geologic periods. For comparison, sea water contains ~10,500 mg/L Na^+ and ~400 mg/L Ca^{2+}. Data compiled from Collins [103].

Geologic Period	Average (Maximum) Metal Ion Concentration (mg/L)	
	Na^+	Ca^{2+}
Oligocene/Paleogene	39,000 (103,000)	2530 (38,800)
Cretaceous	31,000 (88,600)	7000 (37,400)
Jurassic	57,300 (120,000)	25,800 (56,300)
Permian	47,000 (109,000)	8600 (22,800)
Carboniferous (Pennsylvanian)	43,000 (101,000)	9100 (205,000)
Carboniferous (Mississippian)	41,500 (115,800)	8900 (37,800)
Devonian	48,000 (101,000)	18,000 (129,000)

Thus, Equation (16) enables the plot shown in Figure 7c to be constructed, illustrating how the contact angle between the oil and rock surface is affected by the Ca^{2+} concentration change in the example given, from 20,000 mg/L to 5000 mg/L during waterflooding. Here, it is evident that the lowest initial contact angles, indicating high oil affinity with the surface, are influenced most by the change in interfacial tension. In this region, contact angles increase by as much as 50°, potentially improving oil recovery. Such a scenario is depicted in the inset images in Figure 7c, added for illustrative purposes only, but is consistent with the results of the study by Shehata and Nasr-El-Din regarding the dominant role of formation water salinity and composition in improving oil recovery during low salinity waterflooding [98].

Thus, the findings presented herein may provide further insight into potential mechanisms for the low salinity effect. In addition, it is apparent that the ability to predict the effect of water chemistry on oil recovery reservoir requires not only knowledge of the interfacial chemistry of inorganic ions, but also the range of surface-active species present in the crude oil; here, we have only examined the effects of naphthenic acids. For real crude oils, however, the increased complexity and competition resulting from the presence of other interfacially active species, such as asphaltenes and resin components, would be expected to produce a different interfacial behavior [104]. This has been borne out in studies on crude oil fractions to be reported elsewhere.

4. Conclusions

It seems to be a recurring theme that Ca^{2+} plays significant roles in the aqueous colloid and interfacial chemistry of crude oil and, in this respect, is more important than other metal ions. As discussed in the present paper, the characteristics of this ion make it favorable to undergo specific interactions with carboxylic components found in crude oil, to some extent mimicking its unique position in biology [95]. Of the metal ions studied herein, both monovalent and divalent, only Ca^{2+} has been shown to exhibit interfacial coordination with carboxylate groups of a commercial naphthenic acid sample.

The present findings have indicated that, on their own, naphthenic acids (NAs) are only weakly active at the *n*-heptane/water interface at concentrations below ~200 mg/L as determined using the dynamic drop volume technique. However, this behavior is modified in the specific presence of Ca^{2+}, whereas Mg^{2+} and Sr^{2+}, as well as several alkali metal cations, exhibit no comparable effect. We have found that, for a given aqueous Ca^{2+} concentration, there exists an "optimum" NA concentration in the *n*-heptane phase at which a minimum interfacial tension is attained. As shown in Figure 7, the lowest interfacial tension is ~10–15 mN/m, to be compared with a value of ~45 mN/m for the other cations and deionized water. The interaction is not stoichiometric with respect to the respective bulk concentrations, in that the minimum interfacial tension occurs at a bulk Ca^{2+}:NA molar concentration ratio of ~1000:1. The interfacial stoichiometry is unclear at present, but probably relates to the bidentate preference of Ca^{2+}, but further interfacial studies will be necessary in order to provide more definitive evidence.

In fact, there may be no single factor responsible for the results presented in this study. Rather, for Ca^{2+}, the combination of its ionic hydration structure and its thermodynamic and kinetic properties as well as its coordination preferences, particularly relating to carboxylate, may contribute to its behavior at the oil/water interface. Conclusions from studies undertaken some 80 years ago by Alexander et al. also highlighted the specific relevance of Ca^{2+} ions (compared with Na^+, K^+ and Mg^{2+}) to the chemistry of biological films, such as cephalin (kephalin) at the benzene/water interface [105].

Author Contributions: S.E.T. conceptualized and supervised the research and conducted some of the experiments. H.T.C. conducted the majority of the experiments and prepared an initial interpretation of the results. S.E.T. wrote the manuscript.

Funding: This research received no external funding.

Acknowledgments: BP America is acknowledged for initially establishing the Centre for Petroleum and Surface Chemistry at the University of Surrey. S.E.T. is grateful to the Chemistry Department at the University of Surrey for making continued research possible.

Conflicts of Interest: The authors declare no conflicts of interest.

References

1. Speight, J.G. *High Acid Crudes*; Gulf Professional Publishing: Waltham, MA, USA, 2014; Chapter 1.
2. Burrell, G.A. Composition of petroleum and its products. *Ind. Eng. Chem.* **1928**, *20*, 602–608. [CrossRef]
3. Rudzinski, W.E.; Oehlers, L.; Zhang, Y.; Najera, B. Tandem mass spectrometric characterization of commercial naphthenic acids and a Maya crude oil. *Energy Fuels* **2002**, *16*, 1178–1185. [CrossRef]
4. Anderson, W.G. Wettability literature survey—Part 1: Rock/oil/brine Interactions and the effects of core handling on wettability. *J. Pet. Technol.* **1986**, *38*, 1125–1144. [CrossRef]
5. Anderson, W.G. Wettability literature survey—Part 2: Wettability measurement. *J. Pet. Technol.* **1986**, *38*, 1246–1262. [CrossRef]
6. Muller, H.; Pauchard, V.O.; Hajji, A.A. Role of naphthenic acids in emulsion tightness for a low total acid number (TAN)-high asphaltenes oil: Characterization of the interfacial chemistry. *Energy Fuels* **2009**, *23*, 1280–1288. [CrossRef]
7. Acevedo, S.; Gutierrez, X.; Rivas, H. Bitumen-in-water emulsions stabilized with natural surfactants. *J. Colloid Interface Sci.* **2001**, *242*, 230–238. [CrossRef]
8. Arla, D.; Flesisnki, L.; Bouriat, P.; Dicharry, C. Influence of alkaline pH on the rheology of water/acidic crude oil interface. *Energy Fuels* **2011**, *25*, 1118–1126. [CrossRef]
9. Havre, T.E.; Sjöblom, J.; Vindstad, J.E. Oil/water-partitioning and interfacial behavior of naphthenic acids. *J. Dispers. Sci. Technol.* **2003**, *24*, 789–801. [CrossRef]
10. Gallup, D.L.; Curiale, J.A.; Smith, P.C. Characterization of sodium emulsion soaps formed from production fluids of Kutei Basin, Indonesia. *Energy Fuels* **2007**, *21*, 1741–1759. [CrossRef]
11. Kelland, M.A. *Production Chemicals for the Oil and Gas Industry*; CRC Press, Taylor and Francis: Boca Raton, FL, USA, 2009; Chapter 7.
12. Baugh, T.D.; Grande, K.V.; Mediaas, H.; Vindstad, J.E.; Wolf, N.O. The discovery of high molecular weight naphthenic acids (ARN acid) responsible for calcium naphthenate deposits. In Proceedings of the SPE International Symposium on Oilfield Scale, Aberdeen, UK, 11–12 May 2005.
13. Juyal, P.; Mapolelo, M.M.; Yen, A.; Rodgers, R.P.; Allenson, S.J. Identification of calcium naphthenate deposition in South American oil fields. *Energy Fuels* **2015**, *29*, 2342–2350. [CrossRef]
14. Brandal, Ø.; Hanneseth, A.-M.D.; Hemmingsen, P.V.; Sjöblom, J.; Kim, S.; Rodgers, R.P.; Marshall, A.G. Isolation and characterization of naphthenic acids from a metal naphthenate deposit: Molecular properties at oil-water and air-water interfaces. *J. Dispers. Sci. Technol.* **2006**, *27*, 295–305. [CrossRef]
15. Sundman, O.; Simon, S.; Nordgard, E.L.; Sjöblom, J. Study of the aqueous chemical interactions between a synthetic tetra-acid and divalent cations as a model for the formation of metal naphthenate deposits. *Energy Fuels* **2010**, *24*, 6054–6060. [CrossRef]
16. Ge, L.; Vernon, M.; Simon, S.; Maham, Y.; Sjöblom, J.; Xu, Z. Interactions of divalent cations with tetrameric acid aggregates in aqueous solutions. *Colloids Surf. A Physicochem. Eng. Asp.* **2012**, *396*, 238–245. [CrossRef]
17. Andersen, S.I.; Chandra, M.S.; Chen, J.; Zeng, B.Y.; Zou, F.; Mapolelo, M.; Abdallah, W.; Buiting, J.J. Detection and impact of carboxylic acids at the crude oil–water interface. *Energy Fuels* **2016**, *30*, 4475–4485. [CrossRef]
18. Hutin, A.; Argillier, J.-F.; Langevin, D. Influence of pH on oil-water interfacial tension and mass transfer for asphaltenes model oils. Comparison with crude oil behavior. *Oil Gas Sci. Technol.* **2016**, *71*, 58–66. [CrossRef]
19. Arla, D.; Sinquin, A.; Palermo, T.; Hurtevent, C.; Graciaa, A.; Dicharry, C. Influence of pH and water content on the type and stability of acidic crude oil emulsions. *Energy Fuels* **2007**, *21*, 1337–1342. [CrossRef]
20. Alvarado, V.; Wang, X.; Moradi, M. Role of acid components and asphaltenes in Wyoming water-in-crude oil emulsions. *Energy Fuels* **2011**, *25*, 4606–4613. [CrossRef]
21. Wang, X.; Alvarado, V. Effects of aqueous-phase salinity on water-in-crude oil emulsion stability. *J. Dispers. Sci. Technol.* **2012**, *33*, 165–170. [CrossRef]
22. Kumar, N.; Wang, L.; Siretanu, I.; Duits, M.; Mugele, F. Salt dependent stability of stearic acid Langmuir–Blodgett films exposed to aqueous electrolytes. *Langmuir* **2013**, *29*, 5150–5159. [CrossRef] [PubMed]
23. Sheng, J.J. Critical review of low-salinity waterflooding. *J. Pet. Sci. Eng.* **2014**, *120*, 216–224. [CrossRef]
24. Haagh, M.E.J.; Siretanu, I.; Duits, M.H.G.; Mugele, F. Salinity-dependent contact angle alteration in oil/brine/silicate systems: The critical role of divalent cations. *Langmuir* **2017**, *33*, 3349–3357. [CrossRef] [PubMed]

25. Gandomkar, A.; Rahimpour, M.R. The impact of monovalent and divalent ions on wettability alteration in oil/low salinity brine/limestone systems. *J. Mol. Liq.* **2017**, *248*, 1003–1013. [CrossRef]

26. Pooryousefy, E.; Xie, Q.; Chen, Y.; Sari, A.; Saeedi, A. Drivers of low salinity effect in sandstone reservoirs. *J. Mol. Liq.* **2018**, *250*, 396–403. [CrossRef]

27. Ding, H.; Rahman, S. Experimental and theoretical study of wettability alteration during low salinity water flooding-a state of the art review. *Colloids Surf. A Physicochem. Eng. Asp.* **2017**, *520*, 622–639. [CrossRef]

28. Pouryousefy, E.; Xie, Q.; Saeedi, A. Effect of multi-component ions exchange on low salinity EOR: Coupled geochemical simulation study. *Petroleum* **2016**, *2*, 215–234. [CrossRef]

29. Brady, P.V.; Krumhansl, J.L. A surface complexation model of oil–brine–sandstone interfaces at 100 °C: Low salinity waterflooding. *J. Pet. Sci. Eng.* **2012**, *81*, 171–176. [CrossRef]

30. Derkani, M.H.; Fletcher, A.J.; Abdallah, W.; Sauerer, B.; Anderson, J.; Zhang, Z.J. Low salinity waterflooding in carbonate reservoirs: Review of interfacial mechanisms. *Colloids Interfaces* **2018**, *2*, 20. [CrossRef]

31. De Ruiter, R.; Tjerkstra, R.W.; Duits, M.H.G.; Mugele, F. Influence of cationic composition and pH on the formation of metal stearates at oil/water interfaces. *Langmuir* **2011**, *27*, 8738–8747. [CrossRef] [PubMed]

32. Brandal, Ø.; Hanneseth, A.-M.D.; Sjöblom, J. Interactions between synthetic and indigenous naphthenic acids and divalent cations across oil-water interfaces: Effects of addition of oil-soluble non-ionic surfactants. *Colloid Polym. Sci.* **2005**, *284*, 124–133. [CrossRef]

33. Wojciechowski, K.; Buffle, J.; Miller, R. The synergistic adsorption of fatty acid and azacrown ether at the toluene–water interface. *Colloids Surf. A Physicochem. Eng. Asp.* **2005**, *261*, 49–55. [CrossRef]

34. Enever, R.V.; Pilpel, N. Reaction between stearic acid and calcium ions at the air/water interface using surface viscometry. *Trans. Faraday Soc.* **1967**, *63*, 781–792. [CrossRef]

35. Enever, R.V.; Pilpel, N. Reaction between stearic acid and calcium ions at the air/water interface using surface viscometry. Part 2. Mixed films of octadecanol and stearic acid. *Trans. Faraday Soc.* **1967**, *63*, 1559–1566. [CrossRef]

36. Pilpel, N.; Enever, R.V. Reaction between stearic acid and calcium ions at the air/water interface using surface viscometry. Part 3. Mechanism. *Trans. Faraday Soc.* **1968**, *64*, 231–237. [CrossRef]

37. Hindle, R.; Noestheden, M.; Peru, K.; Headley, J. Quantitative analysis of naphthenic acids in water by liquid chromatography–accurate mass time-of-flight mass spectrometry. *J. Chromatogr. A* **2013**, *1286*, 166–174. [CrossRef] [PubMed]

38. Pereira, R.F.P.; Valente, A.J.M.; Fernandes, M.; Burrows, H.D. What drives the precipitation of long-chain calcium carboxylates (soaps) in aqueous solution. *Phys. Chem. Chem. Phys.* **2012**, *14*, 7517–7527. [CrossRef] [PubMed]

39. Chakravarthy, R.; Naik, G.N.; Savalia, A.; Sridharan, U.; Saravanan, C.; Das, A.K.; Gudasi, K.B. Determination of naphthenic acid number in petroleum crude oils and their fractions by mid-Fourier transform infrared spectroscopy. *Energy Fuels* **2016**, *30*, 8579–8586. [CrossRef]

40. Moreira, A.P.D.; Souza, B.S.; Teixeira, A.M.R.F. Monitoring of calcium stearate formation by thermogravimetry. *J. Therm. Anal. Calorim.* **2009**, *97*, 647–652. [CrossRef]

41. Van Hunsel, J.; Joos, P. Study of the dynamic interfacial tension at the oil/water interface. *Colloid Polym. Sci.* **1989**, *267*, 1026–1035. [CrossRef]

42. Deshiikan, S.R.; Bush, D.; Eschenazi, E.; Papadopoulos, K.D. SDS, Brij 58 and CTAB at the dodecane-water interface. *Colloids Surf. A Physicochem. Eng. Asp.* **1998**, *136*, 133–150. [CrossRef]

43. Joos, P.; Rillaerts, E. Theory on the determination of the dynamic surface tension with the drop volume and maximum bubble pressure methods. *J. Colloid Interface Sci.* **1981**, *79*, 96–100. [CrossRef]

44. Harkins, W.D.; Brown, F.E. The determination of surface tension (free surface energy), and the weight of falling drops: The surface tension of water and benzene by the capillary height method. *J. Am. Chem. Soc.* **1919**, *41*, 499–524. [CrossRef]

45. Wilkinson, M. Extended use of, and comments on, the drop-weight (drop-volume) technique for the determination of surface and interfacial tensions. *J. Colloid Interface Sci.* **1972**, *40*, 14–26. [CrossRef]

46. Lee, B.-B.; Ravindra, P.; Chan, E.-S. New drop weight analysis for surface tension determination of liquids. *Colloids Surf. A Physicochem. Eng. Asp.* **2009**, *332*, 112–120. [CrossRef]

47. Robertson, E.J.; Beaman, D.K.; Richmond, G.L. Designated drivers: The differing roles of divalent metal ions in surfactant adsorption at the oil–water interface. *Langmuir* **2013**, *29*, 15511–15520. [CrossRef] [PubMed]

48. Berry, J.D.; Neeson, M.J.; Dagastine, R.R.; Chan, D.Y.C.; Tabor, R.F. Measurement of surface and interfacial tension using pendant drop tensiometry. *J. Colloid Interface Sci.* **2015**, *454*, 226–237. [CrossRef] [PubMed]

49. Dill, K.A.; Bromberg, S. *Molecular Driving Forces: Statistical Mechanics in Chemistry and Biology*; Garland Science, Taylor and Francis: New York, NY, USA, 2003; Chapters 21 and 22.

50. Andersson, M.P.; Olsson, M.H.M.; Stipp, S.L.S. Predicting the pK_a and stability of organic acids and bases at an oil–water interface. *Langmuir* **2014**, *30*, 6437–6445. [CrossRef] [PubMed]

51. Gershevitz, O.; Sukenik, C.N. In situ FTIR-ATR analysis and titration of carboxylic acid-terminated SAMs. *J. Am. Chem. Soc.* **2004**, *126*, 482–483. [CrossRef] [PubMed]

52. Kanicky, J.R.; Shah, D.O. Effect of degree, type, and position of unsaturation on the pK_a of long-chain fatty acids. *J. Colloid Interface Sci.* **2002**, *256*, 201–207. [CrossRef] [PubMed]

53. Kanicky, J.R.; Shah, D.O. Effect of premicellar aggregation on the pK_a of fatty acid soap solutions. *Langmuir* **2003**, *19*, 2034–2038. [CrossRef]

54. Wellen, B.A.; Lach, E.A.; Allen, H.C. Surface pK_a of octanoic, nonanoic, and decanoic fatty acids at the air–water interface: Applications to atmospheric aerosol chemistry. *Phys. Chem. Chem. Phys.* **2017**, *19*, 26551–26558. [CrossRef] [PubMed]

55. Brosseau, Q.; Vrignon, J.; Baret, J.-C. Microfluidic dynamic interfacial tensiometry (µDIT). *Soft Matter* **2014**, *10*, 3066–3076. [CrossRef] [PubMed]

56. Von Szyszkowski, B. Experimental studies on the capillary characteristics of watery solutions of fatty acids. *Z. Phys. Chem.* **1908**, *64*, 385–414.

57. Rosen, M.J. *Surfactants and Interfacial Phenomena*, 3rd ed.; John Wiley & Sons: New York, NY, USA, 2004; p. 45.

58. Braunschweig, B.; Schulze-Zachau, F.; Nagel, E.; Engelhardt, K.; Stoyanov, S.; Gochev, G.; Khristov, K.; Mileva, E.; Exerowa, D.; Miller, R.; et al. Specific effects of Ca^{2+} ions and molecular structure of β-lactoglobulin interfacial layers that drive macroscopic foam stability. *Soft Matter* **2016**, *12*, 5995–6004. [CrossRef] [PubMed]

59. Jones, G.; Ray, W.A. The surface tension of solutions of electrolytes as a function of the concentration. I. A differential method for measuring relative surface tension. *J. Am. Chem. Soc.* **1937**, *59*, 187–198. [CrossRef]

60. Kakati, A.; Sangwai, J.S. Effect of monovalent and divalent salts on the interfacial tension of pure hydrocarbon-brine systems relevant for low salinity water flooding. *J. Pet. Sci. Eng.* **2017**, *157*, 1106–1114. [CrossRef]

61. Moeini, F.; Hemmati-Sarapardeh, A.; Ghazanfari, M.-H.; Masihi, M.; Ayatollahi, S. Toward mechanistic understanding of heavy crude oil/brine interfacial tension: The roles of salinity, temperature and pressure. *Fluid Phase Equilib.* **2014**, *375*, 191–200. [CrossRef]

62. Petersen, P.B.; Johnson, J.C.; Knutsen, K.P.; Saykally, R.J. Direct experimental validation of the Jones–Ray effect. *Chem. Phys. Lett.* **2004**, *397*, 46–50. [CrossRef]

63. Saien, J.; Mishi, M. Equilibrium interfacial tension and the influence of extreme dilutions of uni-univalent salts: An expression of the "Jones–Ray effect". *J. Chem. Thermodyn.* **2012**, *54*, 254–260. [CrossRef]

64. Nguyen, K.T.; Nguyen, A.V.; Evans, G.M. Interactions between halide anions and interfacial water molecules in relation to the Jones–Ray effect. *Phys. Chem. Chem. Phys.* **2014**, *16*, 24661–24665. [CrossRef] [PubMed]

65. Uematsu, Y.; Bonthuis, D.J.; Netz, R.R. Charged surface-active impurities at nanomolar concentration induce Jones–Ray effect. *J. Phys. Chem. Lett.* **2018**, *9*, 189–193. [CrossRef] [PubMed]

66. Langmuir, I. Repulsive forces between charged surfaces in water, and the cause of the Jones-Ray effect. *Science* **1938**, *88*, 430–432. [CrossRef] [PubMed]

67. Okur, H.I.; Chen, Y.; Wilkins, D.M.; Roke, S. The Jones-Ray effect reinterpreted: Surface tension minima of low ionic strength electrolyte solutions are caused by electric field induced water-water correlations. *Chem. Phys. Lett.* **2017**, *684*, 433–442. [CrossRef]

68. Heydweillwer, A. On the physical properties of solutions and their interactions. II. Surface tension and electrical conductivity of aqueous salts. *Ann. Phys.* **1910**, *33*, 145–185.

69. Aveyard, R.; Saleem, S.M. Interfacial tensions at alkane-aqueous electrolyte interfaces. *JCS Faraday Trans. 1* **1976**, *72*, 1609–1617. [CrossRef]

70. Wen, B.; Sun, C.; Bai, B.; Gatapova, E.Y.; Kabov, O.A. Ionic hydration-induced evolution of decane–water interfacial tension. *Phys. Chem. Chem. Phys.* **2017**, *19*, 14606–14614. [CrossRef] [PubMed]

71. Bastos-González, D.; Pérez-Fuentes, L.; Drummond, C.; Faraudo, J. Ions at interfaces: The central role of hydration and hydrophobicity. *Curr. Opin. Colloid Interface Sci.* **2016**, *23*, 19–28. [CrossRef]

72. Persson, I. Hydrated metal ions in aqueous solution: How regular are their structures? *Pure Appl. Chem.* **2010**, *82*, 1901–1917. [CrossRef]
73. Dos Santos, A.P.; Levin, Y. Ions at the water–oil Interface: Interfacial tension of electrolyte solutions. *Langmuir* **2012**, *28*, 1304–1308. [CrossRef] [PubMed]
74. Dos Santos, A.P.; Levin, Y. Surface and interfacial tensions of Hofmeister electrolytes. *Faraday Discuss.* **2013**, *160*, 75–87. [CrossRef]
75. Ikeda, T.; Boero, M.; Terakura, K. Hydration properties of magnesium and calcium ions from constrained first principles molecular dynamics. *J. Chem. Phys.* **2007**, *127*, 074503. [CrossRef] [PubMed]
76. Pavlov, M.; Siegbahn, P.E.M.; Sandström, M. Hydration of beryllium, magnesium, calcium, and zinc ions using density functional theory. *J. Phys. Chem A* **1998**, *102*, 219–228. [CrossRef]
77. Hofmeister, F. Zur Lehre von der Wirkung der Salze. *Arch. Exp. Pathol. Pharmakol.* **1888**, *24*, 247–260. [CrossRef]
78. Kunz, W.; Henle, J.; Ninham, B.W. 'Zur Lehre von der Wirkung der Salze' (about the science of the effect of salts): Franz Hofmeister's historical papers. *Curr. Opin. Colloid Interface Sci.* **2004**, *9*, 19–37. [CrossRef]
79. Collins, K.D.; Washabaugh, M.W. The Hofmeister effect and the behaviour of water at interfaces. *Q. Rev. Biophys.* **1985**, *18*, 323–422. [CrossRef] [PubMed]
80. Lee, Y.; Thirumalai, D.; Hyeon, C. Ultrasensitivity of water exchange kinetics to the size of metal ion. *J. Am. Chem. Soc.* **2017**, *139*, 12334–12337. [CrossRef] [PubMed]
81. Kherb, J.; Flores, S.C.; Cremer, P.S. Role of carboxylate side chains in the cation Hofmeister series. *J. Phys. Chem. B* **2012**, *116*, 7389–7397. [CrossRef] [PubMed]
82. Einspahr, H.; Bugg, C.E. The geometry of calcium-carboxylate interactions in crystalline complexes. *Acta Cryst.* **1981**, *B37*, 1044–1052. [CrossRef]
83. Carrell, C.J.; Carrell, H.L.; Erlebacher, J.; Glusker, J.P. Structural aspects of metal ion-carboxylate interactions. *J. Am. Chem. Soc.* **1988**, *110*, 8651–8656. [CrossRef]
84. Beaman, D.K.; Robertson, E.J.; Richmond, G.L. Metal ions: Driving the orderly assembly of polyelectrolytes at a hydrophobic surface. *Langmuir* **2012**, *28*, 14245–14253. [CrossRef] [PubMed]
85. Papageorgiou, S.K.; Kouvelos, E.P.; Favvas, E.P.; Sapalidis, A.A.; Romanos, G.E.; Katsaros, F.K. Metal–carboxylate interactions in metal–alginate complexes studied with FTIR spectroscopy. *Carbohydr. Res.* **2010**, *345*, 469–473. [CrossRef] [PubMed]
86. Lu, Y.; Miller, J.D. Carboxyl stretching vibrations of spontaneously adsorbed and LB-transferred calcium carboxylates as determined by FTIR internal reflection spectroscopy. *J. Colloid Interface Sci.* **2002**, *256*, 41–52. [CrossRef]
87. Dhavskar, K.T.; Bhargao, P.H.; Srinivasan, B.R. Synthesis, crystal structure and properties of magnesium and calcium salts of *p*-anisic acid. *J. Chem. Sci.* **2016**, *128*, 421–428. [CrossRef]
88. Tang, C.Y.; Huang, Z.; Allen, H.C. Binding of Mg^{2+} and Ca^{2+} to palmitic acid and deprotonation of the COOH headgroup studied by vibrational sum frequency generation spectroscopy. *J. Phys. Chem. B* **2010**, *114*, 17068–17076. [CrossRef] [PubMed]
89. Tang, C.Y.; Huang, Z.; Allen, H.C. Interfacial water structure and effects of Mg^{2+} and Ca^{2+} binding to the COOH headgroup of a palmitic acid monolayer studied by sum frequency spectroscopy. *J. Phys. Chem. B* **2011**, *115*, 34–40. [CrossRef] [PubMed]
90. Vogel, H.J.; Brokx, R.D.; Ouyang, H. Calcium-binding proteins. In *Methods in Molecular Biology, Volume 172: Calcium-Binding Protein Protocols: Reviews and Case Studies*; Vogel, H.J., Ed.; Humana Press Inc.: Totowa, NJ, USA, 2002; Volume 1, Chapter 1.
91. Katz, A.K.; Glusker, J.P.; Beebe, S.A.; Bock, C.W. Calcium ion coordination: A comparison with that of beryllium, magnesium, and zinc. *J. Am. Chem. Soc.* **1996**, *118*, 5752–5763. [CrossRef]
92. Dudev, T.; Lim, C. Monodentate versus bidentate carboxylate binding in magnesium and calcium proteins: What are the basic principles? *J. Phys. Chem. B* **2004**, *108*, 4546–4557. [CrossRef]
93. Nara, M.; Morii, H.; Tanokura, M. Coordination to divalent cations by calcium-binding proteins studied by FTIR spectroscopy. *Biochim. Biophys. Acta* **2013**, *1828*, 2319–2327. [CrossRef] [PubMed]
94. Dudev, T.; Cowan, J.A.; Lim, C. Competitive binding in magnesium coordination chemistry: Water versus ligands of biological interest. *J. Am. Chem. Soc.* **1999**, *121*, 7665–7673. [CrossRef]
95. Williams, R.J.P. Calcium. In *Methods in Molecular Biology, Volume 172: Calcium-Binding Protein Protocols: Reviews and Case Studies*; Vogel, H.J., Ed.; Humana Press Inc.: Totowa, NJ, USA, 2002; Volume 1, Chapter 2.

96. Collins, K.D. Ion hydration: Implications for cellular function, polyelectrolytes, and protein crystallization. *Biophys. Chem.* **2006**, *119*, 271–281. [CrossRef] [PubMed]
97. Kiriukhin, M.Y.; Collins, K.D. Dynamic hydration numbers for biologically important ions. *Biophys. Chem.* **2002**, *99*, 155–168. [CrossRef]
98. Shehata, A.M.; Nasr-El-Din, H.A. Laboratory investigations to determine effect of connate-water composition on Low-salinity waterflooding in sandstone reservoirs. *SPE Reserv. Eval. Eng.* **2017**, *20*, 59–76. [CrossRef]
99. Lashkarbolooki, M.; Ayatollahi, S.; Riazi, M. The impacts of aqueous ions on interfacial tension and wettability of an asphaltenic–acidic crude oil reservoir during Smart Water injection. *J. Chem. Eng. Data* **2014**, *59*, 3624–3634. [CrossRef]
100. Suleimanov, B.A.; Latifov, Y.A.; Veliyev, E.F.; Frampton, H. Comparative analysis of the EOR mechanisms by using low salinity and low hardness alkaline water. *J. Pet. Sci. Eng.* **2018**, *162*, 35–43. [CrossRef]
101. Gonzalez, V.; Taylor, S.E. Physical and chemical aspects of "precursor films" spreading on water from natural bitumen. *J. Pet. Sci. Eng.* **2018**, *170*, 291–303. [CrossRef]
102. Basu, S.; Sharma, M.M. Measurement of critical disjoining pressure for dewetting of solid surfaces. *J. Colloid Interface Sci.* **1996**, *181*, 443–455. [CrossRef]
103. Collins, A.G. *Geochemistry of Oilfield Waters*; Developments in Petroleum Science; Elsevier: New York, NY, USA, 1976; Volume 1, Chapter 7.
104. Sauerer, B.; Stukan, M.; Buiting, J.; Abdallah, W.; Andersen, S. Dynamic asphaltene-stearic acid competition at the oil–water interface. *Langmuir* **2018**, *34*, 5558–5573. [CrossRef] [PubMed]
105. Alexander, A.E.; Teorell, T.; Åborg, G.C. A study of films at the liquid/liquid interface. Part III. A specific effect of calcium ions on kephalin monolayers. *Trans. Faraday Soc.* **1939**, *35*, 1200–1205. [CrossRef]

colloids
and interfaces

MDPI

Addendum

Addendum: Taylor, S.E., et al. Metal Ion Interactions with Crude Oil Components: Specificity of Ca^{2+} Binding to Naphthenic Acid at an Oil/Water Interface. *Colloids Interfaces* 2018, 2, 40.

Spencer E. Taylor [1,*], Hiu Tung Chu [2] and Ugochukwu I. Isiocha [3]

[1] Centre for Petroleum and Surface Chemistry, Department of Chemistry, University of Surrey, Guildford, Surrey GU2 7XH, UK
[2] TÜV Rheinland, 10-16 Pun Shan Street, Tsuen Wan, N.T., Hong Kong; christy.chu@tuv.com
[3] Department of Chemistry, University of Surrey, Guildford, Surrey GU2 7XH, UK; ui00015@surrey.ac.uk
* Correspondence: s.taylor@surrey.ac.uk; Tel.: +44-1483-681999

Received: 25 October 2018; Accepted: 2 November 2018; Published: 6 November 2018

In the article recently published in *Colloids and Interfaces* [1], it was erroneously stated that a hydrated form of calcium chloride, CaCl$_2$·2H$_2$O, was the source of Ca^{2+} ions used to compare with other alkali and alkaline earth ions on the interfacial behavior the metal ions at aqueous/*n*-heptane interfaces containing commercial naphthenic acid exhibit. We have since determined that the real source of Ca^{2+} ions was in fact anhydrous CaCl$_2$ (supplied by Fisher Scientific Ltd., Lot 1496876), which was inadvertently prescribed by the main author. As a consequence, the authors wish to make the readers aware that the interpretations given are not necessarily correct in the original account [1].

In particular, the use of anhydrous CaCl$_2$ causes an increase in pH of the aqueous solutions. In the original paper, an average pH value of the chloride salt solutions of ~6 was reported. This was true for all the salts used, but did not include solutions derived using anhydrous CaCl$_2$. It is correct, however, for the hydrated salt CaCl$_2$·2H$_2$O.

On the other hand, as shown in Figure 1, the pH of aqueous solutions derived from anhydrous CaCl$_2$ increases steadily throughout the concentration range covered in reference [1], that is, 0.01–1 mol/L. As far as we have been able to ascertain, however, there is little quantitative information relating to the solution properties of anhydrous CaCl$_2$ in the literature, except for indications of its alkaline character given in manufacturers' specification sheets (e.g., Sigma–Aldrich state "<1% free alkali as Ca(OH)$_2$" for an equivalent product [2]). This characteristic is most likely a consequence of CaCl$_2$ being obtained commercially as a byproduct of the Solvay process for sodium carbonate manufacture. The purer hydrated forms obtained by crystallization are devoid of such impurities.

The significance of the pH change accompanying the increase in Ca^{2+} concentration is shown in Figure 2, in which the minimum interfacial tension (from Figure 7a in reference [1]) is seen to decrease linearly with the OH$^-$ concentration (determined from the pH, i.e., [OH$^-$] = 10$^{-(14 - pH)}$), indicating that OH$^-$, and not Ca^{2+} ions are responsible for the observed interfacial tension behavior.

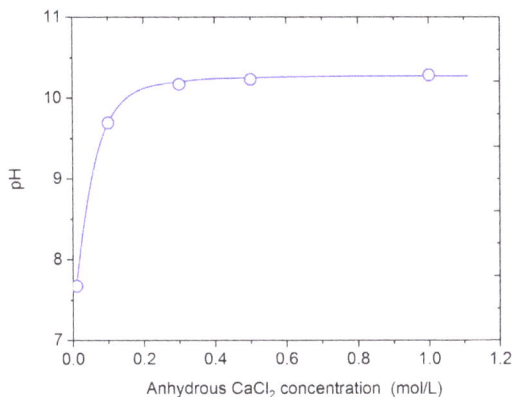

Figure 1. Effect of anhydrous $CaCl_2$ concentration on solution pH.

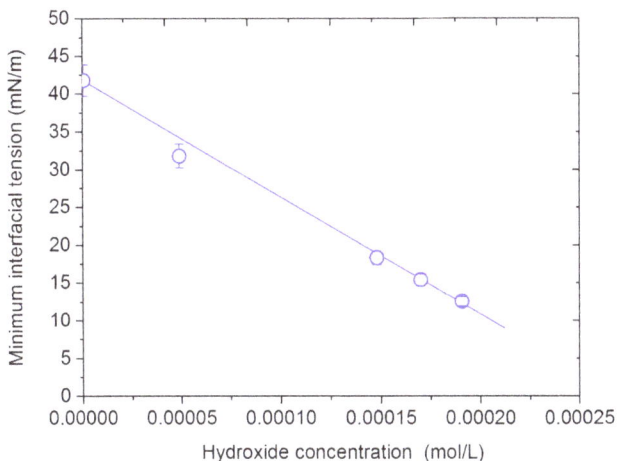

Figure 2. Relationship between the minimum interfacial tension and hydroxide ion concentration.

Notwithstanding the inaccurate reporting of the type of $CaCl_2$ used in the original paper [1], the interfacial tension data were correctly and precisely determined and we believe will make a useful contribution to the literature. However, we appreciate that we can no longer claim that the interpretation and data-fitting regarding the role of Ca^{2+} ions that we presented are unequivocally correct. The original authors would therefore like to apologize for any inconvenience and confusion caused by this disclosure.

References

1. Taylor, S.E.; Chu, H.T. Metal Ion Interactions with Crude Oil Components: Specificity of Ca^{2+} Binding to Naphthenic Acid at an Oil/Water Interface. *Colloids Interfaces* **2018**, *2*, 40. [CrossRef]
2. Available online: http://www.merckmillipore.com/GB/en/product/Calcium-chloride,MDA_CHEM-102083?bd=1#anchor_Specifications (accessed on 17 October 2018).

colloids
and interfaces

MDPI

Review

Interfacial and Colloidal Forces Governing Oil Droplet Displacement: Implications for Enhanced Oil Recovery

Suparit Tangparitkul [1,2], Thibaut V. J. Charpentier [1], Diego Pradilla [3] and David Harbottle [1,*]

[1] School of Chemical and Process Engineering, University of Leeds, Leeds LS2 9JT, UK; suparit.t@cmu.ac.th (S.T.); t.charpentier@leeds.ac.uk (T.V.J.C.)
[2] Department of Mining and Petroleum Engineering, Chiang Mai University, Chiang Mai 50200, Thailand
[3] Grupo de Diseño de Producto y de Proceso (GDPP), Departamento de Ingeniería Química, Universidad de los Andes, Carrera 1 este No. 18A-12, Edificio Mario Laserna, Piso 7, Bogotá 111711, Colombia; d-pradil@uniandes.edu.co
* Correspondence: d.harbottle@leeds.ac.uk; Tel.: +44-113-343-4154

Received: 29 May 2018; Accepted: 16 July 2018; Published: 18 July 2018

Abstract: Growing oil demand and the gradual depletion of conventional oil reserves by primary extraction has highlighted the need for enhanced oil recovery techniques to increase the potential of existing reservoirs and facilitate the recovery of more complex unconventional oils. This paper describes the interfacial and colloidal forces governing oil film displacement from solid surfaces. Direct contact of oil with the reservoir rock transforms the solid surface from a water-wet to neutrally-wet and oil-wet as a result of the deposition of polar components of the crude oil, with lower oil recovery from oil-wet reservoirs. To enhance oil recovery, chemicals can be added to the injection water to modify the oil-water interfacial tension and solid-oil-water three-phase contact angle. In the presence of certain surfactants and nanoparticles, a ruptured oil film will dewet to a new equilibrium contact angle, reducing the work of adhesion to detach an oil droplet from the solid surface. Dynamics of contact-line displacement are considered and the effect of surface active agents on enhancing oil displacement discussed. The paper is intended to provide an overview of the interfacial and colloidal forces controlling the process of oil film displacement and droplet detachment for enhanced oil recovery. A comprehensive summary of chemicals tested is provided.

Keywords: enhanced oil recovery; oil film displacement; colloid and interfacial science; wettability; surfactants; nanoparticle fluids

1. Introduction

The global energy landscape is gradually transitioning towards renewables, however, a reliance on non-renewables, particularly petroleum, will remain for several decades due to its importance as a fuel and chemical feedstock, which is a critical component to the steady improvement in the quality of life of developing countries. While developed countries take the lead on demonstrating the application of non-renewables, their remaining reliance on petroleum as part of the energy matrix remains for the foreseeable future (beyond 2050) [1]. With overall petroleum demand expected to increase [2], demand can only be met by increasing global production, which also coincides with the depletion of 'easy-to-produce' oil.

With few giant oil fields being discovered and new reserves frequently identified in remote/challenging locations, there is a growing need to increase the potential of existing reserves and improve the worldwide average oil recovery factor from as low as 20% to 40%. Methods of enhanced oil recovery (EOR) have the potential to double the produced lifetime of existing proven reserves;

which have a current lifetime without EOR of ~50 years [1]. The growing demand for petroleum is also being met in part by the increased reliance on production from proven unconventional oil reserves, for example, the Canadian oil sands, which has an estimated 300 billion barrels of ultimate potential recoverable reserve (heavy oil), the third largest reserve behind Venezuela and Saudi Arabia [2].

Successful EOR and oil sands operations rely on controlling the process fluid chemistry to favorably affect the mechanisms that govern oil droplet de-wetting and liberation. In this paper, we will provide an overview of the scientific principles influencing oil droplet dynamics on solid surfaces, and extend the discussion to demonstrate how those governing mechanisms can be influenced by the commonly studied surface/interfacially active components namely surfactants and nanoparticles. The step-by-step process by which oil detaches from the solid surface can be summarized in sequence: (i) oil film thinning and rupture; (ii) oil de-wetting (recession) on the solid surface; and (iii) oil-solid surface adhesion and liberation.

Before describing the underlying principles that govern each step, it is worth considering the likely interaction between the oil and solid surface; i.e., the reservoir wettability. The reservoir environment can be characterized as either: (i) water-wet (water droplet contact angle, $\theta = 0°$ to ~70°); (ii) oil-wet ($\theta = $ ~110° to ~180°); and (iii) neutrally-wet ($\theta = $ ~70° to ~110°) exhibiting a similar affinity to both water and oil [3–5]. While it is understood that most reservoir environments were initially water-wet, the reservoir rock can evolve to become more oil-wet due to the deposition/adsorption of several indigenous organic polar species (asphaltenes, resins and naphthenic acids) present in crude oil [6–9]. For oil-wet reservoirs, oil recovery is poor due to no capillary imbibition. Hence, one of the criteria for successful EOR is to enhance capillary imbibition and reverse the wettability change by using chemical additives, although complete reversal to strongly water-wet surfaces is not favored for EOR [10]. An oil layer on a hydrophilic or hydrophobic solid surface is the basis for the following discussions.

2. Background Science

Oil recovery from the reservoir rock occurs by either displacement from squeezing or oil film thinning and rupture to form discrete oil droplets (Figure 1) that are removed by shear; the latter is of interest here.

I. Oil film on substrate   II. Oil film thinning   III. Oil droplet formation   IV. Droplet recession

Figure 1. Schematic showing the four stages of oil film dewetting from a uniform thick film (**I**) to film thinning (**II**); formation of discrete oil patches (**III**); and recession of oil patches to form oil droplets at the new equilibrium wetting condition (**IV**).

The long-time transformation from water-wet to an oil-wet reservoir occurs following the collapse of a thin aqueous layer separating the solid surface and oil layer. The stability of the thin water layer is attributed to the disjoining pressure that accounts for surface forces between the solid-water and water-oil interfaces. The total disjoining pressure (Π) includes contributions from electrostatic (Π_{el}), van der Waals (Π_{vdW}), and structural (Π_{st}) forces [11]

$$\Pi = \Pi_{el} + \Pi_{vdW} + \Pi_{st} \tag{1}$$

with the thin water layer collapsing when Π is negative. The disjoining pressure as a function of aqueous layer thickness has been calculated for a silica/water/oil (bitumen) system of salinity 1 mM KCl and pHs 3, 5, and 9, see Figure 2 (only Π_{el} and Π_{vdW} have been considered). While Π_{vdW} depends on the interaction Hamaker constant, Π_{el} is sensitive to pH and salinity, with the magnitude of the electrostatic force dependent on the surface (zeta) potentials of silica and bitumen, and the Debye length. For crude oil, the pH dependent surface potentials result from ionization and surface activity of natural surfactants (naphthenic acids) [12–14]. At higher pHs, dissociation of the carboxylic-type surfactant increases the surface potential (negative) of the oil-water interface, with the magnitude increasing as more surfactant partitions at the interface. The high surface potentials at pH 9 form very stable thin-water layers, whereas in more acidic conditions, the disjoining pressure maxima decrease, and the thin-water layer in pH 3 is entirely unstable.

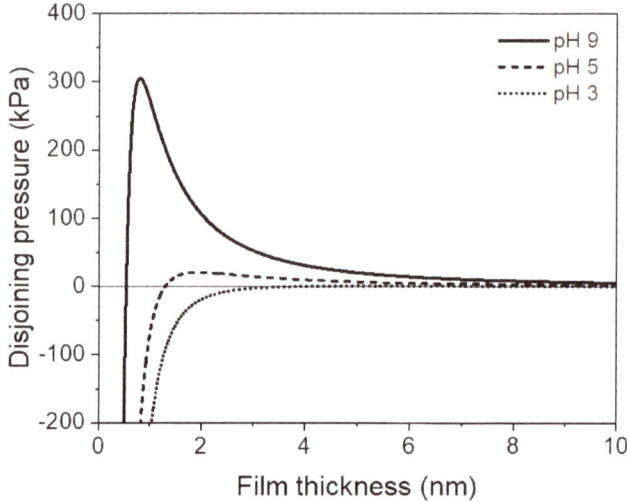

Figure 2. Disjoining pressure ($\Pi = \Pi_{el} + \Pi_{vdW}$) as a function of thin-water layer thickness (h) and pH. Zeta potentials (ψ) at pH 3, 5 and 9 are: 2.5, -55.6 and -78.2 mV for oil (bitumen) [15], and -12, -30 and -38 mV for silica, respectively. The Hamaker constant (A_{SWB}) for silica/water/oil system is 5.7×10^{-21} J [16]. $\Pi_{vdW} = \frac{A_{SWB}}{6\pi h^3}$, $\Pi_{el} = \frac{1}{2}\varepsilon\varepsilon_0\kappa^2 \frac{2\psi_1\psi_2\cos h(\kappa h) - \psi_1^2 - \psi_2^2}{\sin h^2(\kappa h)}$, where ε and ε_0 are the dielectric permittivity of vacuum and relative dielectric permittivity of water, respectively. κ is the Debye length, which accounts for changes in salinity.

The stability of thick oil films is governed by the balance of gravity and capillary forces, with instability and the formation of discrete oil patches having been described analytically by Sharma [17], with the critical film thickness h_{cr} (Equation (2)) dependent on the oil-water interfacial tension ($\gamma_{o/w}$), the three-phase contact angle (θ), and the minimum radius of the hole (r_m)

$$h_{cr} = r_m \ln\left(\frac{2\sin(\pi - \theta)}{r_m[1 + \cos(\pi - \theta)]}\sqrt{\frac{\gamma_{o/w}}{g\rho}}\right) \tag{2}$$

where g is the acceleration due to gravity and ρ the density of oil. For a typical oil-wetted solid surface of three-phase contact angle of 145° and $\gamma_{o/w} = 30$ mN/m, h_{cr} is 0.05 mm for a stable minimum hole radius of 10 μm. Dependence on the fluid and surface properties is rather weak over the range of general applicability, with h_{cr} strongly influenced by the size of the stable hole in the oil film [16]. For very thick films ($h \gg h_{cr}$), thinning of the oil film is needed for dewetting, otherwise holes formed in the film will spontaneously collapse. The mechanisms for film thinning have not been extensively considered but are most likely to result from fluid shear in confined environments. Other factors that can influence the onset of film rupture include gas bubbles trapped in the oil film [18], and surface asperities that lead to non-uniform film thickness.

With the oil film ruptured, the circular hole begins to expand at a rate dependent on the fluid and interfacial properties (to be discussed below) [19]. Away from equilibrium, the process of droplet dewetting is driven by a change in energy following the creation of a new solid-water interface and the loss of oil-solid interface, assuming the change in oil-water interface during droplet recession can be considered negligible, that is

$$\frac{dG}{dA} = \gamma_{S/W} - \gamma_{O/S} \tag{3}$$

where γ is the interfacial tension and subscripts S, W, and O describe the solid, water and oil phases, respectively. Equation (3) can be simplified by the Young's equation for an oil droplet on a solid surface given by

$$\cos\theta = \frac{\gamma_{O/S} - \gamma_{S/W}}{\gamma_{O/W}} \tag{4}$$

to express the energy change during oil recession in terms of the equilibrium contact angle and oil-water interfacial tension (two measurable properties)

$$\frac{dG}{dA} = -\gamma_{O/W}\cos\theta. \tag{5}$$

With $\gamma_{O/W}$ always greater than zero, Equation (5) confirms that oil recession is a spontaneous process when $\theta < 90°$; i.e., the wetted solid surface is more water-wet (hydrophilic). The simple form of Equation (5) provides fundamental insight for effective EOR, highlighting the value of modifying surface wettability and oil-water interfacial tension. The smaller the θ, the more favorable the condition for oil recession. Once the oil droplet has reached equilibrium, the work of adhesion (W_A) between oil and solid surface must be exceeded to liberate the oil droplet. By the reduction in area of oil-solid interface and generation of oil-water and solid-water interfaces, W_A is given by

$$W_A = \gamma_{S/W} + \gamma_{O/W} - \gamma_{O/S} \tag{6}$$

which when combined with the Young's equation leads to

$$W_A = \gamma_{O/W}(1 - \cos\theta) \geq 0. \tag{7}$$

With the unlikely condition of $\theta = 0$ for spontaneous liberation (droplet detachment from the solid surface), Equation (7) confirms the need for energy to detach oil droplets from the wetted surface. In order to detach an oil droplet from the solid surface the hydrodynamic lift force must exceed the contributions from the body and adhesion forces. An approximation of the adhesion force for a partially wetting droplet is, $F_A = \pi r \gamma_{O/W} \sin(\pi - \theta)$, where r is the radius of oil-solid surface contact area [20]. The contour map in Figure 3 indicates the strongest adhesion (red color) when the oil-water interfacial tension and water droplet contact angle are high. Therefore, reducing both the oil-water interfacial tension and oil-water-solid three-phase contact angle leads to more favorable oil droplet liberation.

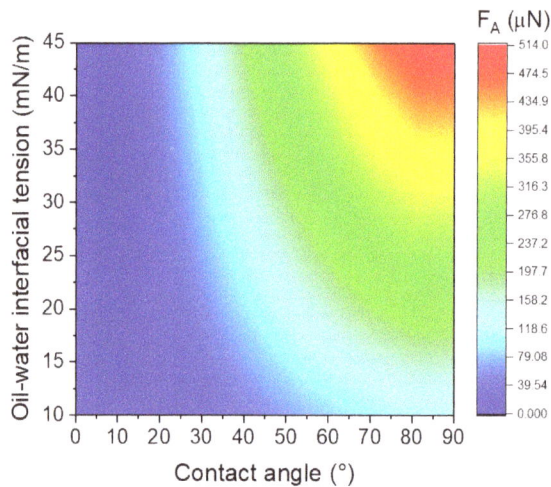

Figure 3. Apparent adhesion force for a partially wetting droplet (droplet volume = 10 μL).

3. Dynamics of oil Film Recession

After creation of a hole on the oil film, the oil film recedes rapidly, governed by the receding force, $F_R = \gamma_{O/W}[\cos(\pi - \theta_d) - \cos(\pi - \theta)]$, acting at the three-phase contact line (θ_d is the dynamic contact angle), with F_R diminishing towards the new equilibrium wetted-state, hence, the velocity of the three-phase contact line decreases with time. The dynamics of oil displacement on a solid surface are frequently described using the (i) hydrodynamic (HD); (ii) molecular-kinetic (MK); or (iii) combined models.

For more viscous fluids, such as crude oil, the hydrodynamic model relies on the solution of creeping flow in the vicinity of the three-phase contact line, with the no slip boundary condition relaxed to allow for finite slipping of the fluid/fluid contact line on a solid surface. Considering an effective slip length (L_S), Cox presented a comprehensive hydrodynamic solution by segmenting the dynamic three-phase contact line into inner, intermediate, and outer regions, and correlated the apparent contact angle to the three-phase contact line displacement velocity, U [21,22]

$$U = \frac{\gamma_{O/W}}{9\mu_o}\left[(\pi - \theta)^3 - (\pi - \theta_d)^3\right]\left[\ln\left(\frac{L}{L_S}\right)\right]^{-1} \tag{8}$$

where μ_o is the oil viscosity, θ is the contact angle measured through the water phase, and L and L_S are the characteristic length of the oil droplet and the slip length, respectively. While determination of the slip length is nontrivial, the term is often used as a fitting parameter of the experimental data.

The molecular-kinetic model accounts for molecular displacements (adsorption/desorption) in the vicinity of the dynamic three-phase contact line. The model assumes that the solid surface behaves as a source of identical adsorption sites, and liquid molecules can detach and attach to neighboring sites by overcoming an energy barrier to molecular displacements [23]. The work to overcome the energy barrier is provided by a driving force governed by $\gamma_{O/W}$ and an imbalance between the equilibrium and dynamic wetting states. The three-phase contact line displacement is described in terms of molecular displacement, defined as the distance between adsorption sites (λ) and a frequency (κ^0) of adsorption/desorption events at equilibrium, as shown in Figure 4.

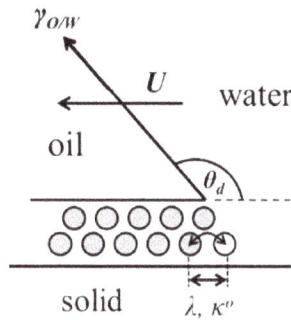

Figure 4. Molecular-kinetic model describes the distance between adsorption sites (λ) and a frequency (κ^0) of adsorption/desorption events. Figure adapted from Blake [24].

The relationship between the dynamic contact angle and the three-phase contact line velocity is given by

$$U = 2\kappa^0 \lambda \sin h \left(\frac{\gamma_{O/W}[\cos(\pi - \theta_d) - \cos(\pi - \theta)]\lambda^2}{2k_B T} \right) \tag{9}$$

where k_B is the Boltzmann constant and T the absolute temperature. The molecular displacement parameters (λ and κ^0) are often combined and treated as the coefficient of contact-line friction, $\zeta = \frac{k_B T}{\kappa^0 \lambda^3}$, to describe the energy dissipated at the three-phase contact line, and neglecting any viscous dissipation in the bulk liquid [24,25]. Similar to the HD model, ζ is treated as an adjustable parameter of the experimental data. Simplification of Equation (9) then follows when the $\sin h$ function is small; i.e., not far from equilibrium—and Equation (9) reduces to the linear form

$$U = \frac{\gamma_{O/W}}{\zeta}[\cos(\pi - \theta_d) - \cos(\pi - \theta)] \tag{10}$$

Since each model neglects a contributing factor, a combined model approach can be considered to account for both the contact-line friction and viscous dissipation. As described by de Gennes and Brochard-Wyart [26,27], the combined model for contact-line displacement is given by

$$U = \frac{\gamma_{O/W}[\cos(\pi - \theta_d) - \cos(\pi - \theta)]}{\zeta + \frac{6\mu_o}{\theta_d} \ln\left(\frac{L}{L_S}\right)} \tag{11}$$

The sequence of images in Figure 5 show the dewetting process for an oil droplet deposited on a solid surface. In this example, a 10 μL droplet of extra heavy oil (13.6° API at 20 °C; SARA: 7.4% saturates, 37.8% aromatics, 15.3% resins, and 39.5% asphaltenes) was deposited on a glass substrate with a water contact angle <5°. Since the oil viscosity was ~6700 mPa·s at 20 °C, the substrate was heated to ~50 °C to promote faster spreading of the oil droplet on the solid surface. With the oil droplet at the equilibrium wetted-state, Milli-Q water was pumped underneath the oil-wetted solid surface at 1400 mL/min to completely submerge the oil droplet. The measurement cell temperature was maintained using a circulating water bath. Since $\frac{dG}{dA} < 0$, oil droplet recession occurs spontaneously and the oil-solid contact area reduced to attain a new equilibrium wetted-state, as described by Equation (4).

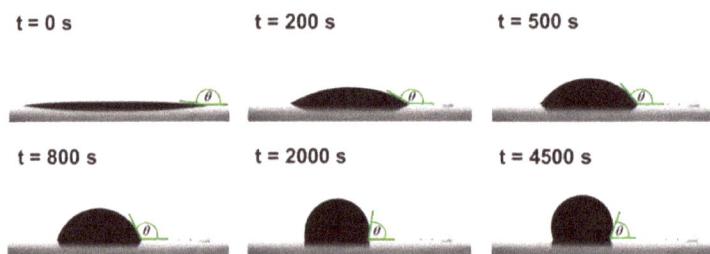

Figure 5. Time-dependent dewetting of an extra heavy oil droplet on a hydrophilic solid surface. The solid surface and water temperature were maintained at 40 °C. Images were captured at 20 fps for the first 15 min and 2 fps thereafter. Data captured using the Theta tensiometer (Biolin Scientific).

The rate of oil film dewetting can be determined from the dynamic contact angle, see Figure 6, with faster dewetting dynamics observed for higher temperature environments. Clearer differentiation between 60 °C and 80 °C is shown in the inset of Figure 6, with the new equilibrium wetted-states (oil-water-solid surface) attained within a few minutes, contrasting the 40 °C sample, which required more than 1 h to reach equilibrium. Moreover, the contact angles at equilibrium were shown to depend on temperature, decreasing from 63.7° to 54.1° and 51.3° with increasing temperature from 40 °C to 60 °C and 80 °C, respectively. Equation (4) shows that changes in the equilibrium wetted-state result from a change in the balance of energies acting on the three interfaces. Measuring $\gamma_{O/W}$ at equivalent temperatures, Figure 7 confirms a small decrease in $\gamma_{O/W}$ with increasing temperature. Hence, if it were assumed that $\gamma_{O/S}$ and $\gamma_{S/W}$ remained independent of temperature, then θ would decrease in good agreement with Equation (4). Previous studies showed variation in the oil-water interfacial tension as a function of pH and temperature [12,13,28–31], with the effect attributed to the partial solubility of naphthenic acids in water [32,33].

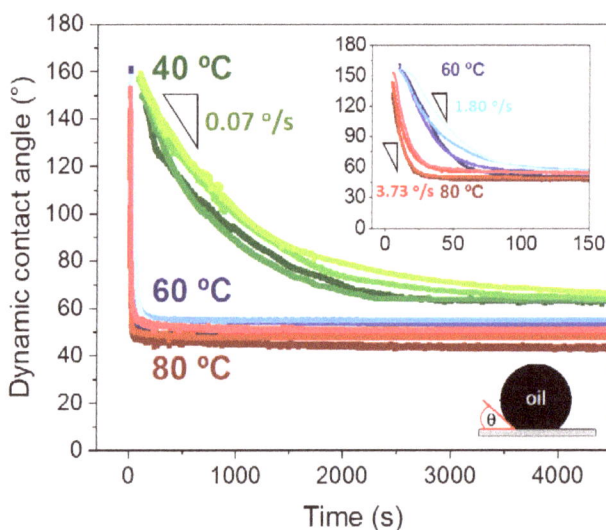

Figure 6. Dewetting dynamics of an extra heavy oil film immersed in Milli-Q water at different temperatures: 40 °C, 60 °C, and 80 °C. Inset is an expanded region of the initial dewetting dynamics to differentiate between the two higher temperatures. Each experimental condition was repeated four times with measurement variability considered to be negligible.

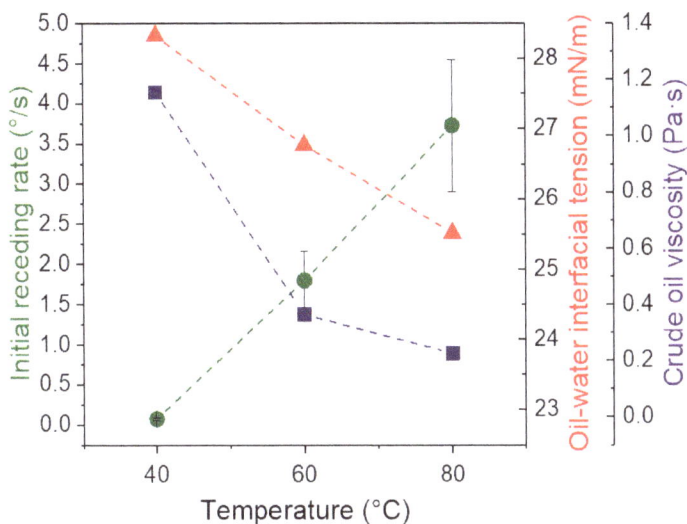

Figure 7. Initial oil droplet receding rate as a function of temperature, correlated to changes in oil viscosity and oil-water interfacial tension (error $\pm$ 0.03 mN/m). Symbols: circle—oil droplet receding rate, triangle—$\gamma_{O/W}$, square—μ_o.

Naphthenic acids are considered to be cyclic carboxylic acids of the general form R–COOH, where R can be any cyclo-aliphatic group [34]. Compared to asphaltenes, naphthenic acids are of lower molecular weight, typically less than 450 g/mol, spanning mainly C_{10} to C_{50} compounds with up to six fused ring structures that are mostly saturated [35]. Naphthenic acids preferentially adsorb on carbonate solid surfaces mainly by chemical interactions to modify the wettability from water-wet to oil-wet as the surface becomes saturated, although the process is reversible at elevated temperatures [36,37].

The initial receding rate ($\frac{d\theta}{dt}$) of the oil film was compared for each temperature with the rates correlated to changes in $\gamma_{O/W}$ and μ_o, see Figure 7. Based on the HD model for contact line displacement (Equation (8)), which includes both parameters, the oil viscosity is the rate dependent parameter since the change in oil viscosity (-80.9%) with temperature is more significant than that of oil-water interfacial tension (-9.9%); between 60 °C and 80 °C the oil viscosity decreased by 38.9% and the initial receding rate increased by 107.2%. The same oil displacement data was fitted to both the HD and MK models (Figure 8). A least-squares difference between the experimental and theoretical θ_d was made

$$\Delta = \sum_{t=0}^{t_e} (\theta_{d,t,m} - \theta_{d,t,e})^2 \tag{12}$$

where $\theta_{d,t,m}$ and $\theta_{d,t,e}$ are the theoretical and experimental dynamic contact angles at time t, respectively, and the model fitting parameters were determined by minimizing the least-squares value.

During the process of oil film dewetting, the model fits appear in reasonable agreement with the experimental data. Slight variation is magnified at higher temperatures when the receding dynamics can be considered rapid for extra heavy crude oil, and experimental variability is more evident. The adjustable fitting parameters for each model (HD—$\ln\left(\frac{L}{L_S}\right)$, MK—$\zeta$) reduced with increasing temperature, suggesting that the slip length of fluid/fluid contact line on a solid surface (L_S) increases and the coefficient of contact-line friction decreases when the oil viscosity is reduced, in good agreement with previous findings [38,39]. While our study only considered dewetting dynamics in Milli-Q water the effect of water chemistry on oil film dewetting has received little attention and is an area for further study. High salinity brine can increase oil-water interfacial tension [40], enhance solvation forces between solid-liquid interfaces [41], bind surfactants to substrate via multicomponent ion exchange [42], and impact the stability of chemical additives used for EOR.

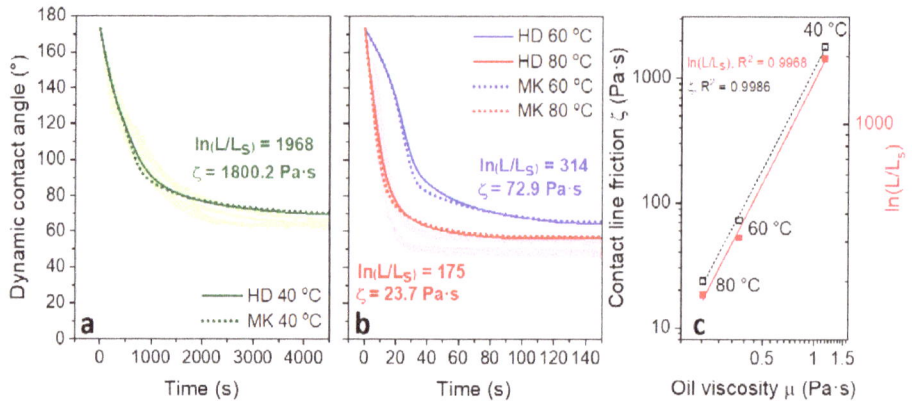

Figure 8. Hydrodynamic (HD) and molecular-kinetic (MK) model fittings of oil film dewetting at 40 °C (**a**), 60 °C and 80 °C (**b**). Shaded lines represent the experimental data and the HD and MK models identified by the solid and dash lines, respectively. (**c**) Optimal fitting parameters, HD (closed symbols)—ln ($\frac{L}{L_s}$), MK (open symbols)—ζ.

4. Surfactant Oil Droplet Displacement

Surfactants are widely used in EOR to reduce $\gamma_{O/W}$ and enhance water-wetting of the solid surface. Surfactants are often described as amphiphilic molecules composed of a hydrophilic head and a hydrophobic tail, thus surfactants favorably partition at solid-liquid and liquid-liquid interfaces. The accumulation of surfactants at an interface is a function of the surfactant concentration in the bulk fluid as described by the general form of the Gibbs' adsorption equation for a binary, isothermal system, $d\gamma = -RT\Gamma_s d\ln C_s$, where Γ_s is the surface excess of surfactant, C_s the surfactant concentration in the bulk fluid, and RT the thermal energy of the system. As a function of concentration, surfactants in solution exist in the monomer-form at low surfactant concentrations, reaching a concentration of maximum solubility of the monomer-form, forming micelles via self-association. This concentration is termed the critical micelle concentration (CMC). Surfactant adsorption and displacement of organic species on solid surfaces and the resultant wettability modification is dependent on the surfactant concentration. At extremely low concentrations, surfactant monomers adsorb as individual molecules with no interaction between the adsorbed molecules. At higher concentrations (<CMC) surfactant molecules associate to form patchy hemi-micelles on the solid surface, with surfactants coordinating in the tail-tail confirmation. Further increases in concentration lead to saturation of all available surface sites and the formation of a surfactant bi-layer at the CMC [43]. Formation of a bi-layer would orientate the surfactant hydrophilic head group away from the solid surface, thus increasing the water-wetting nature of the reservoir rock, favorable for oil droplet displacement (Equations (5) and (7)). Mechanisms for wettability modification by different surfactants are described below.

Composition of the reservoir surface (sandstone, carbonate, and deposited organic species) often dictates the surfactant selection for wettability modification, with surfactants categorized as cationic, anionic and non-ionic, based on the charge characteristics of hydrophilic groups. Surfactant adsorption on the solid surface can occur via electrostatic and van der Waals forces, and hydrogen bonding, with the extent of wettability modification a function of several properties including surfactant adsorption kinetics, surfactant structure, temperature, pH, salinity. A brief summary of surfactants considered for EOR is provided in Table A1, with remarks provided for changes in solid surface wettability and interfacial tension.

While electrostatic interactions are often considered to describe surfactant-solid surface adsorption, such simplicity does not describe the potential for surfactants to modify solid surface wettability, when many other factors such as oil saturation, clay content, divalent cations, pH, and temperature influence the action of the surfactant.

Cationic surfactants are frequently used to treat carbonate reservoirs and include permanently charged ammonium groups (ammonium bromide and ammonium chloride) [44]. Adsorbed polar components of crude oil (i.e., negatively charged naphthenic acids) can be removed from the solid surface by forming ion pairs with cationic surfactants via strong ionic interaction. Removal of contaminants transforms the solid surface wettability to more water-wet [44,45]. The use of cationic surfactants to treat sandstone has also been demonstrated, although the chemical effectiveness in carbonate reservoirs is greater [46].

Anionic surfactants including sulfates, sulfonates, phosphates, and carboxylates, have been shown to modify wettability in both carbonate and sandstone reservoirs. Wettability modification occurs via two mechanisms [44,45]: (i) anionic surfactants interact with the organic species via hydrophobic forces, exposing the surfactant head group to make the solid surface more water-wet (wettability modification for sandstone reservoirs); and (ii) via strong electrostatic forces with carbonate surfaces, anionic surfactants can displace organic species exposing the underlying water-wet surface [46].

Non-ionic surfactants such as alcohols, esters and ethers have been used to modify the wettability of carbonate and sandstone surfaces [47], being highly effective in high salinity water. With no contribution from electrostatic forces, non-ionic surfactants interact via hydrophobic forces with deposited organic species, and hydrogen bonding with hydroxyl groups on the solid surface [48,49]. Research has shown that non-ionic surfactants can modify highly oil-wet carbonate to weakly oil-wet or even water-wet ($\theta < 80°$) following the addition of 0.1 wt % surfactant [50].

Sodium dodecyl sulfate (SDS), an anionic surfactant, was used to displace an oil film deposited on a glass substrate at 60 °C (Figure 9a). Adding SDS to the aqueous phase reduced $\gamma_{O/W}$, and the CMC was measured at ~0.1 wt % (3.5 mM) at 60 °C (Figure 9c). Increasing the SDS concentration from 5×10^{-4} wt % to 5×10^{-3} wt % increased both the rate of oil film displacement and equilibrium oil droplet contact angle (inferred from lower water contact angle, θ). The equilibrium contact angle reduced from 54.1° in the absence of SDS to 48.7° and 36.5° for 5×10^{-4} wt % and 5×10^{-3} wt % SDS, respectively. Fitting the HD and MK models confirmed an increased slip length (L_S) and reduced coefficient of contact-line friction (ζ) at higher SDS concentrations. Figure 9b illustrates the benefit of injecting surfactants at a concentration greater than the CMC. The very low oil-water interfacial tension (~5.75 mN/m) causes the oil film to continually recede and eventually detach from the solid surface when the oil droplet buoyant force (3.77 μN) exceeds the solid surface-oil droplet adhesion force (2.01 μN).

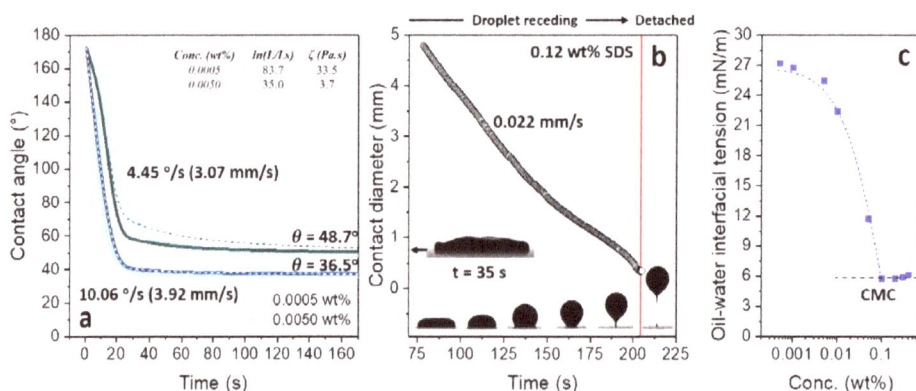

Figure 9. (**a**) Oil film dewetting at 60 °C ($\rho_{oil@60\,°C} = 946$ kg/m^3) with increasing sodium dodecyl sulfate (SDS) concentration (5×10^{-4} wt % and 5×10^{-3} wt %). Shaded lines represent the experimental data with the HD and MK models identified by the solid and dash lines, respectively; (**b**) Oil film dewetting at 60 °C with the SDS concentration above the CMC (SDS = 0.12 wt %). The solid surface-oil droplet contact diameter reaches a minimum of 0.34 mm at the point of oil droplet detachment; (**c**) Extra heavy oil-water interfacial tension as a function of the SDS concentration. The CMC was ~0.1 wt % at 60 °C.

5. Nanoparticle Oil Droplet Displacement

The application of ultra-small particles (nanoparticles) to enhance oil film displacement has been demonstrated. Nanoparticles are typically 1 to 100 nm and are ideal for EOR applications with particle sizes smaller than the pore diameter, hence nanofluids flow through the porous media without obstructing the porous network. In addition, their high surface area to volume ratio increases their effectiveness at low particle concentrations, and promotes their kinetic stability [51]. An overview of nanoparticles (nanofluids) used to displace oil films is provided in Table A2.

For oil film displacement, different nanoparticles have been considered including metal oxides, organic, inorganic, and composite particles. Metal oxides nanoparticles (Al_2O_3, CuO, TiO_2 and Fe_2O_3) have been shown to lower $\gamma_{O/W}$ [52] and increase the disjoining pressure between the solid surface and oil-water interface [53]. The interfacial tension decreases as nanoparticles partition at the oil-water interface from the aqueous phase. Unlike surfactants, nanoparticles are not amphiphilic and their affinity to partition at an oil-water interface is governed by their particle size and surface wettability. The general expression of $E = \pi a^2 \gamma_{O/W}(1 \pm \cos\theta)^2$, describes the particle detachment energy from an oil-water interface ($\pm$describes detachment into either liquid phase), where a is the particle radius [54]. When $\cos\theta = 0$ the particle detachment energy is maximized, although for nanoparticles E can be of the order of a few $k_B T$ which can result in reversible adsorption, similar to a surfactant molecule.

The reduction in oil-water interfacial tension has been shown to correlate to the nanoparticle size and the particle specific surface area [55]. Al_2O_3 nanoparticles of diameter 20 nm and 45 nm were shown to lower the oil-water interfacial tension to ~13.6 mN/m and ~8.6 mN/m, respectively [56]. Compared to surfactants, interfacial tension reduction by nanoparticles is often smaller. For example, silica nanoparticles (7–14 nm) dosed at 0.01–0.10 wt % reduced $\gamma_{O/W}$ to ~10 mN/m from ~15–20 mN/m [57–59], while TiO_2 nanoparticles (58 nm) reduced $\gamma_{O/W}$ from 23 mN/m to 18 mN/m when dosed at 0.01–0.05 wt % [53]. As such, enhanced oil film displacement by nanoparticles is likely to occur via other mechanisms; i.e., structural disjoining pressure and wettability modification.

Structural disjoining pressure is a consequence of nanofluids exhibiting super-spreading behavior. Nanoparticles self-assemble in the vicinity of the three-phase contact line to form a liquid wedge at the de-pinning point, see Figure 10. As nanoparticles accumulate in the liquid wedge a structural disjoining pressure (Equation (14)) gradient is established with the highest pressure at the oil droplet-solid surface vertex, driving the nanofluid to spread and cause the oil film to recede. As explained by Wasan and co-workers [60,61], the spreading coefficient (*S*) of the nanofluid is determined by the sum of the capillary pressures at the equilibrium film thickness ($\Pi_0(h_e)$) and disjoining pressure ($\Pi(h)$)

$$S = \Pi_0(h_e)h_e + \int_{h_e}^{\infty} \Pi(h)dh \tag{13}$$

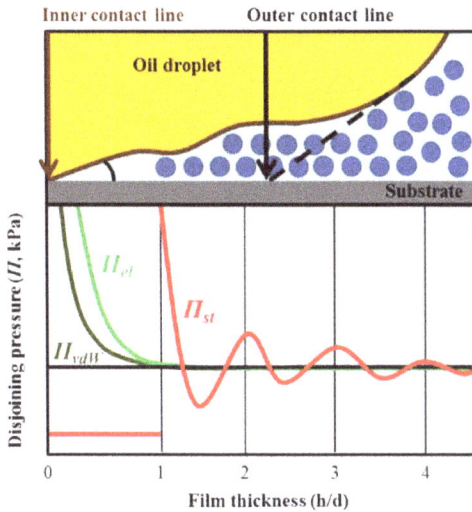

Figure 10. Schematic showing the ordered accumulation of nanoparticles to form a liquid wedge. The structural disjoining pressure increases towards the de-pinning point. The structural disjoining pressure exceeds the Laplace pressure, deforming the meniscus profile as represented by inner and outer contact lines. The contribution from the long range structural disjoining pressure dominates the short range electrostatic and van der Waals forces. The figure has been modified from Zhang et al. [62] and Chengara et al. [11].

When the thickness of the liquid wedge exceeds one particle diameter, nanoparticles accumulate in ordered layers. This layered arrangement of nanoparticles increases the excess pressure in the liquid wedge with the structural disjoining pressure described based on the theory of thin liquid films [63]:

$$\Pi_{st}(h) = \Pi_0 \cos(\omega h + \varnothing_2)e^{-\kappa h} + \Pi_1 e^{-\delta(h-d)}, \ h \geq d$$
$$\Pi_{st}(h) = -P, \ 0 < h < d \tag{14}$$

where *d* is the nanoparticle diameter, *P* the osmotic pressure of nanofluid, and all other parameters (Π_0, Π_1, ω, $\varnothing_2$, κ and δ) are fitted as cubic polynomials varying with particle concentration. Contributions from van der Waals, electrostatic and structural forces have been considered by Chengara et al. [11] (Figure 10). The structural forces are long range and govern the behavior of thick liquid films, with nanoparticle size, concentration, temperature, and fluid salinity, all contributing to the magnitude of the structural disjoining pressure.

Wettability modification by nanoparticles enhances oil droplet displacement when nanoparticles deposit on the solid surface. The deposition/adsorption is influenced by electrostatic forces, with the nanoparticle decorated solid surface more water-wet due to deposition of hydrophilic particles to form a heterogeneous surface and increased nano/micron-scale roughness [64–68]. Wettability of heterogeneous surfaces has been described by Cassie-Baxter [69], with the apparent contact angle on a composite solid surface given by, $\cos\theta_{CB} = f_1\cos\theta_1 + f_2\cos\theta_2$, where f_1 is the fractional area of the surface with contact angle θ_1, f_2 is the fractional area of the surface with contact angle θ_2, and θ_{CB} is the Cassie-Baxter contact angle. The Cassie-Baxter model can be combined with the Wenzel wetting model [70] to account for surface roughness effects, $\cos\theta_W = R'\cos\theta_{CB}$, where R' is the ratio of the true area of the solid to its planar projection. With R' always greater than 1, the Wenzel model confirms nano/micron-scale roughness lowers the contact angle of a water-wet surface, thus increasing the potential for oil droplet displacement. For example, metal oxide nanoparticles (ZrO_2 and NiO < 50 nm) were shown to deposit on an oil-wet surface modifying the contact angle from 152° (untreated surface) to 44° and 86° for ZrO_2 and NiO, respectively. The mean roughness of those surfaces was shown to increase from 70.6 nm (untreated surface) to 2.32 μm (ZrO_2 treated surface) and 330 nm (NiO treated surface) [71].

Oil film displacement can be enhanced when nanoparticles are mixed with surfactants. Fluid blends lower the oil-water interfacial tension below a surfactant only system, with surfactants increasing the interfacial activity of nanoparticles [72]. The decrease in oil-water interfacial tension depends on the surfactant-particle interaction and surfactant concentration [73]. The effect of nanoparticles is lessened at surfactant concentrations greater than the CMC. The use of surfactant blends and composite particles (polymer-coated particles) to enhance oil film displacement have also been considered but such mechanisms are considered outside the scope of this paper [74,75]. Recent studies, which have considered composite fluids (particles), have been summarized in Table A3.

6. Conclusions

While demand for oil continues to rise, challenges in extraction become ever more complex. Extraction from confined, unfavorable environments, and production of unconventional oil is increasing the dependence on alternative extraction methods to deliver enhanced oil recovery. Often the interaction between the oil and solid surface limits recovery with oil strongly adhered to an oil-wet surface. Oil film recession is spontaneous when the solid surface is water-wet, and the adhesion force to be overcome to liberate an oil droplet from a solid surface diminishes with decreasing contact angle and oil-water interfacial tension. The rate of oil film recession and oil droplet equilibrium contact angle can be modified through the careful selection of chemicals. Surfactants have extensively been considered and used in production to lower oil-water interfacial tension and modify the solid surface to more water-wet. An alternative mechanism for oil film displacement has been identified when using nanofluids. Accumulation of nanoparticles in a liquid wedge between oil and solid surface results in a long range structural disjoining pressure gradient causing the three-phase contact line to move (i.e., oil film recede due to super-spreading of the nanofluid).

Controlling interfacial behavior in the reservoir provides a route for enhanced oil recovery. Significant research effort is ongoing to design more effective chemicals that perform in challenging environments (temperature, pressure, salinity, clays), deliver performance at the targeted site (i.e., minimize material loss), and do not impact the environment. Enhanced oil recovery will ensure effective utilization of crude oil resources, and the fundamental mechanisms governing oil film displacement and oil droplet detachment are underpinned by knowledge of interfacial and colloidal forces.

Author Contributions: S.T. performed the experiments; S.T. and D.H. analyzed the data; S.T. and D.H. drafted the manuscript; T.V.J.C. and D.P. reviewed the final manuscript.

Funding: Royal Academy of Engineering Industry-Academia Partnership Program (IAPP1/100150); White-Elephant Academics Scheme of Chiang Mai University.

Acknowledgments: S.T. acknowledges the PhD scholarship from the White-Elephant Academics Scheme of Chiang Mai University. All authors would like to acknowledge the Royal Academy of Engineering and its support through the Industry-Academia Partnership Program (Grant reference: IAPP1/100150).

Appendix

Table A1. Surfactants.

Surfactants	Conc.	Solid Surface	Oil Type	Remarks [a]	Ref.
Cationic surfactants					
n-C$_8$-N(CH$_3$)$_3$Br (C8TAB) in brine	4.0 wt %	Chalk	Crude oil mixed with heptane	Contact angle = 57°, IFT [b] = 2.85 mN/m	[44]
n-C$_{10}$-N(CH$_3$)$_3$Br (C10TAB) in water	0.4 wt %	Calcite	Decane mixed with naphthenic acids	IFT = 2.67 mN/m	[45]
n-C$_{12}$-N(CH$_3$)$_3$Br (C12TAB) in water	0.4 wt %	Calcite	Decane mixed with naphthenic acids	IFT = 0.59 mN/m	[45]
n-C$_{12}$-N(CH$_3$)$_3$Br (C12TAB) in brine	5.0 wt %	Chalk	Crude oil mixed with heptane	Contact angle = 12°, IFT = 0.81 mN/m	[44]
n-C$_{16}$-N(CH$_3$)$_3$Br (C16TAB) in brine	1.0 wt %	Chalk	Crude oil mixed with heptane	Contact angle = 27°, IFT = 0.38 mN/m	[44]
Cetyltrimethylammonium bromide (CTAB) in brine	0.3 wt %	Quartz	Crude oil	Contact angle = 57°	[76]
n-Decyl triphenylphosphonium bromide (C10TPPB) in water	0.4 wt %	Calcite	Decane mixed with naphthenic acids	IFT = 3.56 mN/m	[45]
Cocoalkyltrimethyl ammonium chloride (CAC) in brine	75–2620 ppm (0.0075–0.262 wt %)	Dolomite	Crude oil		[47]
Dodecyltrimethylammonium bromide (DTAB) in brine	0.5 wt %	Calcite	Crude oil	Contact angle = 69°, IFT = 4.8 mN/m	[77]
Dodecyltrimethylammonium bromide (DTAB) in brine	0.06 wt %	Quartz	Crude oil	Contact angle = 95°, IFT = 2.49 mN/m	[78]
n-(C$_8$-C$_{18}$)-N(CH$_3$)$_2$(CH$_2$-Ph)Cl (ADMBACl) in brine	0.5 wt %	Chalk	Crude oil mixed with heptane	Contact angle = 26°, IFT = 0.41 mN/m	[44]
n-C$_8$-Ph-(EO)$_2$-N(CH$_3$)$_2$(CH$_2$-Ph)Cl (Hyamine) in brine	0.2 wt %	Chalk	Crude oil mixed with heptane	Contact angle = 21°, IFT = 0.48 mN/m	[44]
Coconut oil alkyl trimethylammonium chloride (ARQUAD C-50) in water	0.4 wt %	Calcite	Decane mixed with naphthenic acids	IFT = 0.53 mN/m	[45]
Trimethyl tallowalky ammonium choride (ARQUAD T-50) in water	0.4 wt %	Calcite	Decane mixed with naphthenic acids	IFT = 0.69 mN/m	[45]
Methyldodecylbis ammonium tribromide	0.0001–1 mM	Mica	Kerosene mixed with n-decane	Contact angle = 87°, IFT = 0.18 mN/m	[79]

Table A1. *Cont.*

Surfactants	Conc.	Solid Surface	Oil Type	Remarks [a]	Ref.
Anionic surfactants					
n-(C_{12}-C_{15})-(EO)$_{15}$-SO$_3$Na (S-150) in brine	0.5 wt %	Chalk	Crude oil mixed with heptane	Contact angle = 63°, IFT = 2.29 mN/m	[44]
n-C_{13}-(EO)$_8$-SO$_3$Na (B 1317) in brine	0.5 wt %	Chalk	Crude oil mixed with heptane	Contact angle = 40°, IFT = 0.78 mN/m	[44]
n-C_8-(EO)$_3$-SO$_3$Na (S-74) in brine	0.5 wt %	Chalk	Crude oil mixed with heptane	Contact angle = 49°, IFT = 6.72 mN/m	[44]
n-(C_{12}-C_{15})-(PO)$_4$-(EO)$_2$-OSO$_3$Na (APES) in brine	1.0 wt %	Chalk	Crude oil mixed with heptane	Contact angle = 44°, IFT = 0.082 mN/m	[44]
n-(C_8O_2CH$_2$)(n-C_8O_2C)CH-SO$_3$Na (Cropol) in brine	0.5 wt %	Chalk	Crude oil mixed with heptane	Contact angle = 55°, IFT = 8.77 mN/m	[44]
n-C_8-(EO)$_8$-OCH$_2$-COONa (Akypo) in brine	0.5 wt %	Chalk	Crude oil mixed with heptane	Contact angle = 48°, IFT = 2.99 mN/m	[44]
n-C_9-Ph-(EO)$_x$-PO$_3$Na (Gafac) in brine	0.5 wt %	Chalk	Crude oil mixed with heptane	Contact angle = 75°, IFT = 0.42 mN/m	[44]
Sodium dodecyl sulfate (SDS) in brine	0.1 wt %	Chalk	Crude oil mixed with heptane	Contact angle = 39°, IFT = 2.95 mN/m	[44]
Sodium dodecyl sulfate (SDS) in water	0.4 wt %	Calcite	Decane mixed with naphthenic acids	IFT = 4.77 mN/m	[45]
Sodium dodecyl 3EO sulfate in brine	0.05 wt %	Calcite	Crude oil	Contact angle ~45°, IFT = 0.003 mN/m	[80]
Alkyldiphenyloxide disulfonate in Na$_2$CO$_3$/NaCl	0.05 wt %	Calcite	Crude oil	Contact angle ~110°, IFT = 0.0011 mN/m	[50]
Polyether sulfonate in Na$_2$CO$_3$/NaCl	0.30 wt %	Calcite	Crude oil	Contact angle ~80°, IFT = 0.00812 mN/m	[50]
Sodium nonyl phenol ethoxylated sulfate (4EO) in Na$_2$CO$_3$/NaCl	0.05 wt %	Calcite	Crude oil	Contact angle ~60°, IFT = 0.003 mN/m	[50]
C_{12}-C_{13} propoxy sulfate (8PO) in Na$_2$CO$_3$/NaCl	0.05 wt %	Calcite	Crude oil	Contact angle ~40°, IFT = 0.0001 mN/m	[50]
Alkyldiphenyloxide disulphonate + C_{14}T-isofol propoxy sulfate (8PO) in Na$_2$CO$_3$/NaCl	0.075 wt %	Calcite	Crude oil	Contact angle ~70°, IFT = 0.116 mN/m	[50]
Methyl alcohol+Proprietary sulfonate in brine	0.02–0.20 wt %	Shale (siliceous)	Crude oil	Contact angle = 38°, IFT = 0.4 mN/m	[81]
Sodium laureth sulfate in brine	0.02–0.05 wt %	Quartz	Crude oil	Contact angle ~110°, IFT = 2.007 mN/m	[76]
Sodium lauryl monoether sulfate in brine	0.035 wt %	Quartz	Crude oil	Contact angle = 116.1°, IFT = 2.49 mN/m	[78]

Table A1. *Cont.*

Surfactants	Conc.	Solid Surface	Oil Type	Remarks [a]	Ref.
Nonionic surfactants					
Poly-oxyethylene alcohol (POA) in brine	750–1050 ppm (0.075–0.105 wt %)	Dolomite	Crude oil	IFT = 2.0 mN/m	[47]
Ethoxylated C_{11}-C_{15} secondary alcohol (Tergitol 15-S-3) in water	0.4 wt %	Calcite	Decane mixed with naphthenic acids	IFT = 4.44 mN/m	[45]
Ethoxylated C_{11}-C_{15} secondary alcohol (Tergitol 15-S-7) in water	0.4 wt %	Calcite	Decane mixed with naphthenic acids	IFT = 1.39 mN/m	[45]
Ethoxylated C_{11}-C_{15} secondary alcohol (Tergitol 15-S-40) in water	0.4 wt %	Calcite	Decane mixed with naphthenic acids	IFT = 11.5 mN/m	[45]
Nonylphenoxypoly(ethyleneoxy)ethanol (Igepal CO-530) in water	0.4 wt %	Calcite	Decane mixed with naphthenic acids	IFT = 0.33 mN/m	[45]
C_{12}-C_{15} linear primary alcohol ethoxylate (Neodol 25-7) in water	0.4 wt %	Calcite	Decane mixed with naphthenic acids	IFT = 2.02 mN/m	[45]
Secondary alcohol ethoxylate in Na_2CO_3/NaCl	0.10 wt %	Calcite	Crude oil	Contact angle ~20°, IFT = 0.0017 mN/m	[50]
Nonyl phenol ethoxylate in Na_2CO_3/NaCl	0.10 wt %	Calcite	Crude oil	Contact angle ~80°, IFT = 0.0006 mN/m	[50]
Branched alcohol oxyalkylate in brine	0.02–0.20 wt %	Shale (siliceous)	Crude oil	Contact angle = 60°, IFT = 9.8 mN/m	[81]
Polyoxyethylene octyl phenyl ether in brine	0.04 wt %	Quartz	Crude oil	Contact angle = 95°, IFT = 4.05 mN/m	[76]
Alkylpolyglycosides in brine	0.05 wt %	Quartz	Crude oil	Contact angle = 58.8°, IFT = 2.49 mN/m	[28]

[a] not all studies reported contact angle or interfacial tension data. [b] IFT is interfacial tension.

Table A2. Nanoparticles/fluids.

Nanoparticles/Fluids	Solid Surface	Oil Type	Remarks [a]	Ref.
Metal oxides				
TiO_2 (0.01–1 wt %)	Sandstone	Heavy oil	Contact angle = 90°	[82]
TiO_2 (0.01–0.10 wt %)	Sandstone	Heavy crude oil	Slight IFT [b] reduction ~$\Delta\gamma$ = 1 mN/m	[52]
TiO_2 (0.01–0.05 wt %)	Sandstone	Heavy oil	Contact angle change from 127° to 81°, Slight IFT reduction	[53]
Al_2O_3 (0.01–0.10 wt %)	Sandstone	Heavy crude oil	Slight IFT reduction ~$\Delta\gamma$ = 1 mN/m	[52]
NiO (0.01–0.10 wt %)	Sandstone	Heavy crude oil	Slight IFT reduction ~$\Delta\gamma$ = 1 mN/m	[52]
Organic				
Janus nanoparticles (0.0025–0.0004 mM)	NA [c]	Hexane	IFT = 12 mN/m	[83]
Carbon nanotubes (0.05–0.50 wt %)	Glass	Crude oil	IFT reduction ~3 mN/m	[84]
Nanocellulose (0.2–1.0 wt %)	Glass	Crude oil	IFT = 0.7 mN/m	[85]
Inorganic				
SiO_2 (0.1–0.6 wt %)	Carbonate	Crude oil	Contact angle = 51°	[86]
SiO_2 (0.5–4.0 wt %)	Calcite (oil-wet)	n-decane	Contact angle = 20°	[87]
SiO_2 (0.1–5 wt %)	Glass	Crude oil	Contact angle = 0°	[88]
SiO_2 (0.025–0.2 wt %)	Calcite (oil-wet)	n-heptane	Contact angle = 41.7°	[89]
SiO_2 (0.4 effective volume fraction)	Glass	Model oil		[60]
SiO_2 (0.01–0.10 wt %)	Sandstone	Crude oil	Contact angle = 22°, IFT = 7.9 mN/m	[57]
SiO_2 (0.10 wt %)	Sandstone	Light crude oil	Contact angle change from 34° to 32°, IFT reduced from 20 to 10 mN/m	[58]

Table A2. *Cont.*

Nanoparticles/Fluids	Solid Surface	Oil Type	Remarks [a]	Ref.
SiO$_2$ (0.01–0.10 wt %)	Sandstone	Heavy crude oil	Slight IFT reduction ~$\Delta\gamma$ = 1 mN/m	[72]
Hydrophilic silica (0.01–0.10 wt %)	Glass/Sandstone	Light crude oil	Contact angle ~20°, IFT ~8 mN/m	[39]
Hydrophilic, neutralized, and hydrophobic silica (0.2–0.3 wt %)	Sandstone	Crude oil	Contact angle ~35°	[57]
Hydrophobic silica (0.1–0.4 wt %)	Sandstone	Crude oil	Contact angle = 95.4°, IFT = 1.75 mN/m	[90]
Nanostructure particles (0.05–0.50 wt %)	Sandstone	Light crude oil	Wettability index = 0.36 (wettability index = 1 is water-wet)	[91]
Silica colloidal nanoparticles (0.05–0.50 wt %)	Sandstone	Light crude oil	Wettability index = 0.57 (wettability index = 1 is water-wet)	[91]

[a] not all studies reported contact angle or interfacial tension data. [b] IFT is interfacial tension. [c] *NA* is not available.

Table A3. Composite fluids.

Composite Fluids	Solid Surface	Oil Type	Remarks [a]	Ref.
Blend systems				
SDS and SiO$_2$ (Patented nanofluid—No reported concentration)	Glass	Crude oil	Contact angle = 1.2°	[62]
SDS and hydrophilic and hydrophobic SiO$_2$ (Surfactant: 100–6000 ppm, particle: 1000–2000 ppm)	Sandstone	Kerosene	IFT [b] = 1.81 mN/m	[72]
SDS and ZrO$_2$ (Surfactant: 0.001–5 CMC, particle: 0.001–0.050 wt %)	NA [c]	n-heptane	IFT = 10 mN/m	[92]
Composite nanoparticles				
Zwitterionic polymer and SiO$_2$ (coated) (No reported concentration)	Sandstone	n-decane	IFT = 35 mN/m	[74]

[a] not all studies reported contact angle or interfacial tension data. [b] IFT is interfacial tension. [c] *NA* is not available.

References

1. BP. *2017 Energy Outlook*; BP: London, UK, 2017.
2. BP. *BP Statistical Review of World Energy 2017*; BP: London, UK, 2017.
3. Treiber, L.E.; Owens, W.W. *A Laboratory Evaluation of the Wettability of Fifty Oil-Producing Reservoirs*; Society of Petroleum Engineers: Richardson, TX, USA, 1972.
4. Chilingar, G.V.; Yen, T.F. Some Notes on Wettability and Relative Permeabilities of Carbonate Reservoir Rocks, II. *Energy Sources* **1983**, *7*, 67–75. [CrossRef]
5. Pu, W.-F.; Yuan, C.-D.; Wang, X.-C.; Sun, L.; Zhao, R.-K.; Song, W.-J.; Li, X.-F. The Wettability Alteration and the Effect of Initial Rock Wettability on Oil Recovery in Surfactant-based Enhanced Oil Recovery Processes. *J. Dispers. Sci. Technol.* **2016**, *37*, 602–611. [CrossRef]
6. Natarajan, A.; Kuznicki, N.; Harbottle, D.; Masliyah, J.; Zeng, H.; Xu, Z. Understanding Mechanisms of Asphaltene Adsorption from Organic Solvent on Mica. *Langmuir* **2014**, *30*, 9370–9377. [CrossRef] [PubMed]
7. Standal, S.; Haavik, J.; Blokhus, A.M.; Skauge, A. Effect of polar organic components on wettability as studied by adsorption and contact angles. *J. Pet. Sci. Eng.* **1999**, *24*, 131–144. [CrossRef]
8. Kupai, M.M.; Yang, F.; Harbottle, D.; Moran, K.; Masliyah, J.; Xu, Z.H. Characterising rag-forming solids. *Can. J. Chem. Eng.* **2013**, *91*, 1395–1401. [CrossRef]
9. Zeng, H.; Zou, F.; Horvath-Szabo, G.; Andersen, S. Effects of Brine Composition on the Adsorption of Benzoic Acid on Calcium Carbonate. *Energy Fuels* **2012**, *26*, 4321–4327. [CrossRef]
10. Jadhunandan, P.P.; Morrow, N.R. *Effect of Wettability on Waterflood Recovery for Crude-Oil/Brine/Rock Systems*; Society of Petroleum Engineers: Richardson, TX, USA, 1995.
11. Chengara, A.; Nikolov, A.D.; Wasan, D.T.; Trokhymchuk, A.; Henderson, D. Spreading of nanofluids driven by the structural disjoining pressure gradient. *J. Colloid Interface Sci.* **2004**, *280*, 192–201. [CrossRef] [PubMed]
12. Bakhtiari, M.T.; Harbottle, D.; Curran, M.; Ng, S.; Spence, J.; Siy, R.; Liu, Q.X.; Masliyah, J.; Xu, Z.H. Role of Caustic Addition in Bitumen-Clay Interactions. *Energy Fuels* **2015**, *29*, 58–69. [CrossRef]
13. Flury, C.; Afacan, A.; Bakhtiari, M.T.; Sjoblom, J.; Xu, Z. Effect of Caustic Type on Bitumen Extraction from Canadian Oil Sands. *Energy Fuels* **2014**, *28*, 431–438. [CrossRef]
14. Swiech, W.; Taylor, S.; Zeng, H. The Role of Water Soluble Species in Bitumen Recovery from Oil Sands. In Proceedings of the SPE Heavy Oil Conference-Canada, Society of Petroleum Engineers, Calgary, AB, Canada, 10–12 June 2014.
15. Liu, J.; Zhou, Z.; Xu, Z.; Masliyah, J. Bitumen–Clay Interactions in Aqueous Media Studied by Zeta Potential Distribution Measurement. *J. Colloid Interface Sci.* **2002**, *252*, 409–418. [CrossRef] [PubMed]
16. Czarnecki, J.; Radoev, B.; Schramm, L.L.; Slavchev, R. On the nature of Athabasca Oil Sands. *Adv. Colloid Interface Sci.* **2005**, *114*, 53–60. [CrossRef] [PubMed]
17. Sharma, A. Disintegration of macroscopic fluid sheets on substrates—A singular perturbation approach. *J. Colloid Interface Sci.* **1993**, *156*, 96–103. [CrossRef]
18. Srinivasa, S.; Flury, C.; Afacan, A.; Masliyah, J.; Xu, Z. Study of Bitumen Liberation from Oil Sands Ores by Online Visualization. *Energy Fuels* **2012**, *26*, 2883–2890. [CrossRef]
19. Bertrand, E.; Blake, T.D.; de Coninck, J. Dynamics of dewetting. *Colloids Surfaces A Physicochem. Eng. Asp.* **2010**, *369*, 141–147. [CrossRef]
20. Basu, S.; Nandakumar, K.; Masliyah, J.H. A Model for Detachment of a Partially Wetting Drop from a Solid Surface by Shear Flow. *J. Colloid Interface Sci.* **1997**, *190*, 253–257. [CrossRef] [PubMed]
21. Cox, R.G. The dynamics of the spreading of liquids on a solid surface. Part 1. Viscous flow. *J. Fluid Mech.* **1986**, *168*, 169–194. [CrossRef]
22. Cox, R.G. The dynamics of the spreading of liquids on a solid surface. Part 2. Surfactants. *J. Fluid Mech.* **2006**, *168*, 195–220. [CrossRef]
23. Blake, T.D.; Haynes, J.M. Kinetics of liquid/liquid displacement. *J. Colloid Interface Sci.* **1969**, *30*, 421–423. [CrossRef]
24. Blake, T.D. The physics of moving wetting lines. *J. Colloid Interface Sci.* **2006**, *299*, 1–13. [CrossRef] [PubMed]
25. Blake, T.D.; de Coninck, J. The influence of solid–liquid interactions on dynamic wetting. *Adv. Colloid Interface Sci.* **2002**, *96*, 21–36. [CrossRef]
26. De Gennes, P.G. Wetting: Statics and dynamics. *Rev. Mod. Phys.* **1985**, *57*, 827–863. [CrossRef]

27. Brochard-Wyart, F.; de Gennes, P.G. Dynamics of partial wetting. *Adv. Colloid Interface Sci.* **1992**, *39*, 1–11. [CrossRef]

28. Schramm, L.L.; Smith, R.G. The influence of natural surfactants on interfacial charges in the hot-water process for recovering bitumen from the athabasca oil sands. *Colloids Surfaces* **1985**, *14*, 67–85. [CrossRef]

29. Drelich, J.; Miller, J.D. Surface and interfacial tension of the Whiterocks bitumen and its relationship to bitumen release from tar sands during hot water processing. *Fuel* **1994**, *73*, 1504–1510. [CrossRef]

30. Drelich, J.; Bukka, K.; Miller, J.D.; Hanson, F.V. Surface Tension of Toluene-Extracted Bitumens from Utah Oil Sands as Determined by Wilhelmy Plate and Contact Angle Techniques. *Energy Fuels* **1994**, *8*, 700–704. [CrossRef]

31. Long, J.; Drelich, J.; Xu, Z.; Masliyah, J.H. Effect of Operating Temperature on Water-Based Oil Sands Processing. *Can. J. Chem. Eng.* **2007**, *85*, 726–738. [CrossRef]

32. Rogers, V.V.; Liber, K.; MacKinnon, M.D. Isolation and characterization of naphthenic acids from Athabasca oil sands tailings pond water. *Chemosphere* **2002**, *48*, 519–527. [CrossRef]

33. Grewer, D.M.; Young, R.F.; Whittal, R.M.; Fedorak, P.M. Naphthenic acids and other acid-extractables in water samples from Alberta: What is being measured? *Sci. Total Environ.* **2010**, *408*, 5997–6010. [CrossRef] [PubMed]

34. Masliyah, J.H.; Xu, Z.; Czarnecki, J.A. *Handbook on Theory and Practice of Bitumen Recovery from Athabasca Oil Sands: Theoretical Basis*; Kingsley Knowledge Pub.: Cochrane, AB, Canada, 2011.

35. Schramm, L.L.; Stasiuk, E.N.; MacKinnon, M. Surfactants in Athabasca oil sands slurry conditioning, flotation recovery, and tailings processes. In *Surfactants: Fundamentals and Applications in the Petroleum Industry*; Schramm, L.L., Ed.; Cambridge University Press: Cambridge, UK, 2000; pp. 365–430.

36. Kelesoglu, S.; Volden, S.; Kes, M.; Sjoblom, J. Adsorption of naphthenic acids onto mineral surfaces studied by quartz crystal microbalance with dissipation monitoring (QCM-D). *Energy Fuels* **2012**, *26*, 5060–5068. [CrossRef]

37. Ding, L.; Rahimi, P.; Hawkins, R.; Bhatt, S.; Shi, Y. Naphthenic acid removal from heavy oils on alkaline earth-metal oxides and ZnO catalysts. *Appl. Catal. A Gen.* **2009**, *371*, 121–130. [CrossRef]

38. Duvivier, D.; Seveno, D.; Rioboo, R.; Blake, T.D.; de Coninck, J. Experimental Evidence of the Role of Viscosity in the Molecular Kinetic Theory of Dynamic Wetting. *Langmuir* **2011**, *27*, 13015–13021. [CrossRef] [PubMed]

39. Lin, F.; He, L.; Primkulov, B.; Xu, Z. Dewetting Dynamics of a Solid Microsphere by Emulsion Drops. *J. Phys. Chem. C* **2014**, *118*, 13552–13562. [CrossRef]

40. Moeini, F.; Hemmati-Sarapardeh, A.; Ghazanfari, M.-H.; Masihi, M.; Ayatollahi, S. Toward mechanistic understanding of heavy crude oil/brine interfacial tension: The roles of salinity, temperature and pressure. *Fluid Phase Equilib.* **2014**, *375*, 191–200. [CrossRef]

41. Basu, S.; Sharma, M.M. Measurement of Critical Disjoining Pressure for Dewetting of Solid Surfaces. *J. Colloid Interface Sci.* **1996**, *181*, 443–455. [CrossRef]

42. Haagh, M.E.J.; Siretanu, I.; Duits, M.H.G.; Mugele, F. Salinity-Dependent Contact Angle Alteration in Oil/Brine/Silicate Systems: The Critical Role of Divalent Cations. *Langmuir* **2017**, *33*, 3349–3357. [CrossRef] [PubMed]

43. Somasundaran, P.; Zhang, L. Adsorption of surfactants on minerals for wettability control in improved oil recovery processes. *J. Pet. Sci. Eng.* **2006**, *52*, 198–212. [CrossRef]

44. Standnes, D.C.; Austad, T. Wettability alteration in chalk: 2. Mechanism for wettability alteration from oil-wet to water-wet using surfactants. *J. Pet. Sci. Eng.* **2000**, *28*, 123–143. [CrossRef]

45. Wu, Y.; Shuler, P.J.; Blanco, M.; Tang, Y.; Goddard, W.A. *An Experimental Study of Wetting Behavior and Surfactant EOR in Carbonates with Model Compounds*; Society of Petroleum Engineers: Richardson, TX, USA, 2008.

46. Jarrahian, K.; Seiedi, O.; Sheykhan, M.; Sefti, M.V.; Ayatollahi, S. Wettability alteration of carbonate rocks by surfactants: A mechanistic study. *Colloids Surfaces A Physicochem. Eng. Asp.* **2012**, *410*, 1–10. [CrossRef]

47. Xie, X.; Weiss, W.W.; Tong, Z.; Morrow, N.R. *Improved Oil Recovery from Carbonate Reservoirs by Chemical Stimulation*; Society of Petroleum Engineers: Richardson, TX, USA, 2004.

48. Curbelo, F.D.S.; Santanna, V.C.; Neto, E.L.B.; Dutra, T.V.; Dantas, T.N.C.; Neto, A.A.D.; Garnica, A.I.C. Adsorption of nonionic surfactants in sandstones. *Colloids Surfaces A Physicochem. Eng. Asp.* **2007**, *293*, 1–4. [CrossRef]

49. Park, S.; Lee, E.S.; Sulaiman, W.R.W. Adsorption behaviors of surfactants for chemical flooding in enhanced oil recovery. *J. Ind. Eng. Chem.* **2015**, *21*, 1239–1245. [CrossRef]

50. Gupta, R.; Mohanty, K. *Temperature Effects on Surfactant-Aided Imbibition into Fractured Carbonates*; Society of Petroleum Engineers: Richardson, TX, USA, 2010.

51. Vafaei, S.; Borca-Tasciuc, T.; Podowski, M.Z.; Purkayastha, A.; Ramanath, G.; Ajayan, P.M. Effect of nanoparticles on sessile droplet contact angle. *Nanotechnology* **2006**, *17*, 2523–2527. [CrossRef] [PubMed]

52. Alomair, O.A.; Matar, K.M.; Alsaeed, Y.H. *Nanofluids Application for Heavy Oil Recovery*; Society of Petroleum Engineers: Richardson, TX, USA, 2014.

53. Ehtesabi, H.; Ahadian, M.M.; Taghikhani, V. Enhanced Heavy Oil Recovery Using TiO$_2$ Nanoparticles: Investigation of Deposition during Transport in Core Plug. *Energy Fuels* **2015**, *29*, 1–8. [CrossRef]

54. Binks, B.P. Particles as surfactants—Similarities and differences. *Curr. Opin. Colloid Interface Sci.* **2002**, *7*, 21–41. [CrossRef]

55. Fan, H.; Striolo, A. Nanoparticle effects on the water-oil interfacial tension. *Phys. Rev. E* **2012**, *86*, 051610. [CrossRef] [PubMed]

56. Chinnam, J.; Das, D.K.; Vajjha, R.S.; Satti, J.R. Measurements of the surface tension of nanofluids and development of a new correlation. *Int. J. Therm. Sci.* **2015**, *98*, 68–80. [CrossRef]

57. Hendraningrat, L.; Li, S.; Torsæter, O. A coreflood investigation of nanofluid enhanced oil recovery. *J. Pet. Sci. Eng.* **2013**, *111*, 128–138. [CrossRef]

58. Parvazdavani, M.; Masihi, M.; Ghazanfari, M.H. Monitoring the influence of dispersed nano-particles on oil–water relative permeability hysteresis. *J. Pet. Sci. Eng.* **2014**, *124*, 222–231. [CrossRef]

59. Li, S.; Hendraningrat, L.; Torsaeter, O. Improved Oil Recovery by Hydrophilic Silica Nanoparticles Suspension: 2-Phase Flow Experimental Studies. In Proceedings of the International Petroleum Technology Conference, Beijing, China, 26–28 March 2013.

60. Wasan, D.T.; Nikolov, A.D. Spreading of nanofluids on solids. *Nature* **2003**, *423*, 156–159. [CrossRef] [PubMed]

61. Kondiparty, K.; Nikolov, A.; Wu, S.; Wasan, D. Wetting and Spreading of Nanofluids on Solid Surfaces Driven by the Structural Disjoining Pressure: Statics Analysis and Experiments. *Langmuir* **2011**, *27*, 3324–3335. [CrossRef] [PubMed]

62. Zhang, H.; Nikolov, A.; Wasan, D. Enhanced Oil Recovery (EOR) Using Nanoparticle Dispersions: Underlying Mechanism and Imbibition Experiments. *Energy Fuels* **2014**, *28*, 3002–3009. [CrossRef]

63. Trokhymchuk, A.; Henderson, D.; Nikolov, A.; Wasan, D.T. A Simple Calculation of Structural and Depletion Forces for Fluids/Suspensions Confined in a Film. *Langmuir* **2001**, *17*, 4940–4947. [CrossRef]

64. Winkler, K.; Paszewski, M.; Kalwarczyk, T.; Kalwarczyk, E.; Wojciechowski, T.; Gorecka, E.; Pociecha, D.; Holyst, R.; Fialkowski, M. Ionic Strength-Controlled Deposition of Charged Nanoparticles on a Solid Substrate. *J. Phys. Chem. C* **2011**, *115*, 19096–19103. [CrossRef]

65. Darlington, T.K.; Neigh, A.M.; Spencer, M.T.; Guyen, O.T.N.; Oldenburg, S.J. Nanoparticle characteristics affecting environmental fate and transport through soil. *Environ. Toxicol. Chem.* **2009**, *28*, 1191–1199. [CrossRef] [PubMed]

66. Bayat, A.E.; Junin, R.; Samsuri, A.; Piroozian, A.; Hokmabadi, M. Impact of Metal Oxide Nanoparticles on Enhanced Oil Recovery from Limestone Media at Several Temperatures. *Energy Fuels* **2014**, *28*, 6255–6266. [CrossRef]

67. Li, Y.V.; Cathles, L.M.; Archer, L.A. Nanoparticle tracers in calcium carbonate porous media. *J. Nanopart. Res.* **2014**, *16*, 2541. [CrossRef]

68. Li, Y.V.; Cathles, L.M. Retention of silica nanoparticles on calcium carbonate sands immersed in electrolyte solutions. *J. Colloid Interface Sci.* **2014**, *436*, 1–8. [CrossRef] [PubMed]

69. Cassie, A.B.D.; Baxter, S. Wettability of porous surfaces. *Trans. Faraday Soc.* **1944**, *40*, 546–551. [CrossRef]

70. Wenzel, R.N. Resistance of solid surfaces to wetting by water. *Ind. Eng. Chem.* **1936**, *28*, 988–994. [CrossRef]

71. Nwidee, L.N.; Al-Anssari, S.; Barifcani, A.; Sarmadivaleh, M.; Lebedev, M.; Iglauer, S. Nanoparticles influence on wetting behaviour of fractured limestone formation. *J. Pet. Sci. Eng.* **2017**, *149*, 782–788. [CrossRef]

72. Zargartalebi, M.; Barati, N.; Kharrat, R. Influences of hydrophilic and hydrophobic silica nanoparticles on anionic surfactant properties: Interfacial and adsorption behaviors. *J. Pet. Sci. Eng.* **2014**, *119*, 36–43. [CrossRef]

73. Ahualli, S.; Iglesias, G.R.; Wachter, W.; Dulle, M.; Minami, D.; Glatter, O. Adsorption of Anionic and Cationic Surfactants on Anionic Colloids: Supercharging and Destabilization. *Langmuir* **2011**, *27*, 9182–9192. [CrossRef] [PubMed]

74. Choi, S.K.; Son, H.A.; Kim, H.T.; Kim, J.W. Nanofluid Enhanced Oil Recovery Using Hydrophobically Associative Zwitterionic Polymer-Coated Silica Nanoparticles. *Energy Fuels* **2017**, *31*, 7777–7782. [CrossRef]
75. ShamsiJazeyi, H.; Miller, C.A.; Wong, M.S.; Tour, J.M.; Verduzco, R. Polymer-coated nanoparticles for enhanced oil recovery. *J. Appl. Polym. Sci.* **2014**, *131*. [CrossRef]
76. Hou, B.; Wang, Y.; Cao, X.; Zhang, J.; Song, X.; Ding, M.; Chen, W. Surfactant-Induced Wettability Alteration of Oil-Wet Sandstone Surface: Mechanisms and Its Effect on Oil Recovery. *J. Surfactants Deterg.* **2016**, *19*, 315–324. [CrossRef]
77. Karimi, M.; Al-Maamari, R.S.; Ayatollahi, S.; Mehranbod, N. Wettability alteration and oil recovery by spontaneous imbibition of low salinity brine into carbonates: Impact of Mg^{2+}, SO_4^{2-} and cationic surfactant. *J. Pet. Sci. Eng.* **2016**, *147*, 560–569. [CrossRef]
78. Hou, B.; Wang, Y.; Cao, X.; Zhang, J.; Song, X.; Ding, M.; Chen, W. Mechanisms of Enhanced Oil Recovery by Surfactant-Induced Wettability Alteration. *J. Dispers. Sci. Technol.* **2016**, *37*, 1259–1267. [CrossRef]
79. Zhang, R.; Qin, N.; Peng, L.; Tang, K.; Ye, Z. Wettability alteration by trimeric cationic surfactant at water-wet/oil-wet mica mineral surfaces. *Appl. Surf. Sci.* **2012**, *258*, 7943–7949. [CrossRef]
80. Hirasaki, G.; Zhang, D.L. *Surface Chemistry of Oil Recovery from Fractured, Oil-Wet, Carbonate Formations*; Society of Petroleum Engineers: Richardson, TX, USA, 2004.
81. Alvarez, J.O.; Schechter, D.S. *Wettability Alteration and Spontaneous Imbibition in Unconventional Liquid Reservoirs by Surfactant Additives*; Society of Petroleum Engineers: Richardson, TX, USA, 2017.
82. Ehtesabi, H.; Ahadian, M.M.; Taghikhani, V.; Ghazanfari, M.H. Enhanced Heavy Oil Recovery in Sandstone Cores Using TiO_2 Nanofluids. *Energy Fuels* **2014**, *28*, 423–430. [CrossRef]
83. Glaser, N.; Adams, D.J.; Böker, A.; Krausch, G. Janus Particles at Liquid−Liquid Interfaces. *Langmuir* **2006**, *22*, 5227–5229. [CrossRef] [PubMed]
84. Soleimani, H.; Baig, M.K.; Yahya, N.; Khodapanah, L.; Sabet, M.; Demiral, B.M.R.; Burda, M. Impact of carbon nanotubes based nanofluid on oil recovery efficiency using core flooding. *Results Phys.* **2018**, *9*, 39–48. [CrossRef]
85. Wei, B.; Li, Q.; Jin, F.; Li, H.; Wang, C. The potential of a novel nanofluid in enhancing oil recovery. *Energy Fuels* **2016**, *30*, 2882–2891. [CrossRef]
86. Roustaei, A.; Bagherzadeh, H. Experimental investigation of SiO_2 nanoparticles on enhanced oil recovery of carbonate reservoirs. *J. Pet. Explor. Prod. Technol.* **2015**, *5*, 27–33. [CrossRef]
87. Al-Anssari, S.; Barifcani, A.; Wang, S.; Maxim, L.; Iglauer, S. Wettability alteration of oil-wet carbonate by silica nanofluid. *J. Colloid Interface Sci.* **2016**, *461*, 435–442. [CrossRef] [PubMed]
88. Maghzi, A.; Mohammadi, S.; Ghazanfari, M.H.; Kharrat, R.; Masihi, M. Monitoring wettability alteration by silica nanoparticles during water flooding to heavy oils in five-spot systems: A pore-level investigation. *Exp. Therm. Fluid Sci.* **2012**, *40*, 168–176. [CrossRef]
89. Monfared, A.D.; Ghazanfari, M.H.; Jamialahmadi, M.; Helalizadeh, A. Potential Application of Silica Nanoparticles for Wettability Alteration of Oil–Wet Calcite: A Mechanistic Study. *Energy Fuels* **2016**, *30*, 3947–3961. [CrossRef]
90. Shahrabadi, A.; Bagherzadeh, H.; Roostaie, A.; Golghanddashti, H. *Experimental Investigation of HLP Nanofluid Potential to Enhance Oil Recovery: A Mechanistic Approach*; Society of Petroleum Engineers: Richardson, TX, USA, 2012.
91. Li, S.; Genys, M.; Wang, K.; Torsæter, O. *Experimental Study of Wettability Alteration during Nanofluid Enhanced Oil Recovery Process and Its Effect on Oil Recovery*; Society of Petroleum Engineers: Richardson, TX, USA, 2015.
92. Esmaeilzadeh, P.; Hosseinpour, N.; Bahramian, A.; Fakhroueian, Z.; Arya, S. Effect of ZrO_2 nanoparticles on the interfacial behavior of surfactant solutions at air–water and *n*-heptane–water interfaces. *Fluid Phase Equilib.* **2014**, *361*, 289–295. [CrossRef]

colloids
and interfaces

MDPI

Article

Effects of Short-Chain *n*-Alcohols on the Properties of Asphaltenes at Toluene/Air and Toluene/Water Interfaces

Raphael G. Martins, Lilian S. Martins and Ronaldo G. Santos *

Department of Chemical Engineering, Centro Universitário FEI, Av. Humberto de Alencar Castelo Branco, 3972, São Paulo CEP 09850-901, Brazil; raphaelgraciuti@hotmail.com (R.G.M.); lilian_smartins@yahoo.com.br (L.S.M.)
* Correspondence: rgsantos@fei.edu.br; Tel.: +55-11-4353-2900

Received: 5 March 2018; Accepted: 21 March 2018; Published: 23 March 2018

Abstract: Crude oil asphaltenes contain a wide series of chemical species, which includes the most polar compounds and interfacially active agents from the petroleum. Asphaltenes have been considered to be implicated in foam and emulsion formation during the petroleum recovery and production process. In this work, the interfacial activity of organic solutions containing asphaltene and *n*-alcohols was investigated. Asphaltene extraction from a 28° API crude oil produced 2.5 wt % of *n*-pentane precipitated asphaltene (C5I). Dynamic surface and interfacial tensions of asphaltene solutions were assessed by the pendant drop method. Asphaltene films were evaluated at the air-water interface using a Langmuir trough. Results were expressed by means of the interfacial tension time-dependence. Interfacial tension measurements showed alcohols reduce the toluene/water interfacial tension of asphaltene solutions. The interfacial tension was reduced from 23 mN/m to 15.5 mN/m for a 2 g/L solution of asphaltene plus *n*-butanol. Higher asphaltene concentrations did not affect the toluene/air surface tension. The effects of *n*-alcohols on the asphaltene surface activity was dependent on the asphaltene aggregation state. *n*-Alcohols modify the asphaltene film elasticity and the film phase behavior.

Keywords: asphaltene; surface and interfacial tension; alcohols; monolayer; petroleum

---

1. Introduction

Crude oil asphaltenes comprise higher molecular weight polar components that exhibit interfacial activity. Asphaltenes are defined as a solubility class, containing a wide series of chemical species. They constitute the crude oil fraction that is insoluble in low boiling point *n*-alkanes and soluble in toluene, benzene, carbon disulfide, chloroform and other chlorinated hydrocarbon solvents [1]. Strictly, asphaltenes are considered insoluble in *n*-pentane or *n*-heptane and soluble in toluene or benzene. These fractions are commonly abbreviated as C5I and C7I, respectively, according to the precipitant utilized in the extraction [2].

The asphaltene content in the crude oils depends mainly on the nature of the oil. Unconventional (e.g., heavy and extra-heavy) oils usually contain more asphaltene than conventional oils, which directly impacts on the oil recovery, transport and refining processes [3]. Asphaltenes contain polar and nonpolar chemical functionalities, which produce amphiphilic characteristics in solution [4,5]. The asphaltene amphiphilic character gives rise to the colloidal properties, such as the ability to reduce surface tension [1,6,7]. The colloidal aspects of asphaltenes have been studied for a long time [2] because of their impact on emulsion and foam stabilization, wettability alteration, pipeline blockage, damage of reservoir-rock formation and many other impairments.

In solution, asphaltenes exhibit interfacial phenomena that include adsorption and self-association. These phenomena lead to aggregation, flocculation and lastly solid deposition. The formation of asphaltene deposits is a major concern because asphaltene deposition impairs oil recovery and production [2,3]. Deposition of asphaltene can result in the reduction of reservoir-rock permeability and a decrease of the effective pipe diameter, restricting the oil flow.

Despite the assortment of techniques available to avoid asphaltene deposit formation, the addition of dispersants and inhibitors has been preferred because of their effectiveness and low cost [8,9]. The effect of additives on the inhibition of asphaltene precipitation has been investigated [10]. The result showed that ethoxylated nonylphenol and hexadecyl trimethyl ammonium bromide exhibit a positive performance.

Asphaltenes contribute to the formation and stabilization of crude oil emulsions. An interfacial film containing asphaltenes and other petroleum components supports water droplet dispersion into the continuous phase [11,12]. Oil-in-water (o/w) emulsions are pointed out as an emergent technology for moving viscous oils. Heavy oil emulsification can lead to the viscosity decreasing enough to allow a feasible pipeline flow [13,14]. Crude oil emulsions are complex systems containing many chemical structures. Alcohols and electrolytes are recognized additives that modify oil emulsion properties [15]. *n*-Alcohols have been found to be effective co-surfactants in crude oil-in-water emulsions [16]. Short-chain alcohols improve oil-in-water emulsion stability [17]. In addition, the emulsion-stabilizing effect is stressed by increasing the alkyl chain size of the alcohol [18]. This fact is related to the decrease in alcohol solubility in the aqueous phase caused by the alkyl chain length increasing.

Recent work has demonstrated that emulsions containing conventional ethoxylated surfactants and short-chain alcohols present singular properties [11,14]. The authors have shown the occurrence of phase segregation phenomena and the formation of a fluid layer of low viscosity—known as the oil-depleted layer—during stationary rheological flow. As a consequence, the oil flow as o/w emulsions has been described as lubricated flow or slippage. The rheological behavior of the emulsions was described as markedly shear-thinning [14]. Both phase segregation and shear thinning phenomena are supposedly related to the formation of multicomponent interfacial films, which would influence the resistance and the deformability of the dispersed droplet interfaces.

In this work, the effects of the addition of linear alcohols on the interfacial behavior of asphaltenes at toluene/air and toluene/water interfaces has been investigated by means of surface tension and surface pressure measurements. *n*-Alcohols containing 4–8 carbons were blended with petroleum *n*-pentane-precipitated asphaltenes. Surface and interfacial tension were assessed by the pendant drop method. In addition, asphaltenes were spread on the air-water surface to build mixed monolayers with *n*-alcohols.

2. Materials and Methods

2.1. Materials

This study used samples of a Brazilian light oil (28° API) provided as a kind gift by Petrobras. Chemicals were toluene (99.8%), *n*-pentane (>99%), *n*-butanol (>99%), *n*-hexanol (>99%) and *n*-octanol (>99%), purchased from Sigma-Aldrich. Ethoxylated nonylphenol (ULTRANEX) with 10 ethylene oxide units was provided by Oxiteno (Mauá, Brazil) and used as received. Deionized and double-distilled water was used throughout.

2.2. Methods

2.2.1. Crude Oil Fractionation

Asphaltene extraction from crude oil was performed according to its solubility. The *n*-pentane-insoluble asphaltene (C5I) fraction was removed from the crude oil through a modified IP-143/89 methodology. Briefly, 800 mL of flocculant (*n*-pentane) was mixed with 20 g of crude oil in a closed vessel. The mixture was stirred for 24 h at room temperature (20 ± 1 °C). The solid precipitated was removed by vacuum filtration using a Whatman #42 filter paper (Sigma-Aldrich Corp.,

St. Louis, MO, USA) and the soluble phase (deasphaltened oil) was collected. The asphaltenes were separated from the solid precipitate by means of a Soxhlet extraction using *n*-pentane. This procedure was continued until the draining solvent was colorless. The asphaltene was dried before further use.

2.2.2. Surface and Interfacial Tension Measurement

The surface tension (ST) and interfacial tension (IFT) were each measured by the pendant drop method using a Theta Lite optical tensiometer (Biolin Scientific, Gothenburg, Sweden). The equipment captures successive images of the droplet through a charge-coupled device (CCD) camera to analyze shape parameters. The evaluation system applies the axisymmetric drop shape analysis (ADSA) method along with the Laplace equation to determine the tension. Data analyses were supported by OneAttension software, (version 2, 2015-02-04, Gothenburg, Sweden). The equilibrium state was defined through the surface and interfacial tension as a function of time. ST and IFT measurements were carried out at 20 °C (±1 °C).

2.2.3. Surface Pressure Measurement

The surface pressure of spread films on water was evaluated using a Langmuir trough (Biolin Scientific, Sweden) with an area of 242.25 cm^2 (32.3 cm × 7.5 cm). The trough was equipped with two mobile barriers made from hydrophilic Delrin. Deionized and double-distilled water was used as the subphase, which was kept constant at 20 °C. The temperature was controlled within 0.1 °C by a refrigerated circulator (Julabo, Germany). Asphaltenes were spread on the air-water surface from fresh dichloromethane solutions with a concentration of 2 g·L^{-1}. A volume of 50 μL of C5I asphaltene solution was used. The solvent was allowed to evaporate for at least 15 min before starting the measurements. The barrier speed was maintained constant at 10 mm min^{-1}. The surface pressure was zeroed before asphaltene spreading. Tests carried out before the asphaltene addition on the aqueous phase always displayed surface pressure changes lower than 0.05 mN m^{-1}.

3. Results

3.1. Asphaltene Interfacial Behavior

Addition of *n*-pentane to 20 g of oil yields 0.7377 g of solid precipitate. The solid Soxhlet cleaning procedure yields 0.5003 g of C5I asphaltenes, corresponding to a content of 2.5 wt % of asphaltene in the crude oil. The weight difference implies 32 wt % of impurities present in the precipitated solid. The impurities are suggested to be comprised of other oil fractions (saturates, resin and aromatics) precipitated as asphaltene mixed clusters. The C5I asphaltenes were dissolved in toluene at concentrations ranging from 1 g/L to 10 g/L. The surface and interfacial tension of asphaltene solutions as a function of time are shown in Figure 1. ST data are shown in Figure 1a for asphaltene concentrations up to 3 g/L. Surprisingly, surface tension measurements produced a dispersion of data, regardless of the asphaltene concentration. The surface tension of 4–10 g/L solutions displayed similar behavior. At 10 g/L, the surface tension was 28.5 ± 0.2 mN/m, for instance. This result shows that asphaltenes do not exhibit significant surface activity in the air-water interface for the concentration range studied.

Interfacial tension data are presented in Figure 1b. The results indicate that increasing the asphaltene concentration leads to a continuous reduction of the interfacial tension. The equilibrium interfacial tension was evaluated at 900 s. The maximum IFT reduction was achieved at a concentration of 5 g/L, which means an IFT change from 33.5 mN/m for pure toluene to 20.0 mN/m for a 5 g/L asphaltene solution. No further decline of the interfacial tension is seen at concentrations greater than 5 g/L.

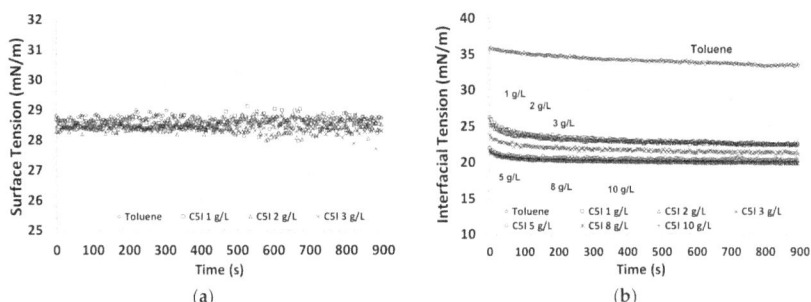

Figure 1. Asphaltene interfacial activity: (a) Surface tension and (b) interfacial tension of C5I in toluene solutions as a function of time.

The equilibrium surface and interfacial tensions were plotted against the solution concentration in order to determine the Critical Aggregation Concentration (CAC) of asphaltene for the toluene-water system. Results are shown in Figure 2. Figure 2a highlights the constant surface tension with increasing asphaltene concentration. This means that no interfacial effects take place at the air-water interface. Figure 2b displays an exponential-type decline of the interfacial tension with increasing concentration. The asphaltene CAC was found at approximately 1.4 g/L.

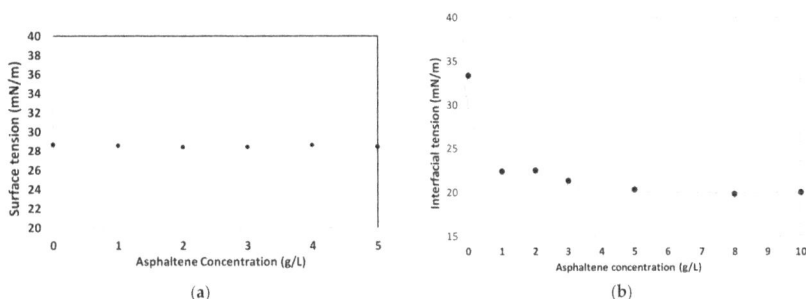

Figure 2. (a) Surface tension and (b) interfacial tension as a function of asphaltene-in-toluene solution concentration.

Figures 1 and 2 describe asphaltene colloidal behavior. Asphaltenes are denoted as single monomers at very low concentrations. As the concentration increases, the monomers adsorb at the interface as stacks, reducing the interfacial tension, as indicated by the reference [2]. At concentrations higher than the CAC, asphaltene is represented as colloidal aggregates.

3.2. Interfacial Behavior for Asphaltene + Alcohol Systems

n-Alcohols were mixed with C5I asphaltenes to evaluate the mixture surface and interfacial tensions. *n*-Butanol, *n*-hexanol and *n*-octanol were mixed with toluene to produce a solvent containing 90 vol % toluene and 10 vol % alcohol. C5I asphaltene was then dissolved in the toluene/alcohol solvent mixtures at concentrations of 0.5 g/L and 2 g/L, in order to evaluate the asphaltene interfacial behavior at concentrations above and below the CAC.

Figure 3 displays the surface and interfacial tension of C5I solutions in a solvent composed of 90% toluene and 10% *n*-butanol. Figure 3a shows that the surface tension of asphaltene solutions was approximately constant over the experimental timescale for this mixture—similar to the pure toluene solutions, although the interfacial tension increases over time (Figure 3b). The time-dependence of the interfacial tension is related to the surface saturation kinetics. The surface saturation rate is governed mainly by two processes: (1) molecular diffusion to the interface and (2) the reorientation of molecules at the interface [19,20]. In addition, *n*-butanol makes the solvent more polar, increasing the water-solvent mutual solubility. As a result, the interfacial tension of the toluene/*n*-butanol mixture against water increases to reach the equilibrium at about 23 mN/m. At a concentration below the CAC, asphaltene did not affect the interfacial tension, exhibiting a behavior similar to the pure solvent. Above the CAC, the solvent interfacial tension was reduced from 23 mN/m to about 15.5 mN/m at equilibrium, indicating asphaltene interfacial activity.

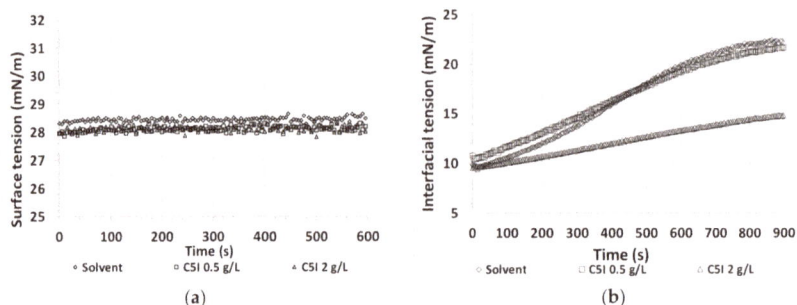

Figure 3. (**a**) Surface tension and (**b**) interfacial tension as a function of time for C5I in 9:1 *v/v* toluene/*n*-butanol solutions.

The surface and interfacial tensions for asphaltene solutions in a 9:1 *v/v* toluene/*n*-hexanol mixture as a function of time are presented in Figure 4. Again, Figure 4a shows behavior analogous to that found in *n*-hexanol, remaining approximately constant over time and across concentration variations. At the oil-water interface, no abrupt change in the solvent interfacial tension was observed, even at a concentration higher than the CAC, as illustrated by Figure 4b. Increasing asphaltene concentration results in a small increase in the interfacial tension. This behavior is suggested to be related to the minor effect of *n*-hexanol on the asphaltene solubility in comparison to *n*-butanol. Besides, the *n*-hexanol solubility in water is much lower than the *n*-butanol solubility.

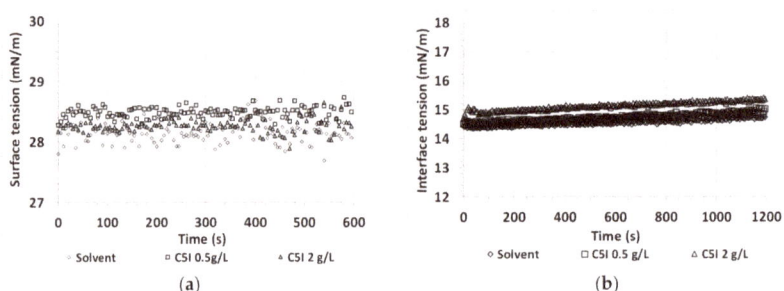

Figure 4. (**a**) Surface tension and (**b**) interfacial tension as a function of time for C5I in 9:1 *v/v* toluene/*n*-hexanol solutions.

Figure 5 illustrates the effects of *n*-octanol on the surface and interfacial tension of asphaltene solutions. The surface tension was about constant over time and with concentration, in agreement with previous results for *n*-butanol and *n*-hexanol (Figure 5a). In the presence of *n*-octanol, the interfacial tension is slightly reduced over time. The IFT curves for the solvent and 0.5 g/L C5I solution are nearly coincident with time, indicating that at concentrations below the CAC, asphaltene demonstrated no effective interfacial activity. For a concentration above the CAC, the IFT is slightly below the solvent IFT. In this case, the diffusion of *n*-octanol molecules must affect the IFT because of the difference in molecular weight.

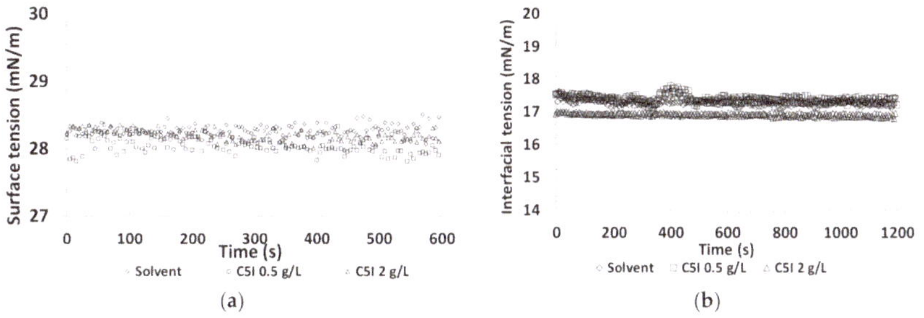

Figure 5. (**a**) Surface tension and (**b**) interfacial tension as a function of time for C5I in 9:1 *v/v* toluene/*n*-octanol solutions.

Surfactants are known to reduce the interfacial tension of oil-water systems (Rudin and Wasan, 1992). Nonylphenol ethoxylated surfactants (NPE) are widely used in crude oil emulsion formulations. A series of tests was therefore carried out to evaluate the combined effects of C5I asphaltene and a conventional surfactant. The surfactant was a NPE with 10 ethylene oxide groups, corresponding to an HLB of 13.3. The asphaltene (C5I) and the surfactant (NPE) were mixed in different proportions and then the C5I + NPE blend was dissolved in the solvent to produce a total concentration of 2 g/L. The solvents were 9:1 *v/v* toluene/alcohol mixtures, as above. The results of interfacial tension measurements are present in Figure 6. The equilibrium interfacial tension was evaluated at 600 s, with the exception of C5I in toluene/*n*-butanol, for which equilibrium was reached at 900 s. The data highlight the influence of the solvent in the asphaltene/surfactant interfacial activity. The greater difference between NPE and C5I interfacial tension was found in toluene, where the IFT was 9 mN/m for NPE and 22.5 mN/m for C5I. For toluene/alcohol mixtures, the interfacial tension was found to be intermediate. The interfacial tension increases with increasing alcohol chain length. The results in Figure 6 underline the variation of the C5I interfacial behavior according to the solvent properties. The addition of alcohol to toluene reduces the interfacial tension. The higher the alcohol polarity, the lower the interfacial tension.

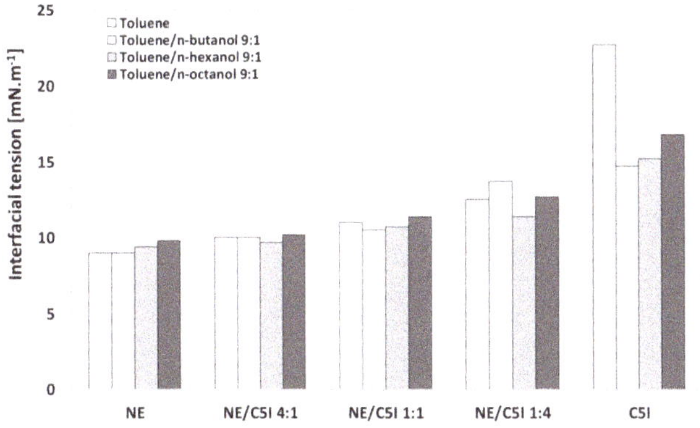

Figure 6. Interfacial tension of C5I + NPE mixtures in different solvents.

3.3. Langmuir Films Containing Asphaltenes and n-Alcohols

Langmuir isotherms resulting from surface pressure measurements for asphaltene films performed at 20 °C are shown in Figure 7 in terms of the surface pressure—surface area (π-A) isotherms. The films were spread at the air-water surface. *n*-Alcohols were added into 2 g/L C5I solutions. The alcohol amounts were: *n*-butanol = 9.5×10^{-5} mol (0.007 g); *n*-hexanol = 8.5×10^{-5} mol (0.0087 g); *n*-octanol = 9.1×10^{-5} mol (0.012 g). The respective resulting C5I:alcohol mass ratios are 70, 87 and 120.

The C5I isotherm shape (see Figure 7) is analogous to asphaltene isotherms previously reported by other authors [11,21–23]. Initially, the isotherm shows a short region indicative of the phase transition from gaseous to liquid expanded states, which is usual for weakly interacting molecules. The film compression produced an expanded liquid region up to about 180 cm^2, at a surface pressure of 18 mN/m. From this point, a second phase transition occurs and the isotherm exhibits behavior corresponding to the coexistence of a two-phase region, containing both an expanded liquid phase and a condensed liquid phase. Further compression up to 30 cm^2 leads the system to a third phase transition, from which only a condensed liquid state is present. At this point, the surface pressure reaches 41 mN/m. The surface pressure reached 46 mN/m at the minimum area. High surface pressure indicates low film compressibility. Film collapse, which is characterized by an abrupt reduction of surface pressure, was not observed. This fact indicates substantial film elasticity.

The interfacial behavior of C5I:alcohol mixed films was dissimilar to the films containing only asphaltenes. First, the gaseous film region was broader for monolayers containing *n*-alcohols. Second, the film elasticity is altered by the addition of alcohols. The presence of *n*-butanol initially leads to a characteristic gaseous film and a smooth phase transition. Films containing *n*-hexanol exhibit only one phase transition, corresponding to the gas-liquid expanded transition. The C5I:*n*-hexanol film displayed a large expanded liquid region, starting with very low surface pressure (5 mN/m). Isotherms obtained from C5I:*n*-octanol films exhibited well-defined phase regions. Initially, behavior representative of a gaseous film is observed, followed by a region corresponding to liquid expanded film. Next, the isotherm shows a clear liquid expanded-liquid condensed phase transition region. Finally, a region containing only liquid condensed states is apparent. *n*-Alcohols produce weaker interactions on the asphaltene film. As a consequence, the adding of alcohols makes the films more compressible.

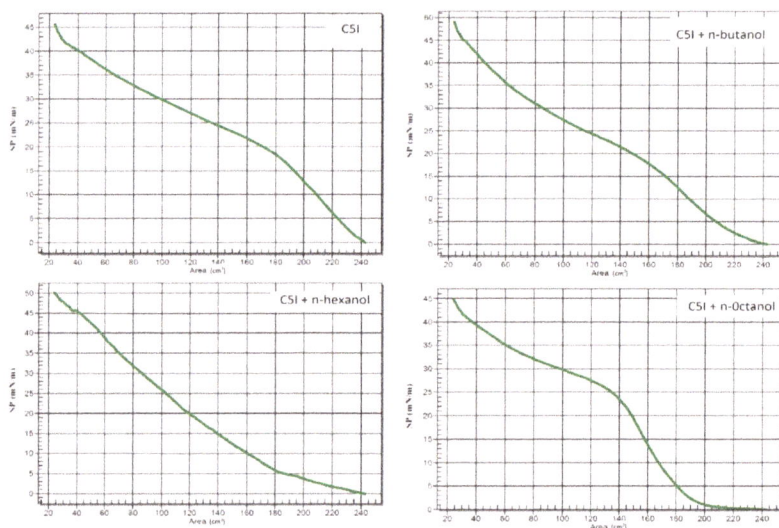

Figure 7. Langmuir isotherms (π-A) for systems containing asphaltenes and different *n*-alcohols. SP: Surface Pressure.

4. Conclusions

Surface/interface tension and surface pressure measurements highlight the influence of linear alcohols on the interfacial properties of crude oil asphaltene. *n*-Pentane asphaltenes extracted from a light oil displayed substantial self-aggregation behavior, verified by tensiometry and interfacial adsorption from surface pressure measurements. Asphaltene did not show an effective reduction in surface tension. *n*-Alcohols promoted a further reduction of the surface and interfacial tension of asphaltene solution. The presence of *n*-alcohols lead to more compressible asphaltene films. It has been reported that oil/water interfacial tension [14] and interfacial film elasticity [11] both regulate key emulsion properties such as stability. In this way, the present results could be useful in the design of crude oil-in-water emulsions with appropriate stability, droplet size and viscosity, by way of adding suitable alcohols into the emulsion formulation. Finally, the results indicate the possibility of *n*-alcohol addition to emulsifying and demulsifying agents, respectively, to heavy oil pipelines and in water-in-oil breaking applications.

Acknowledgments: The authors gratefully acknowledge support from the São Paulo Research Foundation (FAPESP) by means of the Project 2013/25880-3.

Author Contributions: R.G. Santos was responsible for the experimental design, surface pressure measurements, data analysis and data report. R.G. Martins and L.S.M. contribute to the current research by sample preparation and performing surface and interfacial tension measurements.

Conflicts of Interest: The authors declare no conflicts of interest.

References

1. Speight, J.G. *The Chemistry and Technology of Petroleum*, 2nd ed.; Marcel Dekker, Inc.: New York, NY, USA, 1991.
2. Loh, W.; Mohamed, R.S.; Santos, R.G. Crude Oil Asphaltenes: Colloidal Aspects. In *Encyclopedia of Surface and Colloid Science*; Somasundaram, P., Ed.; Taylor & Francis: New York, NY, USA, 2007; Volume 1, pp. 1–18.
3. Santos, R.G.; Loh, W.; Bannwart, A.; Trevisan, O. An overview of heavy oil properties and its recovery and transportation methods. *Braz. J. Chem. Eng.* **2014**, *31*, 571–590. [CrossRef]
4. Ovalles, C.; Garcia, M.D.; Lujano, E.; Aular, W.; Bermudez, R.; Cotte, E. Structure-interfacial activity relationships and thermal stability studies of Cerro Negro crude oil and its acid, basic and neutral fractons. *Fuel* **1998**, *77*, 121–126. [CrossRef]
5. Nenningsland, A.; Simon, S.; Sjoblom, J. Surface properties of basic components extracted from petroleum crude oil. *Energy Fuels* **2010**, *24*, 6501–6505. [CrossRef]
6. Ramos, A.C.S.; Haraguchi, L.H.; Notrispe, F.R.; Loh, W.; Mohamed, R.S. Interfacial and colloidal behavior of asphaltenes obtained from Brazilian crude oils. *J. Pet. Sci. Eng.* **2001**, *32*, 201–216. [CrossRef]
7. Sheu, E.Y. Petroleum asphaltenes—Properties, characterization, and issues. *Energy Fuels* **2002**, *16*, 74–82. [CrossRef]
8. Stephenson, K. Producing asphaltenic crude oils: Problems and solutions. *Pet. Eng. Int.* **1990**, *8*, 24–31.
9. De Boer, R.B.; Leerlooyer, K.; Eigner, M.R.P.; van Berger, A.R.D. Screening of crude oils for asphalt precipitation: Theory, practice and the selection of inhibitors. *Soc. Pet. Eng.* **1992**, *24987*, 259–270. [CrossRef]
10. Mohamed, R.S.; Loh, W.; Ramos, A.C.S.; Delgado, C.C.; Almaeida, V.R. Reversibility and inhibition of asphaltene precipitation in Brazilian crude oils. *Pet. Sci. Technol.* **1999**, *17*, 877–896. [CrossRef]
11. Kabbach, C.B.; Santos, R.G. Effects of pH and temperature on the phase behavior and properties of asphaltene liquid films. *Energy Fuels* **2017**. [CrossRef]
12. Oliveira, P.F.; Santos, I.C.V.M.; Vieira, H.V.P.; Fraga, A.K.; Mansur, C.R.E. Interfacial rheology of asphaltene emulsions in the presence of nanoemulsions based on a polyoxide surfactant and asphaltene dispersant. *Fuel* **2017**, *193*, 220–229. [CrossRef]
13. Langevin, D.; Poteau, S.; Hénaut, I.; Argillier, J.F. Crude oil emulsion properties and their application to heavy oil transportation. *Oil Gas Sci. Technol.* **2004**, *59*, 511–521. [CrossRef]
14. Santos, R.G.; Bannwart, A.C.; Loh, W. Phase segregation, shear thinning and rheological behavior of crude oil-in-water emulsions. *Chem. Eng. Res. Des.* **2014**, *92*, 1629–1636. [CrossRef]

15. Salager, J.L.; Briceño, M.I.; Brancho, C.L. Heavy hydrocarbon emulsions. In *Encyclopedic Handbook of Emulsion Technology*; Sjöblom, J., Ed.; Marcel Dekker: New York, NY, USA, 2001.

16. Santos, R.G.; Briceño, M.I.; Bannwart, A.C.; Loh, W. Physico-chemical properties of heavy crude oil-in-water emulsions stabilized by mixtures of ionic and non-ionic ethoxylated nonylphenol surfactants and medium chain alcohols. *Chem. Eng. Res. Des.* **2011**, *89*, 957–967. [CrossRef]

17. Graciaa, A.; Lachaise, J.; Cucuphat, C.; Bourrel, M.; Salager, J.L. Improving solubilization in microemulsions with additives. 2. Long chain alcohols as lipophilic linkers. *Langmuir* **1993**, *9*, 3371–3374.

18. Bourrel, M.; Chambu, C. The rules for achieving high solubilization of brine and oil by amphiphilic molecules. *Soc. Pet. Eng. J.* **1983**, *23*, 327–338. [CrossRef]

19. Ross, S. The Change of Surface Tension with Time. I. Theories of Diffusion to the Surface. *J. Am. Chem. Soc.* **1945**, *67*, 990–994.

20. Firooz, A.; Chen, P. Surface tension and adsorption kinetics of amphiphiles in aqueous solutions: The role of carbon chain length and temperature. *J. Colloid Interface Sci.* **2012**, *370*, 183–191. [CrossRef] [PubMed]

21. Nordli, K.G.; Sjoblom, J.; Kizling, J.; Stenius, P. Water-in-crude oil-emulsions from the Norwegian continental-shelf. 4. Monolayer properties of the interfacially active crude-oil fraction. *Colloids Surf.* **1991**, *57*, 83–98. [CrossRef]

22. Poteau, S.; Argillier, J.; Langevin, D.; Pincet, F.; Perez, E. Influence of pH on Stability and dynamic properties of asphaltenes and other amphiphilic molecules at the oil-water interface. *Energy Fuels* **2005**, *19*, 1337–1341. [CrossRef]

23. Vieira, V.C.C.; Severino, D.; Oliveira, O.N.; Pavinatto, F.J.; Zaniquelli, M.E.D.; Ramos, A.P.; Baptista, M.S. Langmuir films of petroleum at the air-water interface. *Langmuir* **2009**, *25*, 12585–12590. [CrossRef] [PubMed]

colloids
and interfaces

MDPI

Article

Evaluation of Cyclodextrins as Environmentally Friendly Wettability Modifiers for Enhanced Oil Recovery

Adriana Falconi Telles da Cruz [1], Ramon Domingues Sanches [1], Caetano Rodrigues Miranda [2] and Sergio Brochsztain [1,*]

[1] Centro de Engenharia, Modelagem e Ciências Sociais Aplicadas, Universidade Federal do ABC, Santo André 09210-580, Brazil; adri.falconii@hotmail.com (A.F.T.d.C.); rads_rds@hotmail.com (R.D.S.)
[2] Instituto de Física, Universidade de São Paulo, São Paulo 05508-090, Brazil; caetano.miranda@gmail.com
* Correspondence: sergio.brochsztain@ufabc.edu.br; Tel.: +55-11-4996-8260

Received: 4 February 2018; Accepted: 1 March 2018; Published: 6 March 2018

Abstract: In the present work, the use of Cyclodextrins (CDs) as wettability modifiers for enhanced oil recovery (EOR) was evaluated. Cyclodextrins (CDs) are cyclic oligosaccharides that form inclusion complexes with various organic molecules, including *n*-alkanes. Wettability was evaluated through the contact angle (θ) of an *n*-dodecane drop in contact with a quartz surface and immersed in a 0.6 M NaCl aqueous solution containing the CDs. The quartz surface was functionalized with octadecyltrichlorosilane (OTS), rendering the surface oil-wet (C_{18}-quartz). Here, the *n*-dodecane, the saline solution and the C_{18}-quartz represent the oil, the reservoir brine and an oil-wet rock surface, respectively. In the absence of CDs, the *n*-dodecane drops spread well over the C_{18}-quartz, showing that the surface was oleophilic. In the presence of CDs, remarkable effects on the wettability were observed. The most dramatic effects were observed with α-cyclodextrin (α-CD), in which case the C_{18}-quartz surface changed from oil-wet (θ = 162°) in the absence of CD to water-wet (θ = 33°) in the presence of 1.5% (*w/v*) α-CD. The effects of the CDs can be explained by the formation of surface-active inclusion complexes between the CDs and *n*-dodecane molecules. The CD inclusion complexes can be regarded as pseudo-surfactants, which are less harmful to the environment than the traditional surfactants employed by the petroleum industry.

Keywords: cyclodextrins; inclusion complexes; enhanced oil recovery; contact angles; wettability; Pickering emulsions

1. Introduction

Only about one third of the original oil in place can be recovered using conventional (primary and secondary) recovery methods, such as water-flooding. An incremental amount can be recovered from reservoirs using enhanced oil recovery (EOR) methods [1–3], also known as tertiary recovery methods. According to Lake et al. [1], EOR can be defined as additional oil recovery by the injection of materials not normally present in petroleum reservoirs. The additives used in EOR are usually added to the aqueous phase during water-flooding. An important class of additives for EOR is constituted by compounds that change the wettability of rock surfaces [4–6]. These compounds render the rock surfaces more water-wet and decrease the oil/water interfacial tension, facilitating the displacement of oil by water-flooding processes.

In the present work, we evaluate the use of Cyclodextrins (CDs) as wettability modifiers for EOR. CDs are thoroid-shaped cyclic oligosaccharides constituted by glucose units linked to each other by α-1→4 bonds [7,8]. The most well-known CDs are α-CD, β-CD and γ-CD, with 6, 7 and 8 glucose units, respectively (Figure 1). The macrocyclic CD ring displays a central cavity where various guest molecules can be accommodated, resulting in the formation of inclusion complexes (also known as host-guest complexes) [8]. CDs are environmentally friendly substances, in contrast to many additives employed for EOR. They are prepared from renewable raw materials, and are readily biodegradable and biocompatible [7,8]. In addition to EOR applications, the wettability effects of the CDs could also be exploited for the remediation of oil spills.

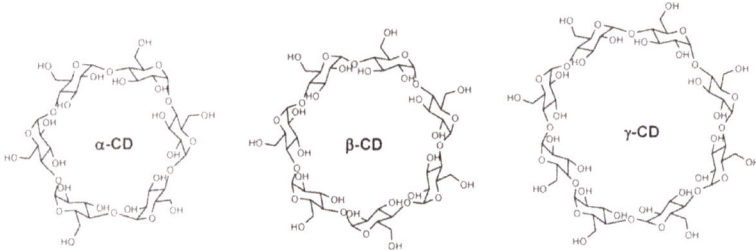

Figure 1. Structure of the most common types of cyclodextrins.

The CDs have been extensively used in the pharmaceutical industry, as well as food and cosmetic additives [7,8]. Nevertheless, there are very few reports on the use of CDs for the petroleum industry [9–13]. Some early patents described processes to extract oil from oil sands using Cyclodextrins [9,10], but no further studies were found in the literature. Leslie et al. [11] used a CD-based polymer as a thickening agent for water permeability reduction in EOR. Similarly, Kjøniksen et al. [12] studied the effect of adding CDs to other polysaccharides for water permeability reduction. In those cases, however, the CD-based additives for EOR were aimed at enhancing the viscosity of the water phase [13], rather than altering the wettability.

The wettability of a triphasic oil/rock/brine system has been mainly evaluated by contact angle measurements [14–22]. In the present work, we used the arrangement shown in Figure 2 for contact angle measurements. A drop of *n*-dodecane was placed under a quartz surface, with the whole system immersed in a saline solution (0.6 M NaCl), where the CDs to be tested were dissolved. In this system, the quartz substrate represents the surface of sandstone rocks, the *n*-dodecane is a model for the oil phase and the saline solution simulates the reservoir brine. The contact angle θ (measured through the water phase) is then related to the interfacial tensions by the Young equation (Equation (1)), where γ_{os}, γ_{ws} and γ_{ow} are the interfacial tensions between the solid (s), oil (o) and water (w) phases, as indicated in Figure 2B. If the oil drop spreads on the substrate (θ > 90°), the surface is considered oil-wet. On the other hand, if the oil has no trend to spread (θ < 90°) the surface is considered water-wet. A contact angle close to 90° means an intermediate-wet surface [14–22].

$$\gamma_{os} - \gamma_{ws} = \gamma_{ow} \cos\theta \tag{1}$$

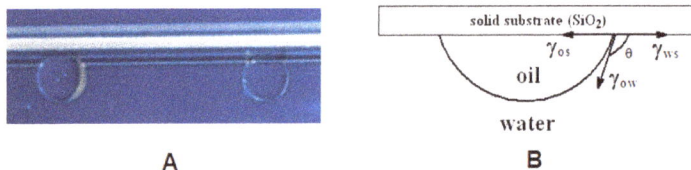

| A | B |

Figure 2. Experimental arrangement employed to measure the effect of cyclodextrins (CDs) on the contact angles. (**A**) Photograph showing *n*-dodecane drops in contact with the quartz surface in a cuvette filled with the aqueous phase; (**B**) Analysis of the drop shape according to the Young equation (Equation (1)), showing the contact-angle (through the water phase) and the interfacial tensions. The solid substrate was a modified quartz surface, the oil drop was *n*-dodecane and the aqueous phase was a brine solution (0.6 M NaCl) containing dissolved CDs.

In a previous report [23], we showed that the CDs can increase the water wettability of quartz surfaces in a three-phase system like that of Figure 2. The observed effects of the CDs, however, were small, because the study started from pristine quartz surfaces, which were already water-wet. Pristine quartz surfaces do not reflect the actual situation found in oil reservoirs, where wettability inversion due to asphaltene deposition renders the rock surfaces oil-wet.

In the present study, we evaluate the potential of CDs for EOR employing quartz surfaces functionalized with octadecyl groups (C_{18}-quartz, Scheme 1). The oleophilic C_{18}-quartz surface is employed as a mimetic system for oil-wet rock surfaces. Remarkable wettability alterations were observed in the presence of the different CDs tested. The most striking effects were observed with α-CD, in which case the C_{18}-quartz changed from oil-wet in the absence of CDs to water-wet in their presence. The effects of CDs on the wettability of quartz surfaces can be attributed to the formation of inclusion complexes between the CDs and *n*-dodecane at the oil/water interface, with the hydrocarbon included in the CD cavity (Scheme 2). Recent studies have shown that the CDs are not surface-active on their own [24,25], but inclusion complexes of the CDs with linear alkanes behave as pseudo-surfactants, decreasing the oil/water interfacial tension [25,26]. Those studies were aiming at applications of the pseudo-surfactants in pharmaceutical and cosmetic products, which means that the present results could also be of interest of researchers in those fields.

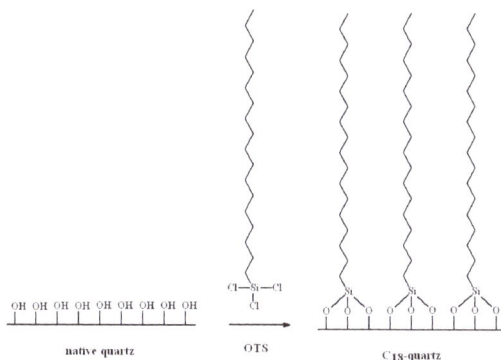

Scheme 1. Functionalization of a quartz surface with octadecyltrichlorosilane (OTS) to give oleophilic C_{18}-quartz.

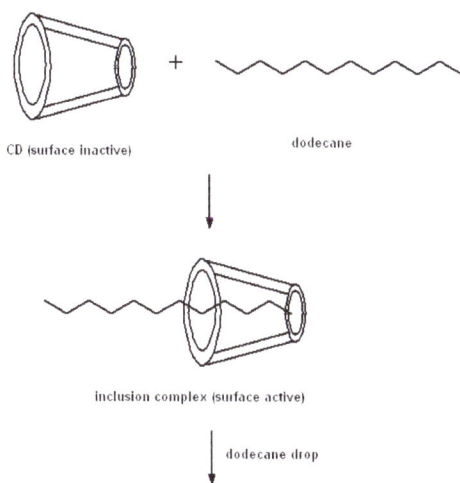

Scheme 2. *Cont.*

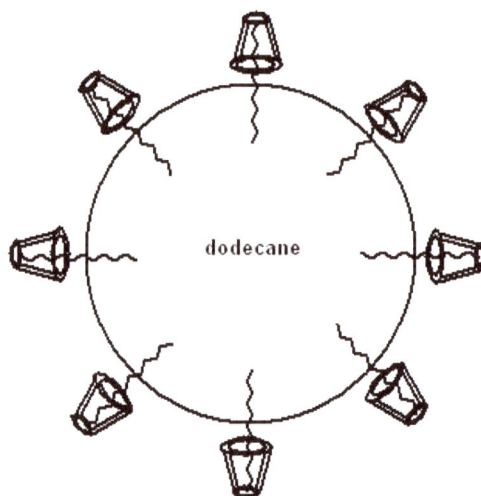

Scheme 2. Formation of inclusion complexes between CDs and *n*-dodecane, and their assembly on the oil drop surface.

2. Materials and Methods

2.1. Materials

Octadecyl trichlorosilane (OTS) was purchased from Sigma-Aldrich (St. Louis, MO, USA). Chloroform, toluene and acetonitrile were HPLC grade from J.T. Baker (Phillipsburg, NJ, USA). α-Cyclodextrin (α-CD), β-cyclodextrin (β-CD) and γ-cyclodextrin (γ-CD) were obtained from Fluka (Buchs, Switzerland). 2-Hydroxypropyl-α-cyclodextrin (HP-α-CD) (average MW~1180; 0.6 mol substitution), 2-hydroxypropyl-β-cyclodextrin (HP-β-CD) (average MW~1380; 0.6 mol substitution) and *n*-dodecane (>99% purity) were purchased from Sigma-Aldrich. NaCl was supplied by Sigma-Aldrich. Aqueous solutions were prepared with deionized water (Barnstead Milli-Q system, Thermo Fisher Scientific, Waltham, MA, USA). A fluorescence cuvette (111-QS, 3600 µL capacity, 10 mm path length) (Hellma Analytics, Müllheim, Germany) was used as the quartz substrate (Figure 2A). According to the supplier, the cuvette windows are made from quartz Suprasil (Heraeus Quarzglas GmbH, Kleinostheim, Germany), which is a synthetic quartz of high purity and homogeneity.

2.2. Instruments

Contact angle measurements were performed with a Theta Optical Tensiometer (Biolin Scientific, Gothenburg, Sweden). The tensiometer records drop images as a function of time and analyses the drop shapes using the included software (OneAttension v2.2, Biolin Scientific). The contact angles are then calculated for individual images using the Young equation (Equation (1), Figure 2B) to fit the entire drop profile. The equipment measures the angles through the *n*-dodecane drop. In order to follow the convention, the values reported in this work are the angles through the aqueous phase (Figure 2B), which are the complementary of the measured angles ($\theta = 180°$—measured angle). The given values are the mean between right and left contact angles for each drop. Advancing and receding contact angles were measured with the tilting cradle accessory of the tensiometer, which allows tilting of the sample stage up to 90°. Interfacial tension measurements were also carried out with the Theta tensiometer, using the pendant drop method, with a dodecane drop hanging from a J-shaped syringe needle inside the brine-containing aqueous CD solutions.

2.3. Methods

2.3.1. Functionalization of the Quartz Surfaces

A solution was prepared with 40 μL OTS in 10 mL *n*-dodecane ([OTS] ≈ 0.01 M). The quartz cuvette was filled with 3 mL of the OTS solution and left standing for 15 min. After that time, the solution was removed and the cuvette was thoroughly rinsed first with *n*-dodecane (8 times), than with toluene (6 times) and finally with chloroform (6 times). After the washing procedure the cuvette was dried with a flow of argon gas. As shown in Figure 3, the quartz surface changed from water-wet to oil-wet after the treatment.

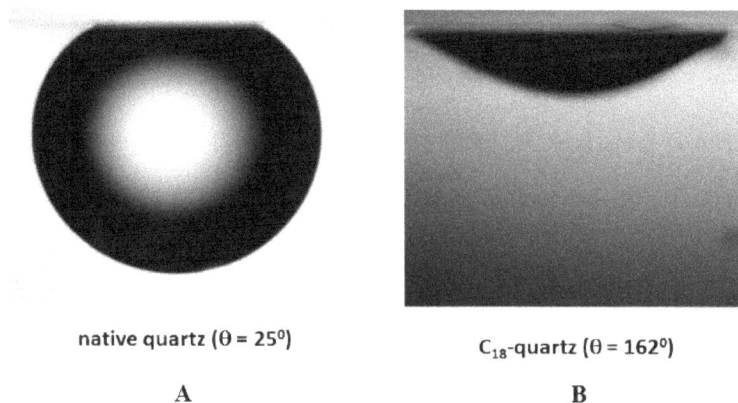

native quartz (θ = 25°)

C_{18}-quartz (θ = 162°)

A   **B**

Figure 3. Drops of *n*-dodecane immersed in the aqueous phase (0.6 M NaCl) and in contact with quartz surfaces. (**A**) Native quartz (before OTS treatment); (**B**) C_{18}-quartz (after OTS treatment).

2.3.2. Contact Angle Measurements

For the measurements, the C_{18}-quartz cuvette was filled up with the aqueous phase and then placed lying down on the sample stage of the equipment. Drops of *n*-dodecane (with ca. 6 μL) were introduced with the help of a microsyringe. Because *n*-dodecane is less dense then water, the drops detached from the needle tip and got attached to the upper cuvette wall (Figure 2A). Typically, at least three drops were deposited and analyzed in each experiment, and the average contact angle (through the aqueous phase) was reported. The experiments were performed with the cuvette open (without the PTFE stoppers), making it easy to inject the drops. Because of the surface tension of the water, the solutions did not leak out of the cuvette. Since the camera was triggered manually, it takes ca. 30 s between injecting the drop, positioning the cuvette and start recording the images. All measurements were performed at 25 °C, with air-equilibrated solutions (pH ≈ 6, not altered by the presence of the CDs). All CD concentrations were expressed as grams of CD per 100 mL (w/v %) (a 1% solution of α-CD or β-CD corresponds to ca. 0.01 mol/L).

The use of fluorescence quartz cuvettes for contact angle measurements was first proposed by Askvik et al. [27], and is a very convenient method, thanks to the optical transparency of the cuvette walls. The cuvette was washed thoroughly between experiments, first with water, then acetonitrile and finally with chloroform, and dried with a flow of argon gas. This cleaning protocol has been repeated several times and the experiments are reproducible after the treatment, indicating that the cuvette remains in its initial oil-wet condition, without OTS desorption.

3. Results and Discussion

3.1. Functionalization of the Quartz Surface

Contact angle measurements confirmed the success of the OTS treatment. When an *n*-dodecane drop was put in contact with a native (water-wet) quartz surface in the presence of saline solution, it was observed a low tendency to spread (Figure 3A), as expected. When the same experiment was performed with C_{18}-quartz, however, the *n*-dodecane drop showed a great affinity to the surface and tended to spread (Figure 3B). Accordingly, the contact angle changed drastically from 25° in native quartz to 162° in C_{18}-quartz, confirming that a layer of OTS was chemisorbed on quartz (Scheme 1), leading to the expected wettability inversion. Grate et al. [28] reported a contact angle of 148° for a similar system, composed of hexadecane drops in contact with silicon wafers modified with dodecyltriethoxysilane, which is consistent with our data, considering the different alkyl chains involved (they modified the surface with dodecyl chains).

Furthermore, *n*-dodecane drops deposited on C_{18}-quartz were not stable. Initially, drops with high contact angle were formed, such as the one in Figure 3B. Such drops remained stable for 1 to 2 min, and then started to collapse, spreading even more on the surface. After about 20 min, the drops reached a final state, forming either very well spread drops with θ > 170° (Figure S2) or even films on the surface, in which case θ = 180° (there was some drop-to-drop variation). These results confirmed that the oleophilic C_{18}-quartz surface had a great affinity for apolar hydrocarbons and therefore can be studied as an analog for reservoir rocks that underwent wettability inversion. Although structural differences exist between octadecyl-covered and asphaltene-covered surfaces, it is important to initiate the studies with a simple mimetic system composed of a pure compound, since asphaltenes isolated from petroleum are complex mixtures containing hundreds of different substances. The next step will be to test the CDs in thin films of deposited asphaltenes.

3.2. Effect of Cyclodextrins on the Wettability of C_{18}-Quartz

The following results were obtained with freshly deposited drops (within 30 s from drop injection) and [CD] fixed at 1.5 g/100 mL (1.5% *w/v*), in order to compare the effects of the different CDs in equal concentration. In a further section the effect of drop aging and varying the [CD] will be studied. As seen in Figure 4, the presence of the CDs in the aqueous phase increased the wettability of C_{18}-quartz by water, decreasing the contact angles (Table 1). The most pronounced effect was found for α-CD, with θ changing from 162° (oil-wet) in the absence of CDs to 33° (water-wet) in the presence of α-CD (Figure 4D). Thus, α-CD completely reversed the wettability inversion effect of the oleophilic OTS layer, resulting in drops with shape and contact angles similar to those observed with the original, unmodified native quartz (compare with Figure 3A). The other CDs tested also showed significant effects (Figure 4, Table 1), although less than in the case of α-CD. In the presence of β-CD, intermediate wettability was observed (θ = 94°), whereas with γ-CD, HP-α-CD and HP-β-CD the surface was slightly oil-wet (θ = 110–111°).

Table 1. Effect of different CDs on the contact angle of *n*-dodecane drops immersed in the aqueous phase and in contact with C_{18}-quartz [1].

CD	θ (Degrees) [2]
none	162
α-CD	33
β-CD	94
γ-CD	111
HP-β-CD	110
HP-α-CD	111

[1] [CD] = 1.5% (*w/v*) in all cases. Aqueous solutions contained 0.6 M NaCl; [2] Measured through the aqueous phase.

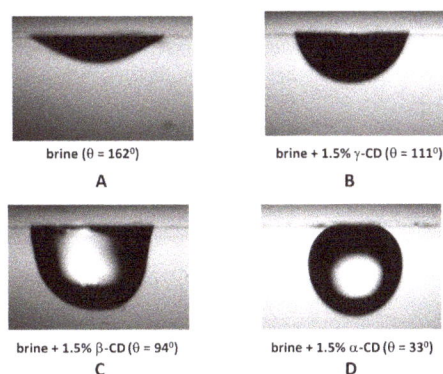

Figure 4. Effects of the different CDs on the shape of *n*-dodecane drops in contact with the C_{18}-quartz surface and immersed in the aqueous phase containing brine (0.6 M NaCl) and the CDs. (**A**) Brine only; (**B**) Brine + γ-CD. (**C**) Brine + β-CD. (**D**) Brine + α-CD.

The effects of CDs can be explained by the formation of inclusion complexes at the water/*n*-dodecane interface (Scheme 2). It is well known that the stability of CD complexes is determined mainly by geometric factors [7,8]. Molecules having a size compatible with the dimensions of the CD cavity are likely to form stable complexes. Linear alkyl chains fit well in the small α-CD cavity (diameter = 5.7 Å), forming more stable complexes than the other CDs [7,8], which explains why α-CD showed the most pronounced effects in Figure 4. β-CD (cavity diameter = 7.8 Å), on the other hand, is known to form stable complexes with simple aromatic molecules such as naphthalene, and γ-CD (cavity diameter = 9.5 Å) can form inclusion complexes with even bulkier aromatic systems [7,8]. Therefore, the larger CDs could in principle perform better with real asphaltenes.

According to the literature [24,25,29–31], the CDs alone are not surface active at air/water interfaces, but inclusion complexes between CDs and linear apolar molecules did present surface-activity, decreasing the air/water surface tension. Bojinova et al. [30] and Machut et al. [31] observed decreases in surface tension from 72 mN/m in pure water to values between 20 and 40 mN/m in the presence of CD complexes with 1-dodecanol and long chain esters (isosorbide dioleate and sorbitan trioleate), respectively. Similar phenomenon has been observed in oil/water interfaces [25,26]. Mathapa and Paunov [25] studied drops of *n*-tetradecane immersed in aqueous Cyclodextrins and reported a decrease of the water/*n*-tetradecane interfacial tension from 44 mN/m in pure water to 37 mN/m in the presence of 1 mM α-CD (ca. 0.1% *w/v*). Inoue et al. [26] observed a decrease in the water/*n*-dodecane interfacial tension from 52 mN/m in pure water to 30 mN/m in the presence of β-CD (0.5% *w/v*). We realized our own measurements (Table 2, Figure S1) and found out that the brine/*n*-dodecane interfacial tension decreased in a nearly linear fashion with [α-CD], from 39 mN/m in the absence of α-CD to 32 mN/m in the presence of 0.75% (*w/v*) α-CD.

Table 2. Interfacial tension (IFT) between dodecane and α-CD solutions of increasing concentrations [1].

[α-CD] (% *w/v*)	IFT (mN/m) [2]
0	39.2
0.125	38.0
0.50	33.4
0.75	32.2

[1] In brine (0.6 M NaCl). [2] Measured after an equilibration time of 400 s (see Figure S1).

According to the Young equation (Equation (1)), a decrease in the oil/water interfacial tension (γ_{ow}) will lead to a decrease in the contact angle, rendering the surface more water-wet, as observed here. The inclusion complexes between the CDs and linear alkanes display amphiphilic character, with the CD as the polar head and a portion of the linear chain protruding out from the cavity as the apolar tail (Scheme 2), and can therefore be regarded as pseudo-surfactants [25,30–32], being an environmentally friendly alternative to the surfactants generally used in EOR. Another possible effect of the CDs is the formation of inclusion complexes with the octadecyl groups anchored at the C_{18}-quartz surface, since in our experimental arrangement the CD was already present prior to the deposition of the hydrocarbon drop.

3.3. Effect of CD Concentration

For EOR purposes, the additive must be active at relatively low concentrations, in order to reduce operation costs. Hence, we investigated here the effect of the CD concentration on the wettability of C_{18}-quartz (Figures 5–7). At low CD concentrations, the initial contact angles (a few seconds after deposition) decreased slightly and in a linear way with CD concentration, up to a critical concentration, above which the contact angles decreased drastically from an oil-wet to a water-wet state. For β-CD, this critical concentration occurred about 1.5% (w/v) (Figures 5 and 7), whereas for α-CD it happened between 0.5% and 1.0% (w/v) (Figures 6 and 7), confirming that α-CD has a more pronounced effect on wettability than β-CD. The higher efficiency of α-CD at lower concentrations as compared to β-CD can be explained by the formation of stable complexes with the dodecyl chains, due to the tight fit of linear alkyl chains within the α CD cavity, as discussed above.

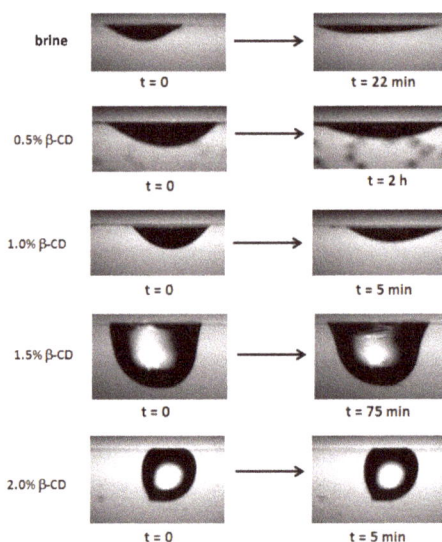

Figure 5. Drops of *n*-dodecane immersed in aqueous saline solutions (0.6 M NaCl) containing β-CD and in contact with C_{18}-quartz. Left column: within 30 s after drop deposition. Right column: the same drop after the time indicated.

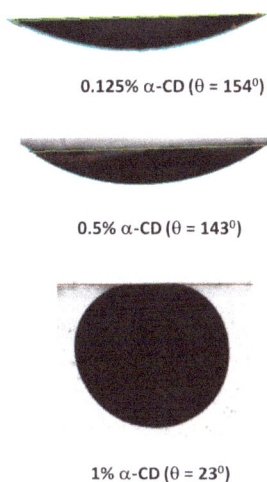

0.125% α-CD (θ = 154⁰)

0.5% α-CD (θ = 143⁰)

1% α-CD (θ = 23⁰)

Figure 6. Drops of *n*-dodecane immersed in aqueous saline solutions (0.6 M NaCl) containing α-CD and in contact with C$_{18}$-quartz.

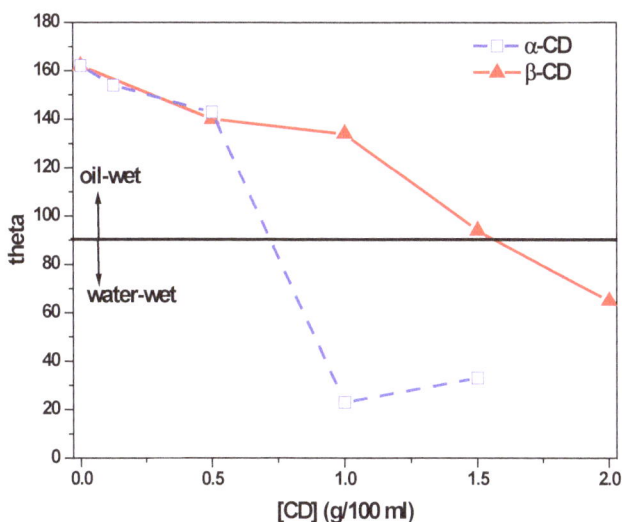

Figure 7. Effect of the CD concentration on the contact angles of an *n*-dodecane drop in contact with C$_{18}$-quartz and immersed in the aqueous phase. The lines connecting the experimental points were added for easy reading.

However, the dynamic behavior of the *n*-dodecane drops at low CD concentrations was quite different from that observed with concentrated CD solutions. In diluted β-CD solutions (<1% *w/v*), the drops behaved similarly to brine without CDs. The initially formed drops were stable for about 1 min, and then relaxed within 10 min to a more spread final state, which remained unchanged for several hours (Figure 5 and Figure S2). As noted above, it takes ca. 30 s in our set up between injecting the drop and start recording the images. In Figure 5 and Figure S2, zero time is the moment the camera was triggered (ca. 30 s after drop deposition). In diluted α-CD solutions, the relaxation process was much faster than with β-CD and could not be recorded, since it was faster than the time needed to trigger the camera.

In concentrated CD solutions (above the transition concentration in Figure 7), on the other hand, the shape of the *n*-dodecane drops changed very little with time, in contrast to the diluted solutions (Figure S2). Instead, membrane-like skins appeared at the oil drop surface within a few minutes as the drop was aged in concentrated CD solutions. Such semi-rigid skins seem to stabilize the drops, preventing them to spread on C_{18}-quartz. This phenomenon can be explained by the precipitation of mycrocrystals of CD inclusion complexes [25], forming a film at the interface, as discussed below. The critical concentration observed in Figure 7 seems to be the onset of film formation.

3.4. Aging of n-Dodecane Drops at High CD Concentrations

When the *n*-dodecane drops were immersed in concentrated CD solutions (above the critical concentration seen in Figure 7), formation of films at the oil/water interface was clearly observed after ca. 5 min (Figure 8 and Figure S3). Similar phenomenon has been observed by other authors, and was attributed to the precipitation of microcrystals of CD/hydrocarbon inclusion complexes at the interface [24,25]. In the case of β-CD, the film had the appearance of a membrane-like wrinkled skin (Figure 8, top), and seemed to be fluid or semi-fluid at the interface. As seen in Figure S3, drops of *n*-dodecane in concentrated β-CD solutions started to spread in the first few minutes after contact (oil started advancing over a previously water-covered surface), but as the skin became more rigid the drop shape got frozen. This shows that the effects of CDs were both kinetic and thermodynamic. Changes in interfacial tensions will affect the wettability according to Equation (1), which is valid for systems in thermodynamic equilibrium. This was likely the case at low [CD]. At high [CD], however, skin formation prevented the system to reach the equilibrium, and the drop was frozen in a non-equilibrium shape.

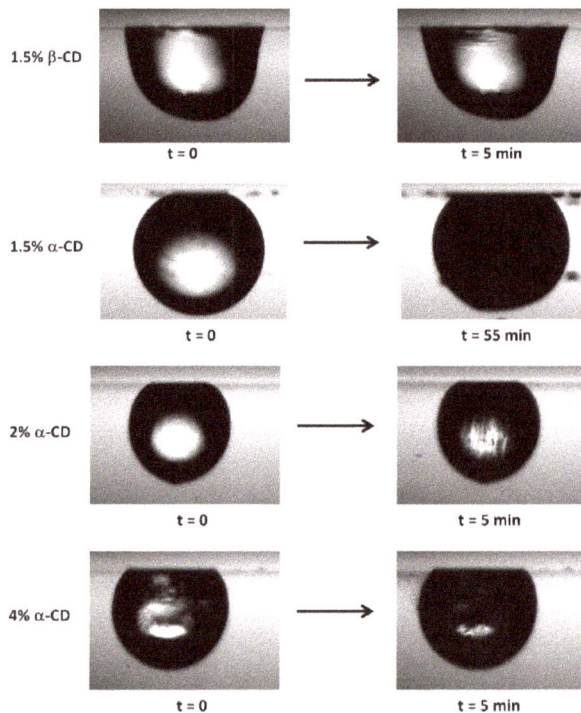

Figure 8. Images of *n*-dodecane drops in contact with a C_{18}-quartz surface and immersed in concentrated CD solutions (containing 0.6 M NaCl). Notice the time-dependent formation of a solid layer of inclusion complexes at the oil/water interface.

The behavior in the presence of concentrated α-CD was quite different. A solid crystalline layer precipitated at the drop surface (Figure 8), as seen by the darkening of the bright areas inside the drops. The images were acquired in light transmission mode, so that the bright areas at $t = 0$ mean that the drops were initially translucent. The crystals formed after some time in the presence of α-CD blocked the light, resulting in the observed darkening. Since α-CD is more soluble in water than β-CD, it was possible to extend the studies to higher concentrations (up to 5% w/v). At very high α-CD concentrations, the interface became saturated with crystals and the excess of solid material was relayed into the aqueous phase as a flow of particles (Figure 8 and Figure S4). The crystals of α-CD/n-dodecane inclusion complexes were less dense that the aqueous phase and flowed up pushed by buoyancy forces (Figure S4).

Hernandez-Pascacio et al. [24] observed the formation of films of α-CD/surfactant inclusion complexes at air/water interfaces. The films could be imaged at the interface using Brewster-angle microscopy and showed a similar aspect as the skins observed in Figure 8 and Figure S3. Mathapa and Paunov observed similar skins around n-tetradecane drops immersed in CD aqueous solutions, and showed that the phenomenon was due to the presence of microcrystals of CD inclusion complexes [25]. They found out that the β-CD complexes with n-tetradecane formed flexible skins, stabilized by stable microcrystals, whereas larger crystals were formed with α-CD, resulting in precipitation at the interface. Our results are in complete agreement with those findings.

The larger crystals verified with α-CD arise because the CD inclusion complexes have a trend to self-assemble in long nanotubes stabilized by hydrogen bonds. The threading of several CD units onto a polymeric chain results in the formation of the so-called pseudo-polyrotaxanes (also known as "molecular necklaces") [24,25]. In the present case several n-dodecane molecules within the CD cavities would play the role of the polymeric chain. Such tubular structures have been seen by microscopy at the interface of n-alkane drops with α-CD solutions [25].

Other studies have shown the formation of CD-based oil-in-water Pickering emulsions (which are emulsions stabilized by solid particles), where the stabilizing particles at the interface were microcrystals of CD inclusion complexes with n-alkanes [26,33,34]. Several other oils were found to form CD-stabilized Pickering emulsions, such as squalane, soybean oil, liquid paraffin [35], medium chain triglycerides [36], olive oil, castor oil and coconut oil [37]. Such stable oil-in-water emulsions could be a mechanism for oil transport within the porous rock media, being an additional effect of CDs in EOR. Pickering emulsions have been often employed to facilitate oil flow in EOR studies [38,39].

The results above showed that α-CD was the most effective in altering the surface wettability among the CDs tested. This can be explained by the tight fit of the alkane chain within the narrower cavity of α-CD, in contrast to a looser fit with β-CD and γ-CD [7,8]. Inoue et al. [26] reported that α-CD formed the most stable inclusion complex with n-dodecane among the native CDs. They also found, however, that complexes of n-dodecane with α-CD were not as good emulsifying agents as those with β-CD and γ-CD. The α-CD/n-dodecane complexes formed unstable emulsions and tended to precipitate as crystals. The β-CD/n-dodecane and γ-CD/n-dodecane complexes, on the other hand, formed stable layers of microcrystals at the oil drop surface, stabilizing the oil-in-water Pickering emulsions [26]. Those results are in agreement with the findings of the present work.

In view of the present results, it can be concluded that the use of α-CD in EOR could be advantageous, since its effects were observed at lower concentrations (Figure 7). However, its use should be restricted to low concentrations, below or near the critical concentration, since the excess of crystals could clog the rock porous system. β-CD, on the other hand, shows a much lower trend to precipitate and therefore could be used for EOR applications above the critical concentration.

3.5. Advancing and Receding Contact Angles

Equilibrium contact angles do not always represent the real contact angles. It is often convenient to measure advancing and receding contact angles. It is generally accepted that advancing contact angles, which are the contact angles obtained when water is advancing over a solid surface previously covered by oil, represent better the reservoir wettability than the equilibrium or receding angles [15,16]. This is because in a water flooding operation, water advances over rock surfaces previously wet by oil. In the tensiometer used here, advancing and receding contact angles were measured by the tilting plate technique (Figure 9), where the stage bearing the drop was tilted using a tilting cradle accessory. The advancing and receding angles were registered in the same image (Figure 9) right after the drop started moving during tilting.

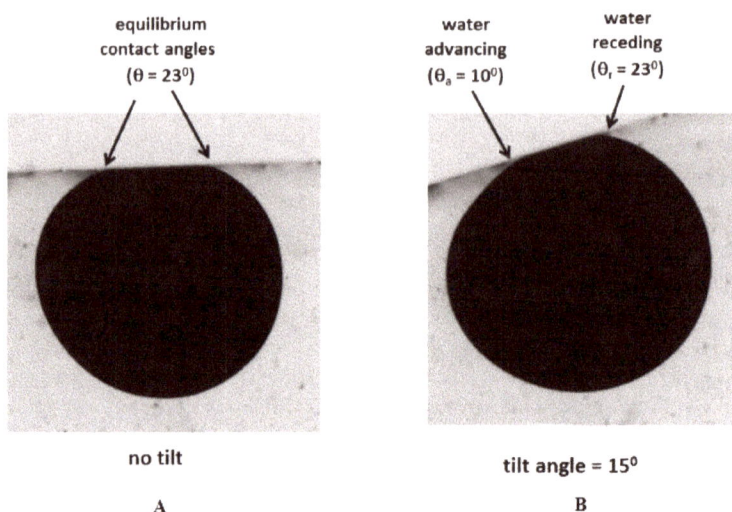

Figure 9. Tilting angle experiment with a drop of *n*-dodecane immersed in aqueous saline solutions (0.6 M NaCl) containing 1% (*w/v*) α-CD and in contact with C_{18}-quartz. (**A**) Initial position (no tilt). (**B**) The same drop just after it started moving on the surface. It can be noticed the water advancing at the contact point on the left and receding at the contact point at the right side.

It was not easy to measure advancing and receding angles in the presence of CDs. With low CD concentrations, time-dependent drop relaxation, as described above, was concomitant with tilting. With high CD concentrations, on the other hand, most drops were strongly adhered to the C_{18}-quartz substrate, and did not detach from the surface even when the quartz cuvette was turned upside down. The reason for the strong oil/C_{18}-quartz adhesion could be an interaction between CD inclusion complexes at the drop interface and surface octadecyl groups on quartz. The same CD molecule could host at the same time a surface octadecyl chain and a *n*-dodecane molecule from the drop surface, resulting in zipper-like attractive interactions. Adhesion of oil to rock surfaces even when those were strongly water-wet has been reported in the literature [15]. In a few favorable cases (near the critical concentration shown in Figure 7), it was possible to make the drop move on the solid surface, and hence advancing and receding contact angles could be obtained, as shown in Figure 9 for 1% (*w/v*) α-CD. It can be seen that the water advancing contact angle was 10°, which is even more water-wet than the equilibrium angle of 23°, confirming the ability of Cyclodextrins in making the C_{18}-quartz surface more water-wet.

4. Conclusions

The results presented here show that the CDs are prospective candidates for pseudo-surfactants with applications in EOR. CD-based pseudo-surfactants are formed upon inclusion of linear hydrocarbon chains within the CD cavities, resulting in inclusion complexes. The pseudo-surfactants were tested on octadecyl-covered surfaces representing oil-wet rock surfaces. Hydrocarbon drops were shown to spread well over these oil-wet surfaces, but the effects were reversed in the presence of the

CDs. The most pronounced effects were found with α-CD, in which case the oil-wet octadecyl-covered surface was changed to water-wet. The pronounced effects of α-CD can be explained by the tight fit of linear alkyl chains within the α-CD cavity, forming stable inclusion complexes, whereas a looser fit is expected for the larger CDs. The effects of α-CD were observed at relatively low concentrations, what is very important for the economic viability in EOR. At higher concentrations, the CD inclusion complexes form a microcrystalline layer around the hydrocarbon drop, resembling Pickering emulsions. The formation of CD-based Pickering emulsions could be explored for oil transport, although care must be taken to avoid pore-clogging by the microcrystals. Many more studies will be needed before the CDs can be introduced in real reservoirs, since it has not been used before for EOR operations. Nevertheless, the present results are encouraging for further investigations on the use of CDs in EOR. The next step in the research will be testing the CDs with asphaltene-covered surfaces.

Supplementary Materials: The following are available online https://www.mdpi.com/2504-5377/2/1/10/s1. Figure S1: Interfacial tension between a drop of dodecane and α-CD solutions of different concentrations in brine (0.6 M NaCl) followed as a function of time. Figure S2: Dynamic behavior of the contact angles followed as a function of time for *n*-dodecane drops immersed in aqueous saline solution (0.6 M NaCl) containing β-CD and in contact with C_{18}-quartz. Figure S3: drops of *n*-dodecane in contact with C_{18}-quartz and immersed in aqueous saline solution (0.6 M NaCl) containing 1.5% (*w/v*) β-CD. Figure S4: Drop of *n*-dodecane in contact with C_{18}-quartz and immersed in aqueous saline solution (0.6 M NaCl) containing 5% (*w/v*) α-CD.

Acknowledgments: S.B. and C.R.M. acknowledge the support of Brazilian agencies FAPESP (grant No 2016/05496-2 and 2017/02317-2) and CNPq (grant No 480189/2011-0). The authors wish to thank FINEP/CTINFRA (call No 01/2010, ref. 1132/10) for the purchase of the Attension Theta Optical Tensiometer.

Author Contributions: S.B. and C.R.M. conceived and designed the experiments and wrote the paper. A.F.T.d.C. and R.D.S. performed the experiments.

Conflicts of Interest: The authors declare no conflict of interest. The funding sponsors had no role in the design of the study; in the collection, analyses, or interpretation of data; in the writing of the manuscript, and in the decision to publish the results.

References

1. Lake, L.W.; Johns, R.; Rossen, B.; Pope, G. *Fundamentals of Enhanced Oil Recovery*; Society of Petroleum Engineers: Richardson, TX, USA, 2014; ISBN 978-1-61399-328-6.
2. Howe, A.M.; Clarke, A.; Mitchell, J.; Staniland, J.; Hawkes, L.; Whalan, C. Visualising surfactant enhanced oil recovery. *Colloids Surf. A Physicochem. Eng. Asp.* **2015**, *480*, 449–461. [CrossRef]
3. Iglauer, S.; Wu, Y.; Shuler, P.J.; Tang, Y.; Goddard, W.A. New surfactant classes for enhanced oil recovery and their tertiary oil recovery potential. *J. Pet. Sci. Eng.* **2010**, *71*, 23–29. [CrossRef]
4. Anderson, W.G. Wettability literature survey. Part 2: Wettability measurement. *J. Petr. Technol.* **1986**, *38*, 1246–1262. [CrossRef]
5. Buckley, J.S. Effective Wettability of Minerals Exposed to Crude Oil. *Curr. Opin. Colloid Interface Sci.* **2001**, *6*, 191–196. [CrossRef]
6. Kumar, M.; Fogden, A. Patterned Wettability of Oil and Water in Porous Media. *Langmuir* **2010**, *26*, 4036–4047. [CrossRef] [PubMed]
7. Szejtli, J. Introduction and General Overview of Cyclodextrin Chemistry. *Chem. Rev.* **1998**, *98*, 1743–1754. [CrossRef] [PubMed]
8. Dodziuk, H. (Ed.) *Cyclodextrins and Their Complexes: Chemistry, Analytical Methods, Applications*; Wiley-VCH: Weinheim, Germany, 2006; ISBN 978-3-527-60844-7.
9. Horikoshi, K.; Shibanai, I.; Nakamura, N. Process for Production of Oil from Oil Sand. Patent CA 1117055 A1, 26 January 1982.
10. Horikoshi, K.; Kato, T.; Shibanai, I. Process for the Oil Extraction from Oil Sand by Using Cyclodextrin and Hydrocarbon Solvents. Patent CA 1208149 A1, 22 July 1986.
11. Leslie, T.; Xiao, H.; Dong, M. Tailor-modified starch/cyclodextrin-based polymers for use in tertiary oil recovery. *J. Pet. Sci. Eng.* **2005**, *46*, 225–232. [CrossRef]
12. Kjøniksen, A.L.; Beheshti, N.; Kotlar, H.K.; Zhu, K.; Nyström, B. Modified polysaccharides for use in enhanced oil recovery applications. *Eur. Polym. J.* **2008**, *44*, 959–967. [CrossRef]

13. Wever, D.A.Z.; Picchioni, F.; Broekhuis, A.A. Polymers for enhanced oil recovery: A paradigm for structure–property relationship in aqueous solution. *Prog. Polym. Sci.* **2011**, *36*, 1558–1628. [CrossRef]
14. Adamson, A.W.; Gast, A.P. *Physical Chemistry of Surfaces*; Wiley-Interscience: New York, NY, USA, 1997.
15. Rao, D.N. Measurements of dynamic contact angles in solid–liquid–liquid systems at elevated pressures and temperatures. *Colloids Surf. Physicochem. Eng. Asp.* **2002**, *206*, 203–216. [CrossRef]
16. Drummond, C.; Israelachvili, J. Fundamental studies of crude oil–surface water interactions and its relationship to reservoir wettability. *J. Pet. Sci. Eng.* **2004**, *45*, 61–81. [CrossRef]
17. Yang, D.; Gu, Y.; Tontiwachwuthikul, P. Wettability determination of the reservoir brine−reservoir rock system with dissolution of CO_2 at high pressures and elevated temperatures. *Energy Fuels* **2008**, *22*, 504–509. [CrossRef]
18. Bera, A.; Kissmathulla, S.; Ojha, K.; Kumar, T.; Mandal, A. Mechanistic study of wettability alteration of quartz surface induced by nonionic surfactants and interaction between crude oil and quartz in the presence of sodium chloride salt. *Energy Fuels* **2012**, *26*, 3634–3643. [CrossRef]
19. Serrano-Saldaña, E.; Domínguez-Ortiz, A.; Pérez-Aguilar, H.; Kornhauser-Strauss, I.; Rojas-González, F. Wettability of solid/brine/n-dodecane systems: Experimental study of the effects of ionic strength and surfactant concentration. *Colloids Surf. Physicochem. Eng. Asp.* **2004**, *241*, 343–349. [CrossRef]
20. Wu, Y.; Shuler, P.J.; Blanco, M.; Tang, Y.; Goddard, W.A., III. An experimental study of wetting behavior and surfactant EOR in carbonates with model compounds. *SPE J.* **2008**, *13*, 26–34. [CrossRef]
21. Hansen, G.; Hamouda, A.A.; Denoyel, R. The effect of pressure on contact angles and wettability in the mica/water/n-decane system and the calcite + stearic acid/water/n-decane system. *Colloids Surf. Physicochem. Eng. Asp.* **2000**, *172*, 7–16. [CrossRef]
22. Freer, E.M.; Svitova, T.; Radke, C.J. The role of interfacial rheology in reservoir mixed wettability. *J. Pet. Sci. Eng.* **2003**, *39*, 137–158. [CrossRef]
23. Lara, L.S.; Voltatoni, T.; Rodrigues, M.C.; Miranda, C.R.; Brochsztain, S. Potential applications of Cyclodextrins in enhanced oil recovery. *Colloids Surf. Physicochem. Eng. Asp.* **2015**, *469*, 42–50. [CrossRef]
24. Hernandez-Pascacio, J.; Garza, C.; Banquy, X.; Díaz-Vergara, N.; Amigo, A.; Ramos, S.; Castillo, R.; Costas, M.; Pinero, A. Cyclodextrin-Based Self-Assembled Nanotubes at the Water/Air Interface. *J. Phys. Chem. B* **2007**, *111*, 12625–12630. [CrossRef] [PubMed]
25. Mathapa, B.G.; Paunov, V.N. Self-assembly of cyclodextrin–oil inclusion complexes at the oil–water interface: A route to surfactant-free emulsions. *J. Mater. Chem. A* **2013**, *1*, 10836–10846. [CrossRef]
26. Inoue, M.; Hashizaki, K.; Taguchi, H.; Saito, Y. Preparation and characterization of n-alkane/water emulsion stabilized by cyclodextrin. *J. Oleo Sci.* **2009**, *58*, 85–90. [CrossRef] [PubMed]
27. Askvik, K.M.; Høiland, S.; Fotland, P.; Barth, T.; Grønn, T.; Fadnes, F.H. Calculation of wetting angles in crude oil/water/quartz systems. *J. Colloid Interface Sci.* **2005**, *287*, 657–663. [CrossRef] [PubMed]
28. Grate, J.W.; Dehoff, K.J.; Warner, M.G.; Pittman, J.W.; Wietsma, T.W.; Zhang, C.; Oostrom, M. Correlation of Oil-Water and Air-Water Contact Angles of Diverse Silanized Surfaces and Relationship to Fluid Interfacial Tensions. *Langmuir* **2012**, *28*, 7182–7188. [CrossRef] [PubMed]
29. Jia, H.; Leng, X.; Zhang, D.; Lian, P.; Liang, Y.; Wu, H.; Huang, P.; Liu, J.; Zhou, H. Facilely control the SDS ability to reduce the interfacial tension via the host-guest recognition. *J. Mol. Liquids* **2018**, *255*, 370–374. [CrossRef]
30. Bojinova, T.; Coppel, Y.; Viguerie, N.L.; Milius, A.; Rico-Lattes, I.; Lattes, A. Complexes between β-Cyclodextrin and Aliphatic Guests as New Noncovalent Amphiphiles: Formation and Physicochemical Studies. *Langmuir* **2003**, *19*, 5233–5239. [CrossRef]
31. Machut, C.; Mouri-Belabdelli, F.; Cavrot, J.-P.; Sayede, A.; Monflier, E. New supramolecular amphiphiles based on renewable resources. *Green Chem.* **2010**, *12*, 772–775. [CrossRef]
32. Dou, Z.-P.; Xing, H.; Xiao, J.-X. Hydrogenated and Fluorinated Host–Guest Surfactants: Complexes of Cyclodextrins with Alkanes and Fluoroalkyl-Grafted Alkanes. *Chem. Eur. J.* **2011**, *17*, 5373–5380. [CrossRef] [PubMed]
33. Mathapa, B.G.; Paunov, V.N. Cyclodextrin stabilised emulsions and cyclodextrinosomes. *Phys. Chem. Chem. Phys.* **2013**, *15*, 17903–17914. [CrossRef] [PubMed]
34. Inoue, M.; Hashizaki, K.; Taguchi, H.; Saito, Y. Emulsion Preparation Using β-Cyclodextrin and Its Derivatives Acting as an Emulsifier. *Chem. Pharm. Bull.* **2008**, *56*, 1335–1337. [CrossRef] [PubMed]

35. Inoue, M.; Hashizaki, K.; Taguchi, H.; Saito, Y. Emulsifying Ability of β-Cyclodextrins for Common Oils. *J. Dispers. Sci. Technol.* **2010**, *31*, 1648–1651. [CrossRef]
36. Li, X.; Li, H.; Xiao, Q.; Wang, L.; Wang, M.; Lu, X.; York, P.; Shi, S.; Zhang, J. Two-way effects of surfactants on Pickering emulsions stabilized by the self-assembled microcrystals of α-cyclodextrin and oil. *Phys. Chem. Chem. Phys.* **2014**, *16*, 14059–14069. [CrossRef] [PubMed]
37. Wu, L.; Liao, Z.; Liu, M.; Yin, X.; Li, X.; Wang, M.; Lu, X.; Lv, N.; Singh, V.; He, Z.; et al. Fabrication of non-spherical Pickering emulsion droplets by Cyclodextrins mediated molecular self-assembly. *Colloids Surf. Physicochem. Eng. Asp.* **2016**, *490*, 163–172. [CrossRef]
38. Sharma, T.; Velmurugan, N.; Patel, P.; Chon, B.H.; Sangwai, J.S. Use of Oil-in-water Pickering Emulsion Stabilized by Nanoparticles in Combination with Polymer Flood for Enhanced Oil Recovery. *Pet. Sci. Technol.* **2015**, *33*, 1595–1604. [CrossRef]
39. Langevin, D.; Poteau, S.; Henaut, I.; Argillier, J.F. Crude Oil Emulsion Properties and their Application to Heavy Oil Transportation. *Oil Gas Sci. Technol.* **2004**, *59*, 511–521. [CrossRef]

colloids
and interfaces

MDPI

Article

Polymer Flow in Porous Media: Relevance to Enhanced Oil Recovery

Arne Skauge [1,2,*], Nematollah Zamani [3], Jørgen Gausdal Jacobsen [1,3], Behruz Shaker Shiran [3], Badar Al-Shakry [1] and Tormod Skauge [2]

[1] Department of Chemistry, University of Bergen, Allegaten 41, N-5020 Bergen, Norway; joja@norceresearch.no (J.G.J.); Badar.Al-Shakry@uni.no (B.A.-S.)
[2] Energy Research Norway, N-5020 Bergen, Norway; Tormod.Skauge@energyresearch.no
[3] Uni Research, N-5020 Bergen, Norway; Nematollah.zamani@uni.no (N.Z.); Behruz.shaker@uni.no (B.S.S.)
* Correspondence: arne.skauge@kj.uib.no; Tel.: +47-5558-3358

Received: 1 June 2018; Accepted: 3 July 2018; Published: 10 July 2018

Abstract: Polymer flooding is one of the most successful chemical EOR (enhanced oil recovery) methods, and is primarily implemented to accelerate oil production by sweep improvement. However, additional benefits have extended the utility of polymer flooding. During the last decade, it has been evaluated for use in an increasing number of fields, both offshore and onshore. This is a consequence of (1) improved polymer properties, which extend their use to HTHS (high temperature high salinity) conditions and (2) increased understanding of flow mechanisms such as those for heavy oil mobilization. A key requirement for studying polymer performance is the control and prediction of in-situ porous medium rheology. The first part of this paper reviews recent developments in polymer flow in porous medium, with a focus on polymer in-situ rheology and injectivity. The second part of this paper reports polymer flow experiments conducted using the most widely applied polymer for EOR processes, HPAM (partially hydrolyzed polyacrylamide). The experiments addressed highrate, near-wellbore behavior (radial flow), reservoir rate steady-state flow (linear flow) and the differences observed in terms of flow conditions. In addition, the impact of oil on polymer rheology was investigated and compared to single-phase polymer flow in Bentheimer sandstone rock material. Results show that the presence of oil leads to a reduction in apparent viscosity.

Keywords: EOR; polymer flooding; in-situ rheology; non-Newtonian flow in porous medium

1. Introduction

The success of polymer flooding depends on the ability of injected solutions to transport polymer molecules deep into a reservoir, thus providing enhanced mobility ratio conditions for the displacement process. In the following sections, we focus on the principal parameters that are crucial in the decision-making process for designing a satisfactory polymer flood design.

The application of polymer flooding to tertiary oil recovery may induce high injection pressures, resulting in injectivity impairment. Since the volumetric injection rate during polymer flooding is constrained by formation fracture pressure, project economics may be significantly affected. Thus, injectivity is a critical parameter and key risk factor for implementation of polymer flood projects.

A large number of injectivity studies, both theoretical and experimental, have been performed in porous media during recent decades, albeit they were mainly studies of linear cores in the absence of residual oil [1–7]. Recently, Skauge et al. [8] performed radial injectivity experiments showing significant reduction in differential pressure compared to linear core floods. This discrepancy in polymer flow in linear cores compared to that in radial disks is partly explained by the of differing pressure conditions that occur when polymer molecules are exposed to transient and semi-transient pressure conditions in radial disks, as opposed to the steady state conditions experienced in linear core floods. In addition, they observed that the onset of apparent shear thickening occurs at significantly higher flux in radial floods. Based on these results, injectivity was suggested to be underestimated from experiments performed in linear core plugs. However, these experiments were performed in the absence of residual oil. If residual oil has a significant effect on polymer propagation in porous media, experiments performed in its absence will not be able to accurately predict polymer performance.

Experimental studies investigating the effects of residual oil on polymer propagation through porous media have been sparse, although they have generally shown decreasing levels of polymer retention in the presence of residual oil [9,10].

The polymer adsorbs to the rock surface and may also block pores due to polymer size (straining) and flow rate (hydrodynamic retention). In addition, different trapping mechanisms may take place. The polymer retention phenomena influence the flow of polymer in porous media, however, these effects are beyond the scope of this paper. The subject has been reviewed in several other books and papers, e.g., Sorbie [11] and Lake [12].

History matches performed in this study aim to highlight the injectivity of partially hydrolyzed polyacrylamides (HPAMs) in radial disks saturated with residual oil, as these conditions best mimic actual flow conditions in oil reservoirs. Results show that the presence of residual oil reduces the apparent viscosity of HPAM in flow through porous media, thus improving injectivity. These results may facilitate increased implementation of polymer EOR (enhanced oil recovery) projects, as previous projects deemed infeasible may now be economically viable.

2. Theory

2.1. In-Situ Rheology

Polymer viscosity as a function of shear rate is usually measured using a rheometer. During the measurement process, polymer solutions are exposed to different shear rates in a stepwise manner. For each shear rate, polymer viscosity is measured after steady state conditions are achieved; at this state, it is referred to as bulk viscosity. However, polymer molecules experience significantly different flow conditions in rheometers compared to porous media. In particular:

(I)   unlike rheometers, porous media exhibit an inherently complex geometry;

(II)   phenomena such as mechanical degradation may change rheological properties;

(III)   although they only demonstrate shear thinning behavior in rheometers, polymer solutions may exhibit apparent shear thickening behavior above a certain critical flow rate;

(IV)   due to the tortuosity of porous media and existence of several contraction-expansion channels, polymer solutions are exposed to a wide range of shear rates at each flow rate and where extensional viscosity becomes more dominant, resulting in significantly different rheology behavior compared to bulk flow.

To account for these contrasting flow conditions, in-situ viscosity has been suggested to describe the fluid flow behavior of polymer solutions in porous media. In-situ viscosity is a macroscopic parameter that can be calculated using Darcy's law for single-phase non-Newtonian fluids:

$$\mu_{app} = \frac{KA}{Q} \frac{\Delta P}{L} \tag{1}$$

It is generally measured in core flood experiments as a function of Darcy velocity. Comparison of in-situ and bulk rheology (Figure 1) shows vertical and horizontal shifts between viscosity curves. Vertical shifts may be due to phenomena such as mechanical degradation, while horizontal shifts are due to a conversion factor between in-situ shear rate and Darcy velocity, shown as α. The red line in Figure 1 shows an increase in apparent viscosity, which is due to polymer adsorption. The adsorbed layer of polymer reduces the effective pore size and blocks smaller pores, both leading to increased resistance to flow e.g., as determined by an increase in pressure at a given rate compared to a non-adsorbing situation. In contrast, a reduction in pressure (and therefore, in apparent viscosity) can be observed in the presence of depleted layers (see e.g., Sorbie [11]) which leads to slip effects.

Due to the time-consuming nature of in-situ measurements, there have been several attempts to investigate in-situ rheology, both analytically and numerically. In spite of extensive studies [13–22], limited success has been achieved to reliably relate in-situ to bulk viscosity based on polymer solution and porous media properties. Most of these models were developed based on analytical solutions of non-Newtonian flow through capillary bundles, which simplifies the complex geometry of porous media.

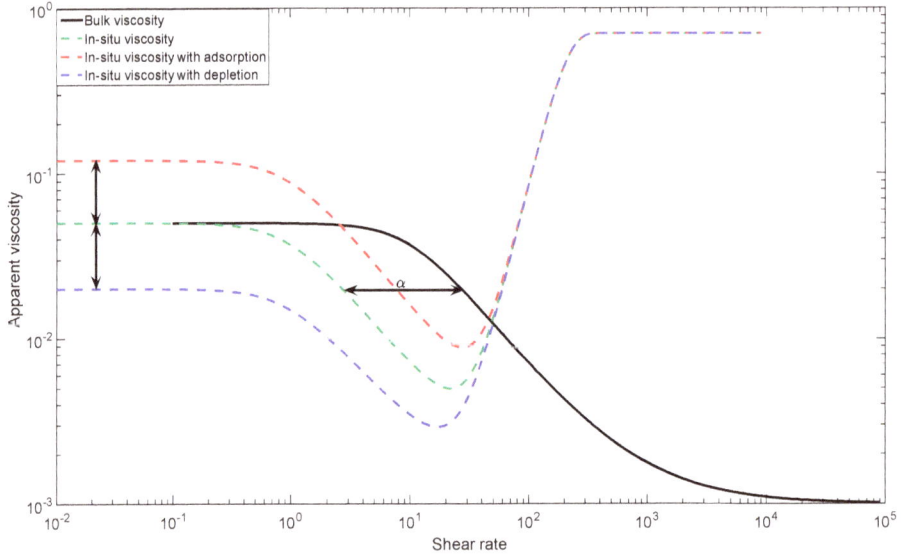

Figure 1. Schematic comparison of in-situ and bulk rheology.

In the following, the calculation procedure of in-situ viscosity is briefly explained:

1. Analytical solutions for a power-law fluid ($\mu = C\,\dot{\gamma}^{n-1}$) at a given flow rate through a capillary tube with an arbitrary radius (R) can be defined by Equation (2). By comparing Equation (2) with the Poiseuille volumetric flow rate for Newtonian fluids in a tube (Equation (3)), an apparent viscosity and shear rate can be obtained from Equations (4) and (5), respectively.

2. The analytical equation in a single tube (Equation (5)) can be extended to account for real porous media by using the capillary bundle approach [23–25]. An equivalent radius of a capillary bundle model for porous media with known porosity (ϕ), permeability (K) and tortuosity (ψ) can be obtained by Equation (6). By calculating the Darcy velocity and substituting the equivalent radius (Equation (6)) into Equation (5), the apparent shear rate as a function of Darcy velocity can be obtained by Equation (7).

$$Q = \frac{\pi n}{3n+1}\left(\frac{\Delta P}{2CL}\right)^{1/n} R^{\frac{3n+1}{n}} \tag{2}$$

$$Q = \frac{\pi}{8\mu}\frac{\Delta P}{L} R^4 \tag{3}$$

$$\mu_{eff} = C\left(\frac{3n+1}{4n}\right)\left(\frac{R\Delta P}{2CL}\right)^{\frac{n-1}{n}} \tag{4}$$

$$\mu_{app} = C\,\dot{\gamma}_{app}^{n-1} \Rightarrow \dot{\gamma}_{app} = \left(\frac{3n+1}{4n}\right)^{\frac{1}{n-1}}\left(\frac{R\Delta P}{2CL}\right)^{1/n} \tag{5}$$

$$R_{eq} = \sqrt{\frac{8K\psi}{\phi}} \tag{6}$$

$$\dot{\gamma}_{app} = 4\left(\frac{3n+1}{4n}\right)^{\frac{n}{n-1}}\frac{U}{\sqrt{8K\phi\psi}} \tag{7}$$

3. The above expressions are considered as an analytical basis for calculating apparent viscosity in porous media. Based on Equation (7), a simplified linear correlation between apparent shear rate and Darcy velocity is generally suggested, i.e., Equation (8), in which the correction factor (α) is the key factor. Some proposed equations for the correction factor are summarized in Table 1. By comparing different coefficients, different values for apparent viscosity may be obtained.

$$\dot{\gamma}_{app} = \alpha\frac{U}{\sqrt{K\phi}} \tag{8}$$

Table 1. Summary of proposed models for correction factor (α).

Model	Equation for Correction Factor (α)	Description
Analytical solution	$4\left(\frac{3n+1}{4n}\right)^{\frac{n}{n-1}}$	n is the power index in power-law region
Hirasaki and Pope [26]	$\frac{12}{\sqrt{150}}\left(\frac{3n+1}{4n}\right)^{\frac{n}{n-1}}$	n is the power index in power-law region
Cannella et al. [16]	$\frac{\beta}{\sqrt{S_w}}\left(\frac{3n+1}{n}\right)^{\frac{n}{n-1}}$	n is the power index in power-law region, S_w is water saturation, β is a constant equal to 6.

Based on the capillary bundle approach, other models were also proposed by Bird et al. [24], Christopher and Middleman [25], and Teeuw and Hesselink [15], in which the modified Blake-Kozeny model is used for power-law fluids (Equation (9)) and apparent viscosity is obtained using Equation (10).

$$U = \left(\frac{K}{\mu_{app}}\frac{\Delta P}{L}\right)^{1/n} \tag{9}$$

$$\mu_{app} = C\left(\frac{3n+1}{4n}\right)^{n}\left(\frac{K\phi}{\beta}\right)^{(1-n)/2} \tag{10}$$

Based on the discussion given by Teeuw and Hesselink [15], tortuosity has a dual effect on both shear rate and shear stress calculations. Christopher and Middleman [25] only incorporated tortuosity in shear stress calculations, while Bird et al. [24] incorporated tortuosity into the shear rate term. The various values of β chosen by different authors are summarized in Table 2.

Table 2. β values applied by different authors where $\psi = 25/12$.

Model	B
Bird et al. [24]	$\sqrt{2\psi}$
Christopher and Middleman [25]	$\sqrt{\frac{2}{\psi}}$
Teeuw and Hesselink [15]	$\sqrt{2}$

Hirasaki and Pope [26] conducted several core flood experiments where permeability was in the range 7–23 mD, porosity in the range 18–20% and residual oil between 20% and 32%. Based on these experiments, they concluded that apparent viscosity could be calculated using the capillary bundle approach and Blake-Kozeny model as follows:

$$\mu_{app} = HU^{n-1} \tag{11}$$

where:

$$H = \frac{C}{12} \left(\frac{9n+3}{n} \right)^n (150K\phi)^{\frac{1-n}{2}} \tag{12}$$

They also included pore size distribution in their calculations:

$$\mu_{app} = \frac{C}{4} \left(\frac{1+3n}{n} \right)^n \frac{\int_0^\infty \sigma(R)R^2 dR}{\left[\int_0^\infty \sigma(R)R^{\frac{1+n}{n}} dR \right]^n} \left(\frac{q}{\phi} \right)^{n-1} \tag{13}$$

Sadowski and Bird [16] used the Ellis model to obtain viscosity from the shear rate. The following equations for apparent viscosity were suggested based on the Blake-Kozeny model and capillary bundle approach:

$$\frac{1}{\mu_{eff}} = \frac{1}{\mu_0} \left(1 + \frac{4}{n+3} \left[\frac{\tau_{RH}}{\tau_{1/2}} \right]^{n-1} \right) \tag{14}$$

$$\tau_{Rh} = \left(\frac{\Delta P}{L} \right) \left[\frac{D_p \phi}{6(1-\phi)} \right] \tag{15}$$

In the above expressions, μ_0, $\tau_{1/2}$ and n are Ellis model parameters that can be measured in rheometers. By applying these equations, they obtained an acceptable match between experimental and predicted results for low to medium molecular weight polymers.

In summary, none of the proposed models for non-Newtonian fluids in porous media based on the capillary bundle approach are in agreement with all experimental results. Therefore, some known limitations of the capillary bundle approach are noted as follows:

- It neglects complex features of porous media such as tortuosity and pore size distribution.
- It assumes unidirectional flow as it neglects interconnectivity between pores.
- It cannot be representative for flow in an anisotropic medium due to its assumption of unique permeability along propagation direction.
- It assumes a single radius along bundles with no variation in cross-sectional area. The contraction-expansion feature of non-Newtonian flow in porous media is of high importance, especially when studying extensional viscosity, yield stress and elasticity.
- It is generally developed based on rheological models in which analytical solutions for velocity profiles are available (e.g., power-law and Ellis model). Analytical solutions for some models (e.g., Carreau model) are quite difficult and the equation for velocity is implicit (Equation (10) for the Carreau model) and needs to be solved iteratively.

$$\frac{\partial v_z}{\partial r} = -\frac{\Delta p\, r}{2L} \left\{ \mu_\infty + \frac{\mu_\infty - \mu_0}{\left[1 + \left(\lambda \frac{\partial v_z}{\partial r} \right)^2 \right]^{\frac{n-1}{2}}} \right\} \tag{16}$$

Duda et al. [27] studied polymer solution rheology inside porous media and reported that experimentally measured pressure drops were greater than those predicted by capillary bundle models, especially at lower values of the Carreau power index. Based on their study, a key reason for underestimating correction factors using the capillary bundle approach is the model's failure to capture either the interconnectivity of pores or non-uniform cross-sections of pore bodies and pore throats (i.e., abrupt contractions and expansions, also known as aspect ratio).

According to the aforementioned limitations of the capillary bundle approach and lack of a universally accepted equation for calculating shear rates in porous media, the application of effective medium theory was eventually suggested. This method was able to remediate certain weaknesses in capillary bundle approach, for example, by incorporating pore interconnectivity and variation in cross-sections. Canella et al. [18] extended this method to account for power-law fluids in porous media. Core floods were conducted using xanthan in the concentration range 300–1600 ppm, rock lithology (sandstone and carbonates) in the permeability range 40–800 mD and various oil residuals (0–29%). Their general assumption was that bulk rheological properties of polymer solutions obey the power-law model, and they suggested the following equation for the relation between shear rate and Darcy velocity based on effective medium theory:

$$\dot{\gamma}_{app} = \beta \left(\frac{3n+1}{4n} \right)^{\frac{n}{n-1}} \frac{q}{\sqrt{K\phi S_w}} \tag{17}$$

Canella et al. achieved a satisfactory match with their experimental results by using a constant value of 6 for β, although this value far exceeds correction factors suggested by other researchers [28–30]. Even though all published results in the literature are not covered by using this correction factor, better agreement between analytical and experimental results was obtained, such as in experiments performed by Teeuw and Hesselink [15] and Gogarty [31].

Canella et al. [18] demonstrated that apparent viscosity depends on both microscopic (connectivity, pore size distribution) and macroscopic properties (permeability, porosity) of porous media. Despite calculation improvements, neither effective medium theory nor the capillary bundle model are able to accurately estimate the correction factor. The great discrepancies in results obtained by the models described above and the wide range of correction factors suggested [17] confirm that a universally accepted model does not yet exist. Insufficiency of these models to predict in-situ viscosity may be attributed to their lack of incorporating time dependence and their use of oversimplified porous media models (e.g., capillary bundle).

To avoid over-simplification of porous media obtained by using the capillary bundle approach, pore network modelling has been suggested. In contrast to the capillary bundle approach, pore network modeling envisages porous media as interconnected bundles with idealized geometries where larger pores (pore bodies) are connected via smaller ones (pore throats). Pore network models have been used by Sorbie et al. [20] to study non-Newtonian fluids that exhibit shear thinning properties; later, several authors studied these phenomena [21,32–35]. Using network modeling, Sorbie et al. [20] showed that in connected (2D) networks of porous media, the average shear rate in the network correlates linearly with the flow rate. This result is not obvious and indeed is rather unexpected. Thus, any formula of the form of Equation (8) which is linear in U, and has a "shift factor", will do well for shear thinning fluids. The paper also shows that a similar argument holds for extensional flow where the extensional rate in the porous medium correlates linearly with flow rate (U). Lopez et al. [21] applied a pore network model to study non-Newtonian fluids using the same approach as for Newtonian fluids, except that viscosity in each bundle was not assumed to be constant and was considered as a function of pressure drop. Therefore, an iterative approach was suggested to calculate pressure drop and apparent viscosity. Although they obtained satisfactory agreement between analytical and experimental results using this approach, Balhoff and Thompson [34] stated that effects of concentration were neglected, and consequently proposed a new model based on CFD calculations to include effects of concentration in calculating conductivity of pore throats. They used pore network modeling to model shear thinning polymer flow with yield stress within a sand-pack.

Zamani et al. [35] studied the effects of rock microstructures on in-situ rheology using digital rock physics and reported that microscopic properties such as aspect ratio, coordination number and tortuosity may affect deviation of in-situ from bulk rheology.

In some experiments [23,27,31,36], in-situ rheology has been reported to deviate significantly from the behavior in bulk flow, such that in-situ rheology may not be calculated directly from bulk rheology using the previously mentioned models. To achieve this, one may use these approaches assuming that either in-situ rheological properties are different from bulk rheological properties (e.g., Hejri et al. [36]) or that the relationship between apparent shear rate and Darcy velocity is non-linear (e.g., Gogarty [31]).

Calculation of in-situ rheology is a controversial subject. Until now, there has been no direct method to obtain it and, generally it has been measured by performing core floods. However, Skauge et al. [37] observed significantly different in-situ rheology for HPAM in linear compared to radial geometry. This discrepancy might be due to differing pressure regimes and flux conditions experienced by polymer solutions flowing through these inherently different flow geometries.

The problem with in-situ rheology calculations extends beyond finding the appropriate correction/shift factor. It also encompasses predicting the onset of extensional viscosity, which is treated as a separate subject in the following section.

2.2. Extensional Viscosity

Several experimental results show that, although polymer solutions (e.g., HPAM) only demonstrate shear thinning behavior in a rheometer, they may exhibit apparent shear thickening behavior above a critical shear rate in porous media (Figure 2) [23,27,31,36]. Generally, polymer flow in porous media may be divided into two distinct flow regimes: shear dominant and extensional dominant flow regimes. Since apparent shear thickening occurs in the extensional flow regime, it may also be referred to as extensional viscosity.

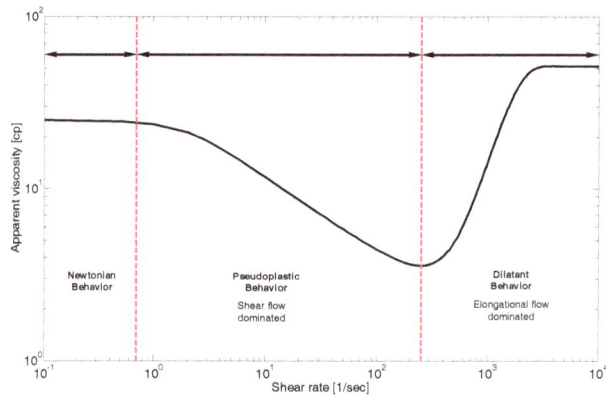

Figure 2. Schematic illustration of apparent viscosity in porous media.

Although its source is poorly understood, extensional viscosity is considered one of the principal aspects of polymer flow in porous media due to its influence on injectivity and oil mobilization. This phenomenon was suggested to be a consequence of elastic properties of polymer solutions (elongational dominated [38] or inertia-dominated flow [39]). As a result, extensional viscosity is often used interchangeably with elongational viscosity, shear thickening behavior, viscosity enhancement, dilatant behavior and viscoelasticity. Two different models are generally used to explain this phenomenon, the transient network model [40–42] and coil stretch model [43]. We adhere to the latter of these models.

Polymer molecules may be envisaged as entangled coils, and when exposed to a flow field, two forces may arise. First, an entropic force that attempts to maintain the existing polymer coil configuration. As coil entanglement increases, higher resistance to deformation is observed. Second the drag force resulting from interactions between solvent fluid and polymer molecules. When shear rate increases beyond a critical rate, molecule configurations change abruptly from coil to stretched states. Therefore, polymer coils start to deform, resulting in anisotropy and stress differences between elongation and compression. Consequently, normal stresses and elastic properties become more dominant.

Choplin and Sabatie [40] suggested that when polymer molecules are exposed to a simple shear flow at a constant shear rate ($\dot{\gamma}$), molecules rotate at a constant angular velocity (ω) proportional to applied shear rate, and in each rotation polymer molecules are stretched and compressed. The time between each rotation can be calculated by Equation (18).

$$t = \frac{\pi}{2} k \, \dot{\gamma} \tag{18}$$

where k is a constant of proportionality, related to viscosity. If t is higher than the Zimm relaxation time, no dilatant behavior occurs. Consequently, the critical shear rate at the onset of dilatant behavior may be calculated based on Zimm relaxation time as follows:

$$\lambda_z = \frac{6}{\pi^2} \frac{M_w}{RT} [\mu]_0 \mu_s \tag{19}$$

$$t \leq \lambda_z, \dot{\gamma} \geq \gamma^* \overset{We = \dot{\gamma}\lambda_z}{\Rightarrow} We^* = \lambda_z * \dot{\gamma}^* = \frac{\pi}{2} k \tag{20}$$

Polymer viscosity behavior in extensional flow may be entirely different from its behavior in pure shear flow, i.e., polymer solution may show simultaneous shear thinning and extension thickening behavior. Theoretically, extensional viscosity can be calculated from Equation (21), where N_1 is normal stress difference and $\dot{\epsilon}$ is stretch rate. The relative importance of extensional viscosity and shear viscosity is defined by a dimensionless parameter known as the Trouton ratio (Equation (22)), initially proposed by Trouton [44]. For non-Newtonian fluids (especially viscoelastic fluids), Tr can reach very large values, such as 10^3 to 10^4 (i.e., when polymer solution demonstrates shear thinning and extension thickening simultaneously).

$$\mu_e = \frac{N_1}{\dot{\epsilon}} \tag{21}$$

$$Tr = \frac{\mu_e}{\mu_s} \tag{22}$$

In Figure 2, the in-situ viscosity of viscoelastic polymers is depicted in both shear and extensional flow regime. At the onset of polymer flow, the generated hydrodynamic force from fluid flow (i.e., drag force) is below the threshold value in terms of overcoming entropic forces. Therefore, polymer configuration persists in a coil shape, and viscosity remains constant and equal to the zero-shear rate viscosity (upper Newtonian plateau). As flow rate increases, polymer molecules are exposed to larger drag forces that disentangle polymer coils and aligns them along the flow direction. This coil alignment reduces resistance to flow (i.e., induces viscosity reduction) and is referred to as shear thinning. When the orientation of polymer molecules is completely aligned, they will start to stretch at increasing flow rates. A change in the deformation of polymer molecules may cause normal stress differences. At low stretch rates ($\dot{\epsilon}$), N_1 is very low and by increasing the stretch rate, N_1 dramatically increases. In other words, beyond the critical shear rate ($\dot{\gamma}_c$), instead of intramolecular interaction, intermolecular interactions will develop which generate amorphous structures much larger than average polymer chain dimensions [28,45].

Within the extensional flow regime, the apparent viscosity generally reaches a maximum value, subsequently followed by a decreasing viscosity interval. This phenomenon may be interpreted as high viscoelastic stresses causing polymer rupture and chain halving, and it has been reported as being more severe in low-permeability porous media [46]. As molecular rupture occurs, new molecular weight distributions emerge (larger molecular weight fractions are distorted) and viscosity behavior of the polymer may be governed by a new molecular weight distribution.

The onset of extensional viscosity-the transition point between shear-dominant and extensional dominant flow-depends on polymer, solvent, and porous media properties. The effects of polymer properties on extensional viscosity can be investigated by using special rheometers that only generate pure extensional flow [47–56]. In the following, the effects of polymer, solvent and porous media properties on the onset of extensional viscosity are explained.

2.2.1. Polymer Concentration

Chauveteau [55] reported that the maximum relaxation time increases with polymer concentration, thus dilatant behavior commences at lower shear rates (Figure 3). He also included the effect of concentration in the expression for Zimm relaxation time, producing Equation (23).

$$\lambda_z = \frac{6}{\pi^2} \mu_s \frac{\mu_{ro} - 1}{C} \frac{M_w}{RT} \tag{23}$$

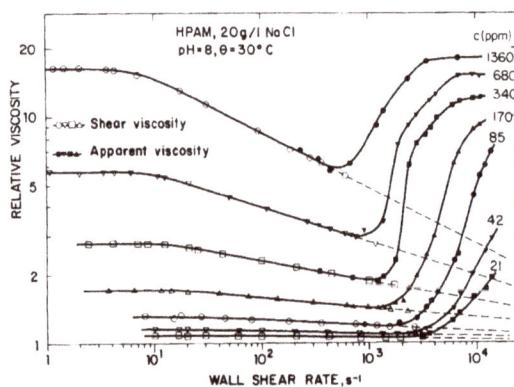

Figure 3. Effect of polymer concentration on the onset of extensional viscosity in a model with 45 successive constrictions. Reproduced with permission from [55].

The effect of concentration on extensional viscosity was also investigated by Lewandowska [56]. In contrast to Chauveteau, he reported that dilatant behavior commences at higher shear rates with increasing polymer concentration. He attributed this observation to the higher degree of entanglement as the concentration increases, thus increasing the extent of the shear thinning region.

Briscoe et al. [57] could not identify a consistent trend between polymer concentration and onset of extensional viscosity. They assumed that only a narrow region of polymer concentrations is able to generate apparent shear thickening behavior. Below a critical concentration limit, defined as the critical overlap concentration (C*), few polymer chains are able to form transient networks. At concentrations above C*, the extent of shear thinning may increase and, consequently, the onset of apparent shear thickening may be delayed. This effect was also studied by Dupuis et al. [58], where they observed that the onset of dilatant behavior decreased with polymer concentration. However, rheological behavior above the critical shear rate deviated among different concentration ranges (low: 30–60 ppm; medium: 120–240 ppm; and high: 480–960 ppm). Jiang et al. [59] also confirmed scattered data for the onset of extensional viscosity as function of polymer concentration. Clarke et al. [60] reported that the onset of extensional viscosity is independent of concentration and only depends on molecular weight.

2.2.2. Molecular Weight

The lengths of polymer chains increase with molecular weight, resulting in higher inter- and intramolecular entanglement. Thus, the extent of the shear thinning region increases and, consequently, delay the onset of dilatant behavior [56]. However, this explanation directly contradicts the expression for the Zimm relaxation time (Equation (23)), where the latter increases with molecular weight and causes critical shear rate to occur at a lower shear rate.

Jiang et al. [59] also studied the effects of molecular weight on the onset of extensional viscosity. They concluded that relaxation time increases with molecular weight, thus the onset of extensional viscosity occurs at lower shear rates. In addition, they observed that this trend was not valid above a critical molecular weight.

Clarke et al. [60] proposed the following correlation for the dependency of the onset of extensional viscosity on polymer molecular weight:

$$\lambda_{ext} \propto MW^2 C_p^0 \tag{24}$$

2.2.3. Salinity Effect

The effect of salinity on polymer rheology may be crucial in some reservoir conditions [11,61,62], and depends on polymer type. For typical EOR polymers (e.g., xanthan, HPAM, or generally non-hydrolyzed polymers), increasing salinity generally reduces coil gyration and hydrodynamic radius. Due to the repulsion between ionic groups in HPAM solutions, increasing salinity compresses the electrical double layer on molecular chains and electrostatic repulsion decreases. In the case of HPAM, the reaction mechanism varies for different metal ions i.e., either monovalent (Na^+) or divalent (Ca^{2+}) cations. In the monovalent case, it may suppress the charge effect and reduce the hydrodynamic radius. In the divalent case, reactions between cations (i.e., Ca^{2+}) can play the role of cross-linkers and influence the conformation and rheological properties of HPAM. In both cases, larger shear rates are required to uncoil polymers and the apparent shear thickening commences at larger shear rates [57,58,63].

2.2.4. Degree of Hydrolysis

When HPAM is dissolved in water, electrostatic repulsion forces cause polymer molecules to expand easily and the shear thinning region is shortened. Therefore, as the degree of hydrolysis increases, the onset of apparent shear thickening decreases [56].

2.2.5. Pressure and Temperature Effect

Although polymers are considered incompressible fluids, they do exhibit some degree of compressibility. Thus, pressure may have an impact on viscosity. By increasing pressure, the free volume between polymer molecules decreases and Brownian motion of polymer chains is inhibited, consequently resulting in viscosity increase of polymer solution. Experimental results [64] indicate that the onset of extensional viscosity decrease significantly with pressure.

The effects of temperature on polymer rheology has also been studied extensively [57,59,65,66] and results show that the critical shear rate and onset of dilatant behavior are retarded with increasing temperature. This behavior may have the following two explanations. Firstly, polymer relaxation time and solvent viscosity should both decrease with increasing temperature, based on Equation (23). Secondly, solvent quality decreases with temperature. By decreasing solvent quality, coil size is reduced, and to compensate for this reduction, a larger shear rate is needed to uncoil and elongate the polymer. Therefore, the onset of extensional viscosity occurs at higher shear rates.

2.2.6. Porous Media Properties

In addition to polymer properties, porous media may also significantly influence the generation of extensional flow, as shown by several experimental [25] and numerical studies [67]. Due to variation in cross-sectional area along its propagation path, polymer molecules are forced to accelerate and decelerate. Consequently, they will experience both stretch and shear flow in porous media, and above a critical flow rate, extensional flow will dominate shear flow.

To envisage polymer flow in porous media, the latter may be considered as a simplified contraction-expansion channel. As polymer molecules enter contractions, they will be compressed and stretched. If the flow is below a critical velocity, deformed polymer molecules have sufficient time to return to their original state. Therefore, when polymer solutions enter subsequent contractions, no stress is stored and no additional resistance to flow is observed. However, if polymer relaxation time is high and polymer molecules are not able to return to their equilibrium state between contractions, stress will be stored and accumulated, thus resulting in steep increases in pressure drop and apparent viscosity. This phenomenon can be interpreted as a memory effect of polymer molecules.

Due to the inherent nature of porous media, polymer molecules are sheared near the wall and elongated at the flow axis. Therefore, molecular momentum is transferred by both tangential and normal stress components in porous media. Seeing that polymer molecules are able to rotate in pore space, molecules are not strained and effective viscosity is only controlled by shear. In contrast, if molecules are exposed to strain for sufficient time, molecule deformation plays a major role and effective viscosity will be defined by strain [25,67–72].

To predict the onset of extensional viscosity in porous media, the dimensionless Deborah number is defined as a ratio between the characteristic relaxation time of a fluid (θ_f) and characteristic time of porous media (θ_p), considered as the average time to travel from one pore body to another (Equation (25)). In other words, the Deborah number may be interpreted as the ratio between elastic and viscous forces. Based on this expression, the Deborah number is zero for Newtonian fluids and infinity for Hookean elastic solids.

$$N_{De} = \frac{\theta_f}{\theta_p} \tag{25}$$

Polymer solutions may have a wide range of molecular weights leading to a large number of relaxation times. Many researchers have used the longest relaxation time as representative of θ_f. However, this may cause the overestimation of Deborah numbers at the onset of extensional viscosity. Relaxation times may also be calculated from normal stress differences [73].

Some experimental observations revealed that the onset of extensional viscosity occurs when N_{De} is larger than 0.5 [74]. However, the Deborah number is not constant in different experiments and a wide range of values has been reported. Marshall and Metzner [73] reported a Deborah number of 0.1 at the onset of extensional viscosity, while Chauveteau [55] reported a relatively high Deborah number of 10. This wide range of reported Deborah numbers at the onset of extensional viscosity is due to difficulties in calculating stretch rates in porous media. To support this idea, Heemskerk et al. [75] reported that by using different polymer types in the same rock sample, critical Deborah numbers (N_{De}) were identical. However, when the same polymer was used in different rock samples, the critical N_{De} varied between 1 and 2. They concluded that measured relaxation times from experimental results can be used to practically define the onset of extensional viscosity, but they acknowledged that equations for calculating stretch rate are not able to capture the exact N_{De} at the onset of extensional viscosity. Zamani et al. [67] proposed that to obtain a more accurate estimation of the critical N_{De}, the stretch rate distribution at the pore scale is required. Metzner et al. [76] concluded that the critical Deborah number might only be used as a first estimation of the onset of extensional viscosity. In Table 3, some suggested equations for the calculation of Deborah number are summarized.

Table 3. Proposed equations for Deborah number calculation.

Model	Equation	Description
Masuda et al. [77]	$N_{De} = \theta_f \, \dot{\gamma}_{eq}$ $\dot{\gamma}_{eq} = \frac{\gamma_c \, |U_w|}{\sqrt{k} \, k_{rw} \phi S_w}$	They used the inverse of the shear rate for θ_p. U_w is the Darcy velocity, k_{rw} is the water relative permeability, S_w is water saturation and $\dot{\gamma}_c$ is a constant equal to 3.97C, where C is an empirical correlation factor to account for the difference between an equivalent capillary model and real porous media
Hirasaki and Pope [26] Haas and Durst [78] Heemskerk et al. [75]	$\frac{1}{\theta_p} = \dot{\varepsilon} = \frac{v}{d} =$ $\frac{U_w}{(1-\phi S_w)\sqrt{150KK_r/(\phi S_w)}}$	

Several experimental results [68,79] show that the Deborah number alone is not sufficient to predict the onset of extensional viscosity. As an explanation, Ranjbar et al. [80] stated that the onset of extensional viscosity highly depends on the elastic properties of polymer solutions and relaxation time alone cannot capture viscoelastic properties. Experimental results reported by Garrouch and Gharbi [79] support this idea. They investigated two different polymer solutions (xanthan and HPAM) in Berea and sand-packs. Calculated Deborah numbers for these two completely and inherently different polymer solutions inside sand-packs were (surprisingly) identical. While xanthan consists of rigid, rod-like molecules that do not show extensional viscosity, HPAM consists of flexible and elastic chain-structured molecules.

Zamani et al. [67] numerically studied the effect of porous media on the onset of extensional viscosity by using real images of porous media obtained from digital rock physics. They confirmed that microscopic features of porous media had significant impact on the onset of extensional viscosity. Furthermore, by increasing the aspect ratio and inaccessible pore volume and decreasing the coordination number, extensional viscosity occurred at lower shear rates, in agreement with several experimental results [55,68,81].

Skauge et al. [37] reported that in radial flow, the onset of extensional viscosity occurred at higher shear rates than at typical core flooding. Since radial flow is more representative of real field conditions, results obtained from radial disks should be more accurate as laboratory data for field implementation.

Briefly summarized, at low shear rates where the amplitude of the elastic component is negligible, flow is controlled by shear forces. In contrast, above a critical shear rate, flow is extensional and governed by elastic forces. Therefore, the response of polymer solutions to imposed stress may be expressed as the sum of shear and elastic components:

$$\Delta P = \Delta P_{shear} + \Delta P_{elastic} \tag{26}$$

$$\mu = \mu_{shear} + \mu_{elastic} \tag{27}$$

The viscosity of polymer solutions under shear flow can be described by empirical equations such as the power-law and Carreau models. To describe viscosity under elongational flow, several models have been suggested, and some of them are summarized in Table 4.

Table 4. Proposed models for calculation of elongational viscosity.

Model	Equation	Description
Hirasaki and Pope [20]	$\mu_{el} = \frac{\mu_{sh}}{[1-N_{De}]}$	
Masuda et al. [77]	$\mu_{elas} = \mu_{sh} C_c (N_{De})^{m_c}$	where C_c and m_c are constant and relate to pore geometry
Delshad et al. [61]	$\mu_{el} = \mu_{max}\left[1 - \exp\left(-(\lambda_2 \tau_r \dot{\gamma})^{n_2-1}\right)\right]$ $\tau_r = \tau_1 + \tau_0 C_p$ $\mu_{max} = \mu_w \left(AP_{11} + AP_{22} \ln C_p\right)$	τ_r is the characteristic relaxation time and can be calculated by dynamic frequency sweep test in the laboratory. Some empirical correlations are also proposed for dependency of different parameters on polymer concentration
Stavland et al. [62]	$\mu_{el} = (\lambda_2 \dot{\gamma})^m$ $\lambda_2 = \left\{ N_{De} \left(\frac{1-\phi}{\phi}\right) \left(\frac{6\alpha \sqrt{\tau}}{\lambda_1}\right) \right\}^{-1}$	m is a non-zero tuning parameter which is known as the elongation exponent and depends on the molecular weight and demonstrates linear correlation with $[\mu]$ C_p. α in the listed formulation is considered 2.5

2.3. Injectivity

Polymer injectivity is a crucial factor governing the economics of polymer flooding projects and its accurate estimation is a prerequisite in terms of optimizing the upper-limit injection rate [82]. Injection well pressure may increase due to one of the following causes: (1) oil bank formation, (2) in-situ polymer viscosity (especially shear thickening due to viscoelasticity) and (3) different types of retention, which cause permeability reduction.

The highest pressure drops observed during polymer flooding are located in the vicinity of the injection wellbore due to dramatic variations in flow rate. Therefore, it is important to include non-Newtonian effects of polymer solutions to accurately predict polymer injectivity. Although both HPAM and xanthan demonstrate shear thinning behavior at low to moderate shear rates, HPAM exhibits apparent shear thickening above a critical flow rate due to its inherent viscoelastic nature. For field applications, injection rates in the vicinity of the injection well may easily exceed the onset of extensional viscosity, and injectivity will then dramatically decrease. In contrast to HPAM, xanthan shows exclusively shear thinning behavior and will attain its highest value of injectivity in the near-wellbore region.

Injectivity investigations at the lab scale are required before implementing field applications, and effects of polymer solution properties, in-situ rheology, temperature, pH, level of retention and the nature of porous media should be accurately measured [83,84]. Furthermore, if screening criteria for polymer type are disregarded, polymer entrapment in narrow pore throats can have significant effects on its injection rate. The salinity of solutions can also affect polymer solubility, resulting in filter cake formation near injection wells or precipitation of polymer molecules in the reservoir. Inaccurate measurement of in-situ rheology and especially the onset of extensional viscosity may lead to either an underestimation or overestimation of injectivity. In some polymer flooding projects, measured injectivity may differ significantly from the simulation or analytical forecast. These unexpected injectivities may be due to the occurrence of mechanical degradation [82,85,86], induced fractures [87–89], or even inaccurate analytical models for calculating in-situ rheology and predicting extensional viscosity.

3. Radial In-Situ Rheology

Injectivity (I) may be defined as the ratio of volumetric injection rate, Q, to the pressure drop, ΔP, associated with polymer propagation between injection well and producer [1]:

$$I = \frac{Q}{\Delta P} \tag{28}$$

As previously mentioned, formation fracture pressure may constrain the value of volumetric injection rate. Due to its significant effect on project economics, accurate determination of differential pressure, and hence injectivity, at a given injection rate is essential. To achieve this, all factors affecting differential pressure during polymer flooding must be quantified. Darcy's law for radial flow may be expressed in terms of differential pressure as follows:

$$\Delta P = \frac{\mu_{app} Q}{2 \pi h k_{e,i}} ln \frac{r_e}{r_w} \tag{29}$$

where μ_{app} is apparent viscosity, h is disk thickness, $k_{e,i}$ is effective permeability to polymer solution, r_e is disk radius and r_w is injection well radius.

In this paper, the ratio of resistance factor (RF) to residual resistance factor (RRF) is used to represent apparent viscosity of polymer solutions propagating through porous media, thus isolating its viscous behavior, i.e.,

$$\mu_{app} = \frac{RF}{RRF} \tag{30}$$

where the resistance factor (RF) represents the pressure increase of polymer relative to brine and the residual resistance factor (RRF) is defined as the ratio of pressure before and after polymer injection (i.e., pressure caused by irreversible permeability reduction induced by retention mechanisms).

Due to their inherent viscoelastic behavior in porous media, synthetic polymers (e.g., HPAM) will exhibit shear-dependent apparent viscosity. Although the common consensus on apparent shear thickening as a phenomenon is accepted, its viscosifying magnitude is still an ongoing topic of debate in scientific communities.

Accurate polymer rheology estimation is a prerequisite for reasonable injectivity estimates due to the proportionality between apparent viscosity and differential pressure. In linear core floods where steady-state pressure conditions exist, polymer flux will remain constant from inlet to outlet, rendering rheology estimation a straightforward task. However, in radial flow, polymer flux is gradually reduced as it propagates from injection well to producer, therefore attaining a range of viscosities rather than one specific value. Since the degree of mechanical degradation generally increases with injection rate, discrepancies in polymer rheology obtained from different injection rates may transpire. Instead of possessing one definite rheology, polymers propagating through radial disks will exhibit both shear-dependent and history-dependent viscosity behavior, thus increasing the complexity of rheology estimation in radial compared to linear models. To date, no correction factor has been suggested to account for this dual nature phenomenon. Even when mechanical degradation is excluded, i.e., when injected and effluent viscosities are approximately equal, this dual nature phenomenon persists, and is suggested to be attributed to non-equilibrium pressure conditions experienced in radial flow and inherent history-dependent nature of polymer molecules.

In addition, synthetic polymers are susceptible to mechanical degradation at high flux, typically in the near-wellbore region, which will impart an irreversible viscosity reduction due to polymer molecule fragmentation. Mechanical degradation induces a pressure drop that improves injectivity. However, since it disrupts the carefully selected viscous properties of the polymer solutions by a non-reversible viscosity decrease, mechanical degradation is not a sought-after phenomenon in polymer flooding. A remediation measure to reduce mechanical degradation is to pre-shear the polymer before injection. Pre-shearing removes the high molecular weight part of the molecular weight distribution, which is believed to be most susceptible to mechanical degradation [6]. Mechanical degradation may also be minimized by shifting to a lower molecular weight polymer. However, this would require higher amounts of polymer to obtain the same concentration, thus potentially influencing polymer project economics.

As mentioned, in radial geometry, high flux causing mechanical degradation occurs principally in the near-wellbore region, as opposed to linear geometry where this high flux persists throughout the entire propagation distance. Therefore, the time that polymer is exposed to high shear is short in radial transient flow pattern, as opposed to that of a steady-state linear core flood, [34]. Based on this time-differing condition between linear and radial flow, it was suggested by Skauge et al. that polymer is degraded to a lesser extent in radial compared to linear flood when injected at the same volumetric flow rate [33].

In summary, there are two principal factors governing injectivity during polymer flooding in linear geometry: (1) viscoelasticity of polymer that induces large injection pressures mainly due to apparent shear thickening behavior at high flux; and (2) mechanical degradation in the near-wellbore region, which causes an entrance pressure drop [1]. In radial disks, two additional factors should be included: (3) non-equilibrium pressure conditions due to kinetic effects; and (4) memory-effects of polymer molecules in non-constant velocity fields.

4. Materials and Methods

Rock: Bentheimer outcrop rock (porosity of ~23%, permeability of about 2.6 Darcy). Based on XRD measurements, Bentheimer consists predominantly of quartz (90.6%) with some feldspar (4.6%), mica (3.2%) and siderite (1.0%).

Polymer: Flopaam 3630S, 30% hydrolyzed, MW = 18 million Da.

Brine: Relatively low salinity with a low content of divalent ions. Brine composition by ions is given in Table 5.

Table 5. Brine ionic composition.

Ion	Concentration (ppm, *w/w*)
Na	1741
K	28
Ca	26
Mg	17
SO₄	160
Cl	2687
TDS	4659
Ionic strength	0.082
Hardness	43

Linear core floods: Core data are summarized in Table 6. All experiments were performed at room temperature and pressure.

Radial core floods: Bentheimer disks were prepared by coating with epoxy resin, vacuuming and saturating with brine. One disk was then drained with an extra heavy oil and aged for 3 weeks at 50 °C to a non-water-wet state. The crude had an initial viscosity of about 7000 cP. The extra heavy oil used for drainage and aging, was then exchanged with a flooding oil of 210 cP. Both experiments were performed at room temperature and pressure. Core data are given in Table 7. The pressure ports were located in the injection and production wells and at radii 0.5, 1.0, 1.5, 2.0, 3.0, 4.0 and 5.0 cm for the disk without oil and at radii 1.1, 2.0, and 5.0 cm for the disk containing oil.

The Bentheimer cores show a pore-throat distribution function similar to other outcrop sandstone material, Figure 4. All porous media have local pore-size variation, involving continuous contraction and expansion of pore-scale transport.

Table 6. Core data for linear core floods.

Experiment	Conc.	L (cm)	D (cm)	ϕ (-)	K_{wi} (Darcy)	K_{wf} (Darcy)	RRF (-)	η_i (cP)	η_e (cP)
No oil	500 ppm	9.54	3.77	0.24	2.48	1.35	1.84	6.81	6.62
No oil	1500 ppm	4.89	3.79	0.24	1.99	0.32	6.29	33.76	32.87
With oil, not aged	500 ppm	10.44	3.78	0.23	1.83	0.36	5.08	6.65	6.77
With oil, aged	500 ppm	9.85	3.78	0.23	2.27	0.27	8.41	6.99	5.90

Table 7. Core data for radial core floods.

Experiment	Diameter (cm)	Thickness (cm)	Well Radius (cm)	ϕ (-)	PV (mL)	Soi (frac)	Sorw (frac)	$K_{w,abs}$ (Darcy)	$K_{w,Sorw}$ (Darcy)	K_{wf} (Darcy)
No oil	30.00	2.20	0.15	0.24	373	n.a.	n.a.	2.600	n.a.	0.056
With oil	29.90	2.21	0.30	0.23	352	0.91	0.22	1.551	0.041	0.039

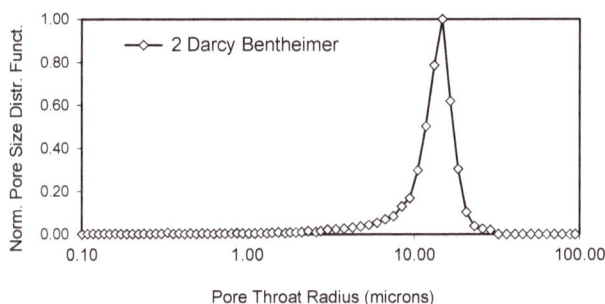

Figure 4. Mercury injection derived pore throat distribution for Bentheimer core material used in the polymer flow experiments.

Simulation: The experimental set-up enabled detailed monitoring of pressure by internal pressure ports located at various distance from the injection well. Differential pressure as function of radial distance was history matched using the STARS simulator, developed by Computer Modeling Group (CMG). The simulation model encompassed a radial grid with 360 sectors, each consisting of 150 grid block cells in radial direction, where the grid block cell size is 1 mm. Porous media permeability (tuning parameter) was obtained by history matching water floods prior to polymer flooding. Local permeability variation improved the history match compared to analytical solution (Darcy's law for radial flow). Permeability data obtained from water floods were used in subsequent polymer floods to isolate the effects of polymer apparent viscosity on differential pressure. In polymer floods, as the permeability obtained from the precursor water flood was held constant, apparent viscosity could be quantitatively investigated as a function of velocity and was used as the tuning parameter to history match differential pressure. The STARS simulation tool can include both shear thinning and thickening behavior of viscoelastic fluids.

Due to the inherent grid averaging calculation method of the simulation tool, the velocity in the first grid block after the injection well was below its analytical value. Because of a rapid velocity decrease with distance in radial models, this phenomenon was addressed by decreasing the injection well radius, thus effectively parallel shifting the position of the first grid block towards the injection well until the correct velocity was attained. This was a necessary step, since the tuning parameter is apparent viscosity as a function of velocity.

5. Polymer In-Situ Rheology in Linear Cores

Four Bentheimer outcrop cores were used to study polymer in-situ rheology in linear systems. Petro-physical properties of core samples as well as properties of polymer solutions are given in Table 6. Two experiments were carried out to examine the effect of polymer concentration on in-situ rheology of the polymer solution. Partially-hydrolyzed Flopaam 3630S at 500 ppm and 1500 ppm was injected into the cores and the in-situ rheology of the polymer solutions was measured. The two concentrations were chosen to give viscosities representative of the upper and lower limit of what would be economically viable for polymer flooding in an oil field. Both concentrations are above the polymer critical overlap concentration, C^*. The results are presented in Table 6 and Figure 5. The bulk viscosity of 1500 ppm 3630S is about 34 cP which is about 5 times that of 500 ppm 3630S. Comparing in-situ rheology of 500 ppm and 1500 ppm 3630S shows that the onset and degree of apparent shear thickening behavior are fairly similar for both concentrations. This is in line with observations by Skauge et al. [8] and Clarke et al. [60] that the onset of extensional viscosity is independent of polymer concentration and only depends on polymer molecular weight. It is noted that this is generally only true for $C^* < C < C_{lim}$, where C_{lim} is the economic limit for polymer concentration, typically between 1500 and 2500 ppm. Table 6 and Figure 5 show that the magnitude of resistance factor (RF) and residual resistance factor (RRF) are about 4 and 3 times higher for 1500 ppm compared to 500 ppm, respectively. This implies that polymer injectivity is a function of polymer concentration, and better injectivity is achieved with lower polymer concentrations.

A series of experiments was also performed to study the effect of the presence of residual oil on polymer in-situ rheology. In these experiments, Bentheimer cores at residual oil saturations of about 22% and different initial wettability states were flooded with polymer and the in-situ rheology behavior was compared to that of single-phase polymer injection in absence of residual oil. Prior to polymer injection, the cores containing oil were water flooded to residual oil saturation. At the end of the water flood, the flow rates were increased to generate pressures higher than that expected for the subsequent polymer flood. This was performed in order to avoid oil mobilization during the polymer flood and, indeed, no oil production was observed during the subsequent polymer flood. The results are presented and compared in Figure 6. As this figure shows, the onset of apparent shear thickening is not affected by the presence of residual oil or the wettability state of the cores. However, the slope of apparent shear thickening and magnitude of resistance factor is significantly affected by oil presence in the cores. That is, although onset of apparent shear thickening is independent of oil presence in porous media and its wettability condition, the results show that the degree of apparent shear thickening is lower when oil is present in the porous media.

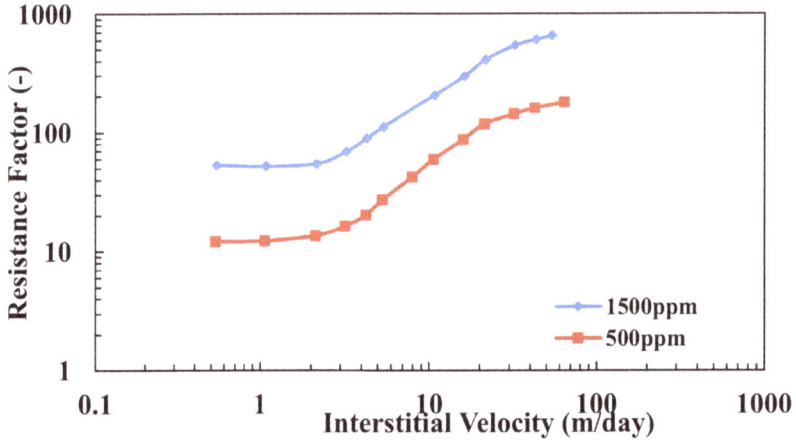

Figure 5. Resistance factor versus interstitial velocity of pre-filtered Flopaam 3630S HPAM polymer dissolved in 1 wt% NaCl brine.

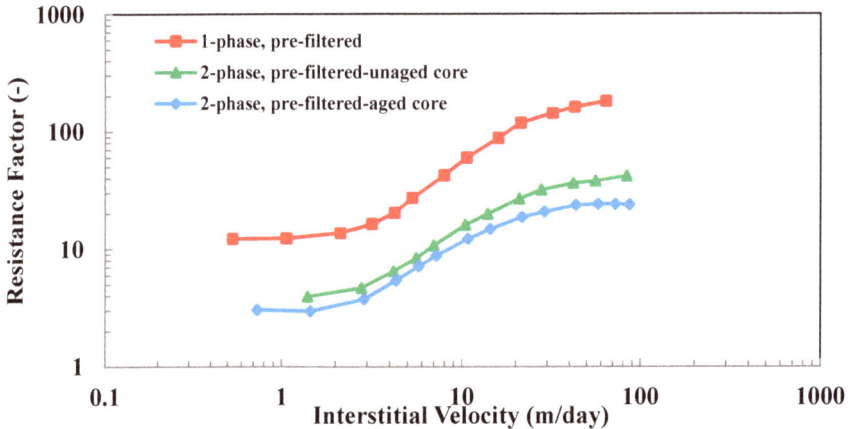

Figure 6. Resistance factor versus interstitial velocity of 500 ppm pre-filtered Flopaam 3630S partially hydrolyzed polyacrylamide (HPAM) polymer dissolved in 1 wt% NaCl brine, single-phase polymer flow and polymer flow at residual oil saturation.

It is important to note that a lower resistance factor in the presence of oil is achieved while porous media is partially occupied by residual oil. Therefore, unlike the single-phase system, in which the pore volume (assuming no inaccessible pore volume) is available for polymer flow, only PV*(1-Sor) is available for polymer flow in two-phase system. This influences and reduces permeability and therefore an even higher resistance factor is expected in the presence of oil. However, the results do not show such an effect, and a lower resistance factor and polymer injectivity is observed with the presence of oil in porous media, which supports the significance of the positive effect of oil on polymer injectivity. The effluent polymer viscosity is reduced by 18% compared to the injected polymer solution for the single-phase, water-wet case, while there is no reduction in effluent viscosity for the two-phase experiments (water-wet and non-water-wet). Shiran and Skauge [90] studied wettability using the same crude oil for aging and found that intermediate wettability was achieved. The end-point water relative permeability confirms that a similar condition was obtained.

Polymer injection in cores with residual oil results in a lower resistance factor which means better polymer injectivity. Furthermore, the resistance factor in the aged core with the non-water-wet state is lower than the resistance factor in the water-wet core. The lower resistance factor in the presence of oil could be attributed to lower adsorption/retention of polymer molecules on rock surface, as reported by Broseta et al. [10]. The rock surface in the presence of an oil film, and especially in less water-wet conditions is partially covered by crude oil polar components during flooding. Therefore, in comparison to single-phase systems, the rock surface has fewer adsorption sites to adsorb polymer molecules. The analysis of reduced apparent viscosity in the presence of oil, assumes that end-point water relative permeability remains constant for polymer as it does for water. The RRF measured with brine after the polymer injection is assumed constant for all rate variation of polymer flow. Under these assumptions a lower resistance factor and better polymer injectivity is expected.

6. Polymer In-Situ Rheology in Radial Flow

Recently, polymer injectivity was analyzed by matching field injectivity tests [5,6,91,92]. In addition to history matching, modification of equations to incorporate fractures and polymer degradation in the near-wellbore zone were reported. The laboratory experiment simplifies the analysis as additional complications like fractures and strong heterogeneity can be avoided.

In earlier studies of radial flow experiments, Skauge et al. [37] used local pressure taps as a function of radial distance from the injection well to derive in-situ rheology. These experiments demonstrated both shear thinning and strong apparent shear thickening behavior.

Two radial flow experiments were performed on circular Bentheimer sandstone disks of 1.6 and 2.6 D permeability with 30 cm diameter and 2.2 cm thickness, see Table 7. The first experiment was performed on a disk that was drained with crude oil and aged to non-water-wet conditions. The second experiment was performed in the absence of oil on a water-wet disk. For the first experiment, the disk was flooded extensively with brine to reach residual oil saturation, Sorw = 0.22. Bump rates were applied to avoid oil mobilization by viscous forces during the subsequent polymer flood. The polymer flood was performed by first saturating the disk with polymer at a low rate to avoid mechanical degradation due to shearing. Thereafter, rate variations were performed to determine in-situ rheology of the polymer. A brine flush was performed between concentration slugs to remove non-adsorbed polymer.

Concentrations of 800 and 2000 ppm were chosen to represent lower and upper boundaries of the semi-dilute region. The second experiment included the same steps, except for water flooding to Sorw. In this case, the water flood was performed to obtain a pressure reference for the subsequent polymer flood. No oil production was detected during polymer floods.

Differential pressure was measured by internal pressure ports located at different radii from the injection well. The 800-ppm HPAM solution was injected in the presence of residual oil at flow rates of 2.2 and 2.8 mL/min, and in the absence of residual oil at 2.0 and 4.0 mL/min. Differential pressure decay as a function of radial distance from injector is shown in Figure 7. The pressure transition zone from semi-steady-state to steady-state is extended compared to the case without oil. Most notable is the difference in pressure in the injection well. While differential pressures measured from internal pressure ports are higher for the two-phase system (as expected), well injection pressure is significantly lower in the presence of residual oil. Taking the pressure ratio of pressure ports at ~1 cm from injection well as a reference, injection pressure should be 5–6 times higher for the disk with oil, compared to the one without. Instead, the injection pressure is 25% lower. There may be several reasons for this observed result. One reason may be that the presence of oil reduces the effective pore volume, thereby leading to higher flow velocities for the polymer in the near-well region. This would subsequently lead to higher effective shear forces on the polymer, producing mechanical degradation. If mechanical degradation occurs, it has only a minor effect on the shear viscosity. The shear viscosity is 15.1 mPas for the effluent sample taken at 2.0 mL/min, while it was 16.0 mPas for the injected solution (measured at 22 °C, 10 1/s). However, as discussed in Section 2.2, it is the extensional viscosity that is the determining factor for high pressures in near-well region. Changes in extensional viscosity are intrinsically hard to measure and were not performed here. It is still possible that the increase in shear forces for the case with residual oil lead to a reduction in extensional viscosity but not for the case without oil where the effective pore volume was larger. The two other reasons are related to the wetting state of the porous media. If the oil is located in smaller pores, polymer flow is diverted to larger pores where it flows at higher velocities (higher flux). Since the velocity increase takes place in larger pores, only minor degradation would be expected. A third reason may be that porous media is fractionally oil-wet and that there is a difference in the slip conditions for the water-wet and the oil–wet surfaces. This may reduce effective shear for the oil-wet surfaces leading to reduced mechanical degradation. Although there have been speculations on the "lubricating" effect of oil-wet surfaces, no clear evidence of the effect on apparent viscosity or injectivity for core material have been shown to date. It is not possible to differentiate between the three phenomena based on the pressure data alone.

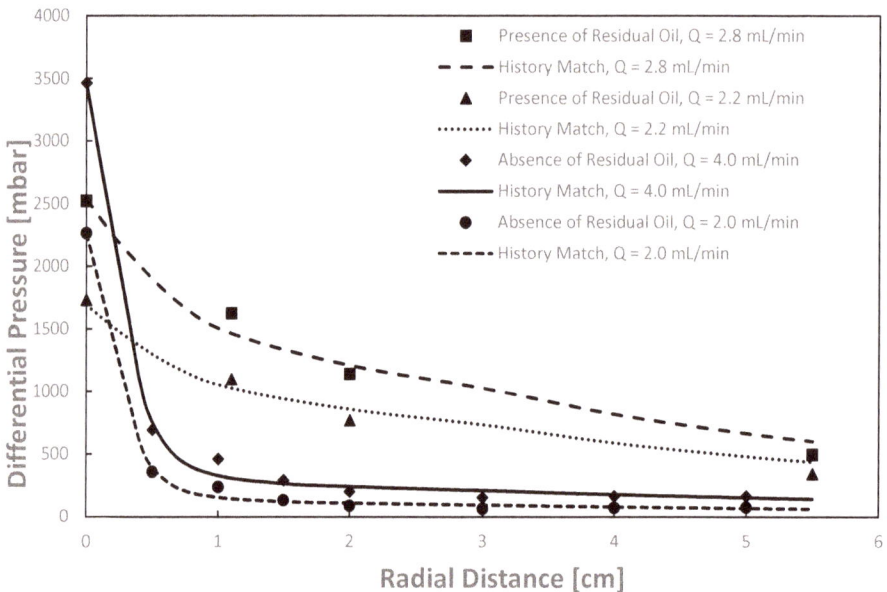

Figure 7. Differential pressure profiles for 800-ppm HPAM floods in the presence and absence of residual oil in radial geometry as a function of distance from injector to producer for four flow rates.

Each of the polymer floods were history matched using STARS (CMG). The measured differential pressures as a function of distance from injection well were used as history match parameters, while polymer apparent viscosity was used as a tuning variable. History matches and polymer rheology from both experiments for 800-ppm HPAM floods are shown in Figures 7 and 8, respectively. It is evident from Figure 8 that the polymer rheology is significantly influenced by the presence of residual oil. In terms of absolute values, the apparent viscosity is between a factor of 5 and 10, and it is higher in the absence, compared to the presence of residual oil. Furthermore, the onset of apparent shear thickening shifts to lower velocities in the presence of residual oil. This occurrence is suggested to result from reduced propagation cross-section caused by the residual oil saturation. When flow channels in porous media become narrower, the extensional flow regime is reached at a lower flux, and HPAM exhibits viscoelastic behavior at an earlier stage, thus the onset of apparent shear thickening commences at a lower flux. The effect of shifting the onset of apparent shear thickening to a lower flux may be detrimental for injectivity. However, since the apparent shear thickening seems to be much more extensive in the absence of residual oil, the rheology shows that overall injectivity is significantly improved in presence of residual oil.

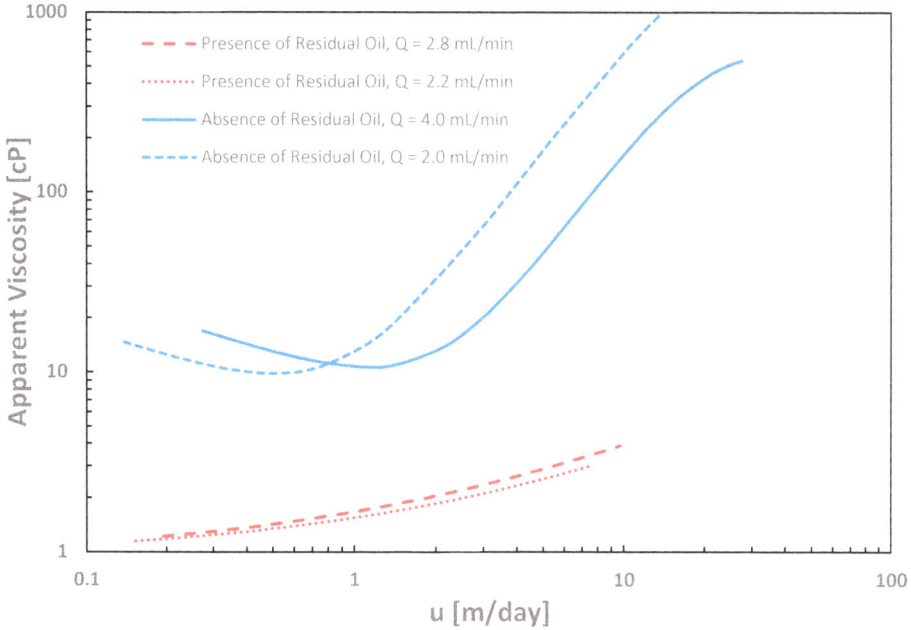

Figure 8. Apparent viscosity from history match of differential pressure for 800 ppm HPAM in presence and absence of residual oil in radial geometry.

History matches and polymer rheology in the presence and absence of residual oil for 2000 ppm floods are shown in Figures 9 and 10, respectively. In order to evaluate the influence of polymer concentration on in-situ rheology, 2000 ppm HPAM was injected in both disks. The differential pressures are shown in Figure 9. In this case, the injection rates were Q = 2.0 and 5.0 mL/min for the disk with no oil, and Q = 1.4 and 1.6 mL/min for the disk with oil. These data show the same trend as for the 800 ppm injection: strong reduction in injection well pressure in the presence of residual oil and extension of the transition zone.

In accordance with the 800 ppm solution, polymer viscosity was significantly higher in the absence compared to presence of residual oil, and ranged between a factor of 6 and 16 in their joint velocity interval, Figure 10. In addition, the 2000 ppm solution also showed a decrease in the onset of apparent shear thickening in the presence of residual oil, consistent with the lower concentration solution investigated. Similar to the 800 ppm solution, apparent shear thickening is observed to be much more extensive in absence of residual oil, thus improved injectivity in the presence of residual oil is further corroborated.

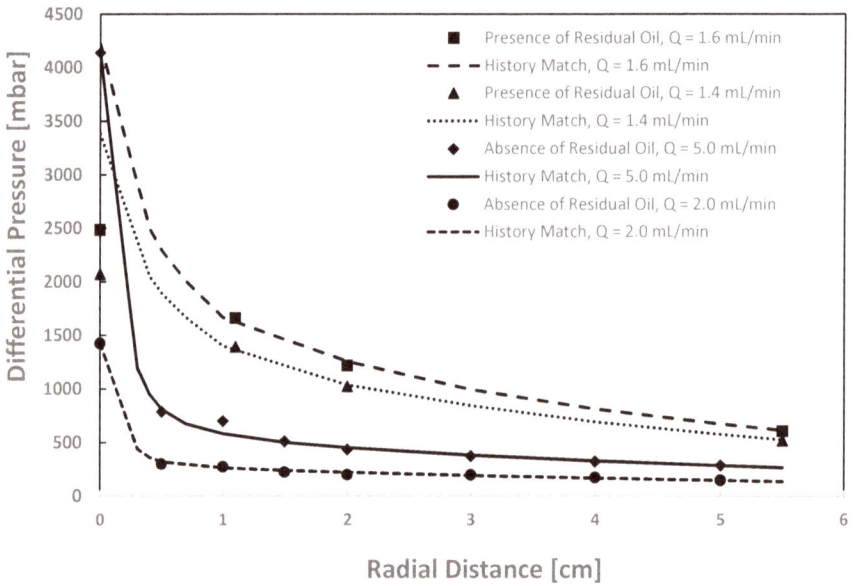

Figure 9. Differential pressure profiles for 2000 ppm 3630S HPAM floods in presence and absence of residual oil in radial geometry as a function of distance from injector to producer for four flow rates.

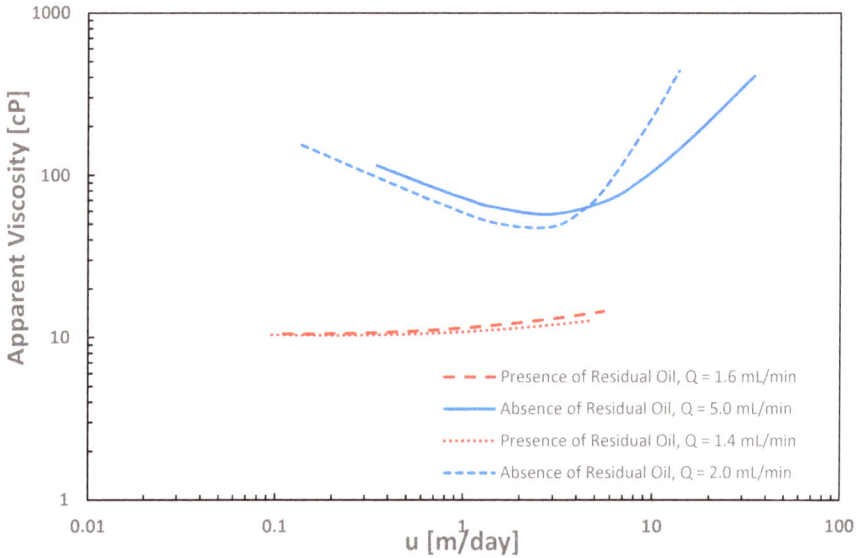

Figure 10. Apparent viscosity from the history match of differential pressure for 2000 ppm 3630S HPAM in the presence and absence of residual oil in radial geometry.

7. Conclusions

A review of polymer flow in a porous medium was presented. The available EOR analytical models we evaluated have limitations in accurately describing flow of polymer at high shear rates, e.g., near injector, and this leads to underestimating or overestimating of polymer injectivity.

The experimental results presented expand our insight into polymer flow in a porous medium. Shear thinning behavior may be present in core floods while bulk rheology is predominant from rheometer measurements. Linear polymer flow experiments are dominated by apparent shear thickening which is not measured in standard rheometers. The extensional viscosity, which is the main cause of the apparent shear thickening behavior, occurs at flow velocities strongly influenced by the porous media.

Linear core floods are commonly used for evaluating polymer in-situ rheology and injectivity, but they suffer from steady-state conditions throughout the core as opposed to the well injection situation where both pressure and shear forces are nonlinear gradients.

In the linear core floods, the onset of apparent shear thickening is independent of polymer concentration, when polymer type, brine composition and porous media are held constant. It is also independent of the presence of oil and wettability for the three cases evaluated here.

Radial flow injections show more complex in-situ rheology. The in-situ rheology shows a much higher degree of apparent shear thickening in the presence of oil. This may be due to restrictions in the pore space available. In the absence of oil, high concentration polymer (2000 ppm) showed shear thinning behavior. The onset of apparent shear thickening was shifted to higher flow velocities. There is a need for further development of numerical models that incorporate memory effects and possible kinetic effects for high polymer flow rates in the near-well region.

Both linear and radial experiments confirm lower apparent viscosity when oil is present in the porous medium. This conclusion is based on the assumption that brine end-point relative permeability is unchanged for polymer injection compared to two-phase flow by water injection. No extra oil was produced during polymer injection and this support the lowering of in-situ polymer viscosity in the presence of oil.

Author Contributions: A.S., B.S.S., and T.S. were responsible for the experimental program and simulation study. N.Z. structured the review contribution, while B.A.-S. and J.G.J. are Ph.D. students working on polymer injectivity, including the experimental study and simulation work. All authors contributed in writing the paper.

Funding: This research received no direct external funding.

Acknowledgments: The authors gratefully acknowledge support from the Norwegian Research Council, Petromaks 2 program. Badar Al-Shakry acknowledge support from, PDO, Petroleum Development Oman. Arne Skauge acknowledge support from Energi Simulations, Canada as the Energi Simulation EOR chair at University of Bergen.

Conflicts of Interest: The authors declare no conflicts of interest.

Nomenclature

A	Cross section area
C	Power-law constant
C_p	Polymer concentration
D_p	Grain size diameter
De	Deborah number
h	Disk thickness
H	Constant, equation 11
k	Constant, equation 18
K_{ei}	Effective permeability to polymer
K	Permeability
L	Length of model
M_w	Polymer molecular weight
N_1	Normal stress difference

n	Ellis, Carreau or power-law constant
P	Pressure
Q	Flow rate
R	Radius
r_e	Disk radius
r_w	Injection well radius
RF	Resistance factor
RRF	Residual resistance factor
R_{eq}	Equivalent radius obtained from Blake-Kozeny model
S_w	Water saturation
T	Temperature
Tr	Trouton ratio
U	Darcy velocity
Wi	Weissenberg number
I	Injectivity
α	Correction factor
β	Constant, equation 10
ω	Angular velocity
$\dot{\varepsilon}$	Stretch rate
ΔP	Pressure drop
$\dot{\gamma}$	Shear rate
$\tau_{1/2}$	Ellis model parameter
$\dot{\gamma}_{eff}$	Effective shear rate
$\dot{\gamma}_{app}$	Apparent shear rate
$\dot{\gamma}_c$	Critical shear rate
λ	Polymer relaxation time
λ_z	Zimm relaxation time
μ	Viscosity
μ_{app}	Apparent viscosity
μ_{eff}	Effective viscosity
μ_0	Upper Newtonian plateau
μ_s	Solvent viscosity
μ_{sh}	Shear rate viscosity
μ_e	Elongational viscosity
μ_∞	Lower Newtonian plateau
ϕ	Porosity
ψ	Tortuosity
θ_f	Characteristic relaxation time of fluid
θ_p	Characteristic time of porous media

References

1. Seright, R.S. The Effects of Mechanical Degradation and Viscoelastic Behavior on Injectivity of Polyacrylamide Solutions. *Soc. Pet. Eng. J.* **1983**, *23*, 475–485. [CrossRef]
2. Shuler, P.J.; Kuehne, D.L.; Uhl, J.T.; Walkup, G.W., Jr. Improving Polymer Injectivity at West Coyote Field, California. *Soc. Pet. Eng. Reserv. Eng.* **1987**, *2*, 271–280. [CrossRef]
3. Southwick, J.G.; Manke, C.W. Molecular Degradation, Injectivity, and Elastic Properties of Polymer Solutions. *Soc. Pet. Eng. Reserv. Eng.* **1988**, *3*, 1193–1201. [CrossRef]
4. Yerramilli, S.S.; Zitha, P.L.J.; Yerramilli, R.C. Novel Insight into Polymer Injectivity for Polymer Flooding. Presented at the SPE European Formation Damage Conference and Exhibition, Noordwijk, The Netherlands, 5–7 June 2013. [CrossRef]
5. Lotfollahi, M.; Farajzadeh, R.; Delshad, M.; Al-Abri, K.; Wassing, B.M.; Mjeni, R.; Awan, K.; Bedrikovetsky, P. Mechanistic Simulation of Polymer Injectivity in Field Tests. Presented at the SPE Enhanced Oil Recovery Conference, Kuala Lumpur, Malaysia, 11–13 August 2015. [CrossRef]
6. Glasbergen, G.; Wever, D.; Keijzer, E.; Farajzadeh, R. Injectivity Loss in Polymer Floods: Causes, Preventions and Mitigations. Presented at the SPE Kuwait Oil & Gas Show and Conference, Mishref, Kuwait, 11–14 October 2015. [CrossRef]

7. Al-Shakry, B.; Shiran, B.S.; Skauge, T.; Skauge, A. Enhanced Oil Recovery by Polymer Flooding: Optimizing Polymer Injectivity. Presented at the SPE Kingdom of Saudi Arabia Technical Symposium and Exhibition, Dammam, Saudi Arabia, 23–26 April 2018.

8. Skauge, T.; Kvilhaug, O.A.; Skauge, A. Influence of Polymer Structural Conformation and Phase Behaviour on In-situ Viscosity. Presented at the 18th European Symposium on Improved Oil Recovery, Dresden, Germany, 14–16 April 2015. [CrossRef]

9. Hughes, D.S.; Teeuw, D.; Cottrell, C.W.; Tollas, J.M. Appraisal of the Use of Polymer Injection To Suppress Aquifer Influx and To Improve Volumetric Sweep in a Viscous Oil Reservoir. *Soc. Pet. Eng.* **1990**, *5*, 33–40. [CrossRef]

10. Broseta, D.; Medjahed, F.; Lecourtier, J.; Robin, M. Polymer Adsorption/Retention in Porous Media: Effects of Core Wettability on Residual Oil. *Soc. Pet. Eng.* **1995**, *3*, 103–112. [CrossRef]

11. Sorbie, K.S. *Polymer-Improved Oil Recovery*; Blackie and Son Ltd.: Glasgow, UK, 1991.

12. Lake, L.W. *Enhanced Oil Recovery*; Prentice Hall: Upper Saddle River, NJ, USA, 1989.

13. Savins, J.G. Non-Newtonian Flow through Porous Media. *Ind. Eng. Chem.* **1969**, *61*, 18–47. [CrossRef]

14. Sadowski, T.J. Non-Newtonian Flow through Porous Media. II. Experimental. *J. Rheol.* **1965**, *9*, 251–271. [CrossRef]

15. Sochi, T. Non-Newtonian flow in porous media. *Polymer* **2010**, *51*, 5007–5023. [CrossRef]

16. Sadowski, T.J.; Bird, R.B. Non-Newtonian Flow through Porous Media. I. *J. Rheol.* **1965**, *9*, 243–250. [CrossRef]

17. Teeuw, D.; Hesselink, F.T. Power-Law Flow And Hydrodynamic Behaviour of Biopolymer Solutions In Porous Media. Presented at the SPE Fifth International Symposium on Oilfield and Geothermal Chemistry, Stanford, CA, USA, 28–30 May 1980. [CrossRef]

18. Cannella, W.J.; Huh, C.; Seright, R.S. Prediction of Xanthan Rheology in Porous Media. Presented at the 63rd Annual Technical Conference and Exhibition of the Society of Petroleum Engineers, Houston, TX, USA, 2–5 October 1988. [CrossRef]

19. Fletcher, A.J.P.; Flew, S.R.G.; Lamb, S.P.; Lund, T.; Bjornestad, E.; Stavland, A.; Gjovikli, N.B. Measurements of Polysaccharide Polymer Properties in Porous Media. Prepared for presentation at the SPE International Symposium on Oilfield Chemistry, Anaheim, CA, USA, 20–22 February 1991. [CrossRef]

20. Sorbie, K.S.; Clifford, P.J.; Jones, E.R.W. The Rheology of Pseudoplastic Fluids in Porous Media Using Network Modeling. *J. Colloid Interface Sci.* **1989**, *130*, 508–534. [CrossRef]

21. Lopez, X.; Valvatne, P.H.; Blunt, M.J. Predictive network modeling of single-phase non-Newtonian flow in porous media. *J. Colloid Interface Sci.* **2003**, *264*, 256–265. [CrossRef]

22. Pearson, J.R.A.; Tardy, P.M.J. Models for flow of non-Newtonian and complex fluids through porous media. *J. Non-Newton. Fluid Mech.* **2002**, *102*, 447–473. [CrossRef]

23. Willhite, G.P.; Uhl, J.T. Correlation of the Flow of Flocon 4800 Biopolymer with Polymer Concentration and Rock Properties in Berea Sandstone. In *Water-Soluble Polymers for Petroleum Recovery*; Springer Publishing: Manhattan, NY, USA, 1988.

24. Bird, R.B.; Stewart, W.E.; Lightfoot, E.N. *Transport Phenomena*; John Wiley and Sons, Inc.: New York, NY, USA, 1960.

25. Christopher, R.H.; Middleman, S. Power-Law Flow through a Packed Tube. *Ind. Eng. Chem. Fundam.* **1965**, *4*, 422–426. [CrossRef]

26. Hirasaki, G.J; Pope, G.A. Analysis of Factors Influencing Mobility and Adsorption in the Flow of Polymer Solution Through Porous Media. *Soc. Pet. Eng. J.* **1974**, *14*, 337–346. [CrossRef]

27. Duda, J.L.; Hong, S.-A.; Klaus, E.E. Flow of Polymer-Solutions in Porous-Media—Inadequacy of the Capillary Model. *Ind. Eng. Chem. Fundam.* **1983**, *22*, 299–305. [CrossRef]

28. Kishbaugh, A.J.; McHugh, A.J. A rheo-optical study of shear-thickening and structure formation in polymer solutions. Part I: Experimental. *Rheol. Acta* **1993**, *32*, 9–24. [CrossRef]

29. Pope, D.P.; Keller, A. Alignment of Macromolecules in Solution by Elongational Flow—Study of Effect of Pure Shear in a 4 Roll Mill. *Colloid Polym. Sci.* **1977**, *255*, 633–643. [CrossRef]

30. Binding, D.M.; Jones, D.M.; Walters, K. The Shear and Extensional Flow Properties of M1. *J. Non-Newton. Fluid Mech.* **1990**, *35*, 121–135. [CrossRef]

31. Gogarty, W.B. Rheological Properties of Pseudoplastic Fluids in Porous Media. *Soc. Pet. Eng. J.* **1967**, *7*, 149–160. [CrossRef]

32. Perrin, C.L.; Tardy, P.M.J.; Sorbie, K.S.; Crawshaw, J.C. Experimental and modeling study of Newtonian and non-Newtonian fluid flow in pore network micromodels. *J. Colloid Interface Sci.* **2006**, *295*, 542–550. [CrossRef] [PubMed]

33. Sochi, T.; Blunt, M.J. Pore-scale network modeling of Ellis and Herschel–Bulkley fluids. *J. Pet. Sci. Eng.* **2008**, *60*, 105–124. [CrossRef]

34. Balhoff, M.T.; Thompson, K.E. A macroscopic model for shear-thinning flow in packed beds based on network modeling. *Chem. Eng. Sci.* **2006**, *61*, 698–719. [CrossRef]

35. Zamani, N.; Bondino, I.; Kaufmann, R.; Skauge, A. Computation of polymer in-situ rheology using direct numerical simulation. *J. Pet. Sci. Eng.* **2017**, *159*, 92–102. [CrossRef]

36. Hejri, S.; Willhite, G.P.; Green, D.W. Development of Correlations to Predict Biopolymer Mobility in Porous Media. *SPE Reserv. Eng.* **1991**, *6*, 91–98. [CrossRef]

37. Skauge, T.; Skauge, A.; Salmo, I.C.; Ormehaug, P.A.; Al-Azri, N.; Wassing, L.M.; Glasbergen, G.; Van Wunnik, J.N.; Masalmeh, S.K. Radial and Linear Polymer Flow—Influence on Injectivity. Prepared for presentation at the SPE Improved Oil Recovery Conference, Tulsa, Oklahoma, 11–13 April 2016. [CrossRef]

38. Faber, T.E. *Fluid Dynamics for Physicists*; Cambridge University Press: Cambridge, UK, 1995.

39. Brown, E.; Jaeger, H.M. The role of dilation and confining stresses in shear thickening of dense suspensions. *J. Rheol.* **2012**, *56*, 875–923. [CrossRef]

40. Choplin, L.; Sabatie, J. Threshold-Type Shear-Thickening in Polymeric Solutions. *Rheol. Acta* **1986**, *25*, 570–579. [CrossRef]

41. Indei, T.; Koga, T.; Tanaka, F. Theory of shear-thickening in transient networks of associating polymer. *Macromol. Rapid Commun.* **2005**, *26*, 701–706. [CrossRef]

42. Odell, J.A.; Müller, A.J.; Keller, A. Non-Newtonian behaviour of hydrolysed polyacrylamide in strong elongational flows: A transient network approach. *Polymer* **1988**, *29*, 1179–1190. [CrossRef]

43. Degennes, P.G. Coil-Stretch Transition of Dilute Flexible Polymers under Ultrahigh Velocity-Gradients. *J. Chem. Phys.* **1974**, *60*, 5030–5042. [CrossRef]

44. Trouton, F.T. On the coefficient of viscous traction and its relation to that of viscosity. *R. Soc.* **1906**, *77*, 426–440. [CrossRef]

45. Edwards, B.J.; Keffer, D.J.; Reneau, C.W. An examination of the shear-thickening behavior of high molecular weight polymers dissolved in low-viscosity newtonian solvents. *J. Appl. Polym. Sci.* **2002**, *85*, 1714–1735. [CrossRef]

46. Hatzignatiou, D.G.; Moradi, H.; Stavland, A. Experimental Investigation of Polymer Flow through Water- and Oil-Wet Berea Sandstone Core Samples. Presented at the EAGE Annual Conference & Exhibition incorporationg SPE Europec, London, UK, 10–13 June 2013. [CrossRef]

47. McKinley, G.H.; Sridhar, T. Filament-stretching rheometry of complex fluids. *Annu. Rev. Fluid Mech.* **2002**, *34*, 375–415. [CrossRef]

48. Fuller, G.G.; Cathey, C.A.; Hubbard, B.; Zebrowski, B.E. Extensional Viscosity Measurements for Low-Viscosity Fluids. *J. Rheol.* **1987**, *31*, 235–249. [CrossRef]

49. Meadows, J.; Williams, P.A.; Kennedy, J.C. Comparison of the Extensional and Shear Viscosity Characteristics of Aqueous Hydroxyethylcellulose Solutions. *Macromolecules* **1995**, *28*, 2683–2692. [CrossRef]

50. Anna, S.L.; McKinley, G.H.; Nguyen, D.A.; Sridhar, T.; Muller, S.J.; Huang, J.; James, J.F. An interlaboratory comparison of measurements from filament-stretching rheometers using common test fluids. *J. Rheol.* **2001**, *45*, 83–114. [CrossRef]

51. Shipman, R.W.G.; Denn, M.M.; Keunings, R. Mechanics of the Falling Plate Extensional Rheometer. *J. Non-Newton. Fluid Mech.* **1991**, *40*, 281–288. [CrossRef]

52. Sridhar, T.; Tirtaatmadja, V.; Nguyen, D.A.; Gupta, R.K. Measurement of Extensional Viscosity of Polymer-Solutions. *J. Non-Newton. Fluid Mech.* **1991**, *40*, 271–280. [CrossRef]

53. Tirtaatmadja, V.; Sridhar, T. A Filament Stretching Device for Measurement of Extensional Viscosity. *J. Rheol.* **1993**, *37*, 1081–1102. [CrossRef]

54. James, D.F.; Chandler, G.M.; Armour, S.J. A Converging Channel Rheometer for the Measurement of Extensional Viscosity. *J. Non-Newton. Fluid Mech.* **1990**, *35*, 421–443. [CrossRef]

55. Chauveteau, G. Moluecular interpretation of several different properties of flow of coiled polymer solutions through porous media in oil recovery conditions. Presented at the 56th Annual Fall Technical Conference and Exhibition of the society of Petroleum Engineers of AIME, San Antonio, TX, USA, 5–7 October 1981. [CrossRef]
56. Lewandowska, K. Comparative studies of rheological properties of polyacrylamide and partially hydrolyzed polyacrylamide solutions. *J. Appl. Polym. Sci.* **2007**, *103*, 2235–2241. [CrossRef]
57. Briscoe, B.; Luckham, P.; Zhu, S.P. Pressure influences upon shear thickening of poly(acrylamide) solutions. *Rheol. Acta* **1999**, *38*, 224–234. [CrossRef]
58. Dupuis, D.; Lewandowski, F.Y.; Steiert, P.; Wolff, C. Shear Thickening and Time-Dependent Phenomena—The Case of Polyacrylamide Solutions. *J. Non-Newton. Fluid Mech.* **1994**, *54*, 11–32. [CrossRef]
59. Jiang, B.; Keffer, D.J.; Edwards, B.J.; Allred, J.N. Modeling shear thickening in dilute polymer solutions: Temperature, concentration, and molecular weight dependencies. *J. Appl. Polym. Sci.* **2003**, *90*, 2997–3011. [CrossRef]
60. Clarke, A.; Howe, A.M.; Mitchell, J.; Staniland, J.; Hawkes, L.A. How Viscoelastic-Polymer Flooding Enhances Displacement Efficiency. *Soc. Pet. Eng. J.* **2016**, *21*, 675–687. [CrossRef]
61. Delshad, M.; Kim, D.H.; Magbagbeola, O.A.; Huh, C.; Pope, G.A.; Tarahhom, F. Mechanistic Interpretation and Utilization of Viscoelastic Behavior of Polymer Solutions for Improved Polymer-Flood Efficiency. Prepared for presentation at the 2008 SPE/DOE Improved Oil Recovery Symposium, Tulsa, OK, USA, 19–23 April 2008. [CrossRef]
62. Stavland., A.; Jonsbråten, H.C.; Lohne, A.; Moen, A.; Giske, N.H. Stavland. A.; Jonsbråten, H.C.; Lohne, A.; Moen, A.; Giske, N.H. Polymer Flooding—Flow Properties in Porous Media Versus Rheological Parameters. Presented at the SPE EUROPEC/EAGE Annual Conference and Exhibition, Barcelona, Spain, 14–17 June 2010. [CrossRef]
63. Aitkadi, A.; Carreau, P.J.; Chauveteau, G. Rheological Properties of Partially Hydrolyzed Polyacrylamide Solutions. *J. Rheol.* **1987**, *31*, 537–561. [CrossRef]
64. Lee, K.; Huh, C.; Sharma, M.M. Impact of Fractures Growth on Well Injectivity and Reservoir Sweep during Waterflood and Chemical EOR Processes. Presented at the SPE Annual Technical Conference and Exhibition, Denver, CO, USA, 30 October–2 November 2011. [CrossRef]
65. Hu, Y.; Wang, S.Q.; Jamieson, A.M. Rheological and Rheooptical Studies of Shear-Thickening Polyacrylamide Solutions. *Macromolecules* **1995**, *28*, 1847–1853. [CrossRef]
66. Cho, Y.H.; Dan, K.S.; Kim, B.C. Effects of dissolution temperature on the rheological properties of polyvinyl alchol solutions in dimethyl sulfoxide. *Korea-Aust. Rheol. J.* **2008**, *20*, 73–77.
67. Zamani, N.; Bondino, I.; Kaufmann, R.; Skauge, A. Effect of porous media properties on the onset of polymer extensional viscosity. *J. Pet. Sci. Eng.* **2015**, *133*, 483–495. [CrossRef]
68. Gupta, R.K.; Sridhar, T. Viscoelastic Effects in Non-Newtonian Flows through Porous-Media. *Rheol. Acta* **1985**, *24*, 148–151. [CrossRef]
69. Smith, F.W. Behavior of Partially Hydrolyzed Polyacrylamide Solutions in Porous Media. *J. Pet. Technol.* **1970**, *22*, 148–156. [CrossRef]
70. Kemblowski, Z.; Dziubinski, M. Resistance to Flow of Molten Polymers through Granular Beds. *Rheol. Acta* **1978**, *17*, 176–187. [CrossRef]
71. Wissler, E.H. Viscoelastic Effects in the Flow of Non-Newtonian Fluids through a Porous Medium. *Ind. Eng. Chem. Fundam.* **1971**, *10*, 411–417. [CrossRef]
72. Vossoughi, S.; Seyer, F.A. Pressure-Drop for Flow of Polymer-Solution in a Model Porous-Medium. *Can. J. Chem. Eng.* **1974**, *52*, 666–669. [CrossRef]
73. Marshall, R.J.; Metzner, A.B. Flow of Viscoelastic Fluids through Porous Media. *Ind. Eng. Chem. Fundam.* **1967**, *6*, 393–400. [CrossRef]
74. Durst, F.; Haas, R.; Interthal, W. Laminar and Turbulent Flows of Dilute Polymer-Solutions—A Physical Model. *Rheol. Acta* **1982**, *21*, 572–577. [CrossRef]
75. Heemskerk, J.; Rosmalen, R.; Janssen-van, R.; Holtslag, R.J.; Teeuw, D. Quantification of Viscoelastic Effects of Polyacrylamide Solutions. Presented at the SPE/DOE Fourth Symposium on Enhanced Oil Recovery, Tulsa, OK, USA, 15–18 April 1984. [CrossRef]

76. Metzner, A.B.; White, J.L.; Denn, M.M. Constitutive equations for viscoelastic fluids for short deformation periods and for rapidly changing flows: Significance of the deborah number. *AIChE J.* **1966**, *12*, 863–866. [CrossRef]

77. Masuda, Y.; Tang, K.-C.; Miyazawa, M.; Tanaka, S. 1D Simulation of Polymer Flooding Including the Viscoelastic Effect of Polymer Solution. *SPE Reserv. Eng.* **1992**, *7*, 247–252. [CrossRef]

78. Haas, R.; Durst, F. Viscoelastic Flow of Dilute Polymer-Solutions in Regularly Packed-Beds. *Rheol. Acta* **1982**, *21*, 566–571. [CrossRef]

79. Garrouch, A.A.; Gharbi, R.C. A Novel Model for Viscoelastic Fluid Flow in Porous Media. Presented at the 2006 SPE Annual Technical Conference and Exhibition, San Antonio, TX, USA, 24–27 September 2006. [CrossRef]

80. Ranjbar, M.; Rupp, J.; Pusch, G.; Meyn, R. Quantification and Optimization of Viscoelastic Effects of Polymer Solutions for Enhanced Oil Recovery. Presented at the SPE/DOE Eight Symposium on Enhanced Oil Recovery, Tulsa, OK, USA, 22–24 April 1992. [CrossRef]

81. Kawale, D.; Marques, E.; Zitha, P.L.J.; Kreutzer, M.T.; Rossen, W.R.; Boukany, P.E. Elastic instabilities during the flow of hydrolyzed polyacrylamide solution in porous media: effect of pore-shape and salt. *Soft Matter* **2017**, *13*, 765–775. [CrossRef] [PubMed]

82. Seright, R.S.; Seheult, J.M.; Talashek, T. Injectivity Characteristics of EOR Polymers. *SPE Reserv. Eval. Eng.* **2009**, *12*, 783–792. [CrossRef]

83. Kulawardana, E.U.; Koh, H.; Kim, D.H.; Liyanage, P.J.; Upamali, K.; Huh, C.; Weerasooriya, U.; Pope, G.A. Rheology and Transport of Improved EOR Polymers under Harsh Reservoir Conditions. Presented at the Eighteenth SPE Improved Oil Recovery Symposium, Tulsa, OK, USA, 14–18 April 2012. [CrossRef]

84. Sharma, A.; Delshad, M.; Huh, C.; Pope, G.A. A Practical Method to Calculate Polymer Viscosity Accurately in Numerical Reservoir Simulators. Presented at the SPE Annual Technical Conference and Exhibition, Denver, CO, USA, 31 October–2 November 2011. [CrossRef]

85. Manichand, R.N.; Moe Soe Let, K.P.; Gil, L.; Quillien, B.; Seright, R.S. Effective Propagation of HPAM Solutions Through the Tambaredjo Reservoir During a Polymer Flood. *SPE Prod. Oper.* **2013**, *28*, 358–368. [CrossRef]

86. Zaitoun, A.; Makakou, P.; Blin, N.; Al-Maamari, R.S.; Al-Hashmi, A.-A.R.; Abdel-Goad, M.; Al-Sharji, H.H. Shear Stability of EOR Polymers. *SPE J.* **2011**, *17*, 335–339. [CrossRef]

87. Suri, A.; Sharma, M.M.; Peters, E. Estimates of Fracture Lengths in an Injection Well by History Matching Bottomhole Pressures and Injection Profile. *SPE Reserv. Eval. Eng.* **2011**, *14*, 405–417. [CrossRef]

88. Zechner, M.; Clemens, T.; Suri, A.; Sharma, M.M. Simulation of Polymer Injection under Fracturing Conditions—A Field Pilot in the Matzen Field, Austria. *SPE Reserv. Eval. Eng.* **2014**, *18*, 236–249. [CrossRef]

89. Van den Hoek, P.; Mahani, H.; Sorop, T.; Brooks, D.; Zwaan, M.; Sen, S.; Shuaili, K.; Saadi, F. Application of Injection Fall-Off Analysis in Polymer flooding. Presented at the 74th EAGE Conference & Exhibition incorporating SPE EUROPEC 2012, Copenhagen, Denmark, 4–7 June 2012. [CrossRef]

90. Shiran, B.S.; Skauge, A. Wettability and Oil Recovery by Polymer and Polymer Particles. Presented at the SPE Asia Pacific Enhanced Oil Recovery Conference, Kuala Lumpur, Malaysia, 11–13 August 2015. [CrossRef]

91. Al-Abri, K.; Al-Mjeni, R.; Al-Bulushi, N.K.; Awan, K.; Al-Azri, N.; Al-Riyami, O.; Al-Rajhi, S.; Teeuwisse, S.; Ghulam, J.; Abu-Shiekha, I.; et al. Reducing Key Uncertainities Prior to a Polymer Injection Trial in a Heavy Oil Reservoir in Oman. Presented at SPE EOR Conference at OGWA, Muscat, Oman, 31 March–2 April 2014. [CrossRef]

92. Li, Z.; Delshad, M. Development of an Analytical Injectivity Model for Non-Newtonian Polymer Solutions. Presented at the SPE Reservoir Simulation Symposium, Woodlands, TX, USA, 18–20 February 2014. [CrossRef]

colloids
and interfaces

MDPI

Article

Waterflooding of Surfactant and Polymer Solutions in a Porous Media Micromodel

Hsiang-Lan Yeh and Jaime J. Juárez *

Department of Mechanical Engineering, Iowa State University, 2529 Union Drive, Ames, IA 50011, USA; hsiang@iastate.edu
* Correspondence: jjuarez@iastate.edu; Tel.: +1-515-294-3298

Received: 1 May 2018; Accepted: 11 June 2018; Published: 12 June 2018

Abstract: In this study, we examine microscale waterflooding in a randomly close-packed porous medium. Three different porosities were prepared in a microfluidic platform and saturated with silicone oil. Optical video fluorescence microscopy was used to track the water front as it flowed through the porous packed bed. The degree of water saturation was compared to water containing two different types of chemical modifiers, sodium dodecyl sulfate (SDS) and polyvinylpyrrolidone (PVP), with water in the absence of a surfactant used as a control. Image analysis of our video data yielded saturation curves and calculated fractal dimension, which we used to identify how morphology changed the way in which an invading water phase moved through the porous media. An inverse analysis based on the implicit pressure explicit saturation (IMPES) simulation technique used mobility ratio as an adjustable parameter to fit our experimental saturation curves. The results from our inverse analysis combined with our image analysis show that this platform can be used to evaluate the effectiveness of surfactants or polymers as additives for enhancing the transport of water through an oil-saturated porous medium.

Keywords: porous media; optical video microscopy; microfluidics; waterflooding; surfactants; polymers

---

1. Introduction

Understanding multiphase flows in porous media is critical for enhancing the recovery of oil from porous bedrock [1]. When a reservoir is tapped through a wellbore in the bedrock for the first time, a pressure differential between the reservoir and the wellbore can be used as a primary form of recovery. Once the pressure differential reaches hydrostatic equilibrium, secondary forms of recovery such as waterflooding (i.e., water injection) can be used to extract the remaining oil [2]. Approximately 40% of the total available oil is extracted using these two methods [3].

Chemical methods, including the addition of surfactants, polymers, or alkali to the water, are used to enhance the recovery of oil beyond the 40% limit encountered with hydrostatic pressure and water [4,5]. Surfactants in water reduce the interfacial tension between the water and the oil, which also reduce the capillary forces and enhance the oil displacement efficiency [6,7]. Polymers act as viscosifying agents, where the inherent viscoelasticity of the polymer flood generates localized non-Newtonian shear stress that pulls oil threads [8,9]. A modification of the relative difference in phase viscosities, known as mobility, is known to enhance oil displacement [10,11]. A combination of chemical methods has been proposed as an effective method for enhancing oil recovery [12].

Understanding how these mechanisms act in isolation to each other and how they can be combined is critical for improving oil displacement efficiency. However, transport through oil-bearing porous media occurs at a rate of one linear foot per day [13], and the direct observation of large-scale oil displacement phenomena can take weeks [14], making the study of multiphase processes slow and difficult. The slow rate of observation is a direct result of transport through the interconnected network of grains that make up the porous media [15]. Micromodels based on lab-on-a-chip platforms offer one possible approach to experimentally investigating multiphase processes in porous media micromodels at shorter time scales [16].

Micromodels enable the direct observation of flow through porous media. In this approach, an optically transparent flow cell is constructed with a uniform distribution of glass or quartz beads dispersed inside to act as the porous grain structure, and direct visualization of the flow is then performed using optical microscopy techniques [17,18]. Although advances in microfabrication technology allow for the manufacturing of complex pore structures [19], most micromodels used to study multiphase fluid flow through pore media have been done in rectangular pore bodies and throats [16,20–25]. Computer-aided design of microchannels [26,27] can be used to mimic heterogeneous porous media structure. However, this method produces 2D pore structures that are not representative of 3D pore space. Micro-computed tomography (micro-CT) [28,29] or focused ion beam-scanning electron microscopy (FIB-SEM) [30] can be used to reconstruct the 3D pore structure for incorporation into porous media micromodels. Reproducing 3D pore space using these approaches is expensive and limited in scale [31–33], which led to a recent proposal to utilize microfluidic devices with patterned channels as a 2.5D structure to mimic the transport characteristics of 3D pore space [34].

This article presents a microfluidic platform to evaluate chemical methods for enhanced oil recovery through the waterflooding process. The platform consisted of a 750-µm-wide microfluidic channel, fabricated by soft lithography, into which glass beads saturated by silicone oil were packed to form a random porous bed. This allowed us to bridge the gap between 2D and 3D pore structures by utilizing soft lithography to create 2D microchannels and glass beads to form a porous 3D network. In this way, our platform helps to fill in the critical gap of 2.5D porous media micromodels by presenting an alternate pathway towards the fabrication of these types of devices.

We used optical fluorescence microscopy to track the invasion of an aqueous fluorescent dye to the main channel through a side channel. Tracking the dye enabled us to evaluate the saturation of the aqueous phase relative to the oil phase. The injection of pure water was compared to the injection of water containing two different kinds of chemical modifiers: sodium dodecyl sulfate (SDS) and polyvinylpyrrolidone (PVP). SDS is an anionic surfactant and PVP is a polymer. Our analysis based on optically tracking the invasion of the aqueous phase demonstrated that chemical modifiers significantly improved the displacement of oil from the microfluidic channel, although image analysis of fractal dimension morphology illustrated differences in aqueous phase invasion.

Micromodels offer qualitative information on flow characteristics, but quantitative tools for comparison to experiment are not well-developed [35]. Our work also seeks to address this gap by introducing an inverse analysis based on implicit pressure explicit saturation (IMPES) to determine the mobility of our surfactant and polymer phases. The analysis presented here offers an approach to interpreting two-phase flow data in a porous microfluidic channel and obtaining parameters such as saturation and mobility that can be used to compare effectiveness of different chemical methods for enhancing oil recovery. Optical measurements of waterflooding can also guide the assessment of other multiphase flow problems, such as the transport of sequestered carbon dioxide in porous bedrock [36,37], filtration of contaminants [38,39], and the additive manufacturing of complex fluid networks [40] or thermal management [41].

2. Materials and Methods

2.1. Device Fabrication

Our device was designed using AutoCAD and printed as a mask onto a transparent plastic sheet (CAD/Art Services, Bandon, OR, USA). The channel consisted of a main channel (750 µm wide and 17 mm long) along with a side channel (20 µm wide and 3 mm long) for injection of the fluorescent dye. The porous media was assembled from glass beads packed within the main channel. The fabrication process, shown in Figure 1A, began with conventional photolithography used to transfer the pattern from the mask to a 4-inch silicon wafer (University Wafer, South Boston, MA, USA) using photoresist (SU8 2050, Microchem, Westborough, MA, USA). The photoresist thickness was approximately 54 µm based on profilometry measurements.

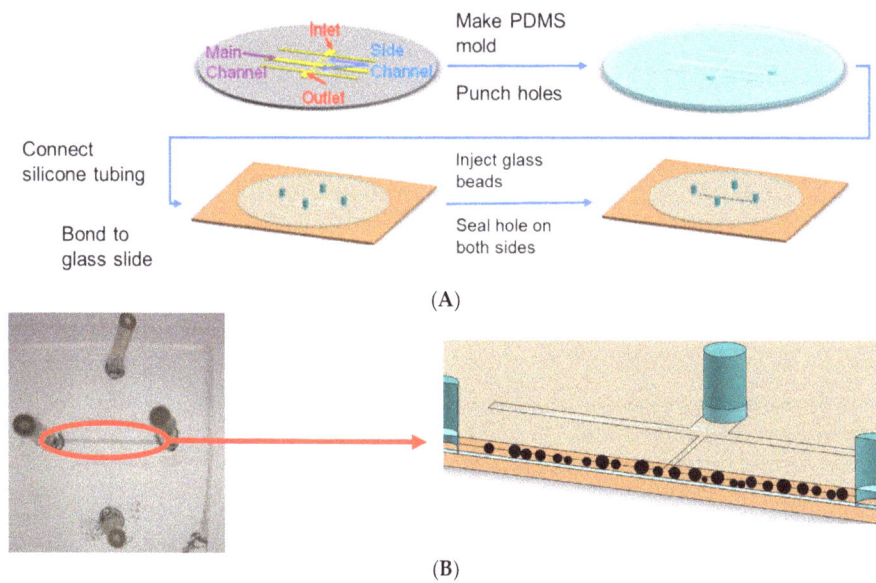

(A)

(B)

Figure 1. (**A**) The design of the microfluidic device used in this work along with the procedures used to fabricate the channel; (**B**) A photograph of the microfluidic device with a schematic that illustrates where glass beads are packed in to form a porous structure. PDMS: polydimethylsiloxane.

Once the pattern was produced onto a silicon substrate through photolithography, it was transferred to a polystyrene petri dish into which polydimethylsiloxane (PDMS, Sylgard 184, Dow Corning, Midland, MI, USA) was poured to form a mold. The PDMS was mixed with a ratio of 10:1 monomer to curing agent by weight. After pouring the uncured PDMS over the pattern, the mold was placed in a desiccator, where air bubbles were removed from the mold using a vacuum pump. The mold was placed into an oven to cure for 2 h at 60 °C. The cured mold was then removed from the silicon master using a razor, and 3-mm-diameter holes were opened at the channel ends with a biopsy punch to create injection points for the fluids.

The PDMS microfluidic device was bonded to a microscope slide (Fisher Scientific, Catalog# 12-550C, Pittsburgh, PA, USA) that had been washed in acetone and methanol to remove organic debris, after which the solvent was washed off the slide with deionized water (ARIES High Purity Water System, Aries Filterworks, Berlin, NJ, USA) and the slide was then dried with nitrogen. Dust debris was removed from the PDMS using adhesive office tape, after which the microscope slide and the PDMS were placed inside a plasma cleaner (Harrick Plasma, PDC-32G, Ithaca, NY, USA) and treated for 2 min. The PDMS and the microscope slide were then placed in contact with one another before being heated at 60 °C for around 2 h to improve sealing. The microfluidic device was then removed from the oven and silicone tubing (Saint-Gobain, Version SPX-50, Product#ABX00001, Courbevoie, France) was inserted through the 3-mm holes, with a small amount of PDMS placed around the tubing perimeter to ensure a firm seal. The device was finally placed back in the oven for 1 h to cure.

The porous media was formed using randomly close-packed soda lime glass microspheres (P2050SL-2.5 35–45 µm—1 kg, Cospheric, Santa Barbara, CA, USA). Sieves with 38 µm, 45 µm, and 63 μm meshes were used to separate the glass microspheres and achieve a diameter range of ~48–63 μm after several rounds of sieving. Since the diameter of our microspheres was larger than the side channels used to inject our aqueous phase, we expected the microspheres to remain in place. A laboratory scale (LW Measurements, Model# HRB224, Rohnert Park, CA, USA) was used to weigh samples of 0.15 g of glass microspheres, which were dispersed in 20 mL of DI water and pumped into the large microchannel using a syringe pump (GenieTouch, Kent Scientific, Torrington, CT, USA) until the channel was filled with glass microspheres. After introducing the glass microspheres, the device was placed in an oven for 8 h to evaporate excess water, making it ready for use. The device was weighed both before and after the injection process to determine the amount of glass microspheres injected into the device. This mass measurement was used with an estimate of the average microsphere radius (~55.5 μm) to calculate the total volume occupied by the glass microspheres. This measurement provides an estimate of the porosity of the structure formed by the packed glass microspheres, which we found to be comparable to calculations of porosity based on image analysis. The porosity value that we obtained from these two measurements was small (Supplementary Tables S1 and S3). Multiple devices were fabricated using this technique, and we selected devices based on image analysis that were within 10% of the target average porosities (0.063, 0.113, and 0.143).

2.2. Image Capture and Processing

Experimental observations of water injection were accomplished using an Olympus IX70 microscope with a 10× objective lens. A scientific CMOS camera (Optimos, QImaging, Surrey, BC, Canada) was used to record video and capture images (480 × 270) for porosity analysis. An LED light source (wLS, QImaging) was used to illuminate the sample and excite the fluorescent dye. Video of the dye being injected into the fluid was captured at a rate of five frames per second. The scale for these images was 3.7281 microns per pixel.

After capturing experimental videos, we used ImageJ (National Institutes of Health, Version 1.51, Bethesda, MD, USA) to crop the videos and retain the main channel where the oil displacement occurred. These videos were then analyzed using a program written in MATLAB (Mathworks, Version 2017B, Natick, MA, USA) to track changes in fluorescence due to aqueous phase invasion. The program applies a boxcar filter to reduce the image noise. A threshold value for each frame of the video was calculated to retain the area occupied by the fluorescent dye. This area was compared to unoccupied area to obtain the degree of water saturation. This result was also used to obtain the fractal dimension of the aqueous phase.

2.3. Device Characterization

MATLAB code [42], initially developed to measure porosity and pore radius for thin rock samples, was adapted to provide a measure of the porosity of our glass bead packed bed. The MATLAB code used in this work was used to accurately determine the porosity of 3D rock samples using 2D images obtained through micro-CT scans [43]. Before performing an experiment, an optical image was taken of the point where the main channel met with the dye injection port (Figure 2, top-left). After the image is converted to black and white (Figure 2, top-right), the MATLAB algorithm draws a series of test lines across the image to determine the average sizes of pore space (white) and grain size (black), with the ratio of empty pore space to total image area representing the porosity for a thin sample. The average porosity measured within the device was consistent with measurements made by weighing the device. The observed pore size was 20 μm for the test sample (Supplementary Table S2), while the distribution of pore sizes is shown in Figure 2. The average porosities for all of our packed beds were 0.063, 0.113, and 0.143. The porosity values fell within a reasonable range based on predictions for randomly packed polydisperse spheres in 3D, with 0.03 being a theoretical lower limit [44].

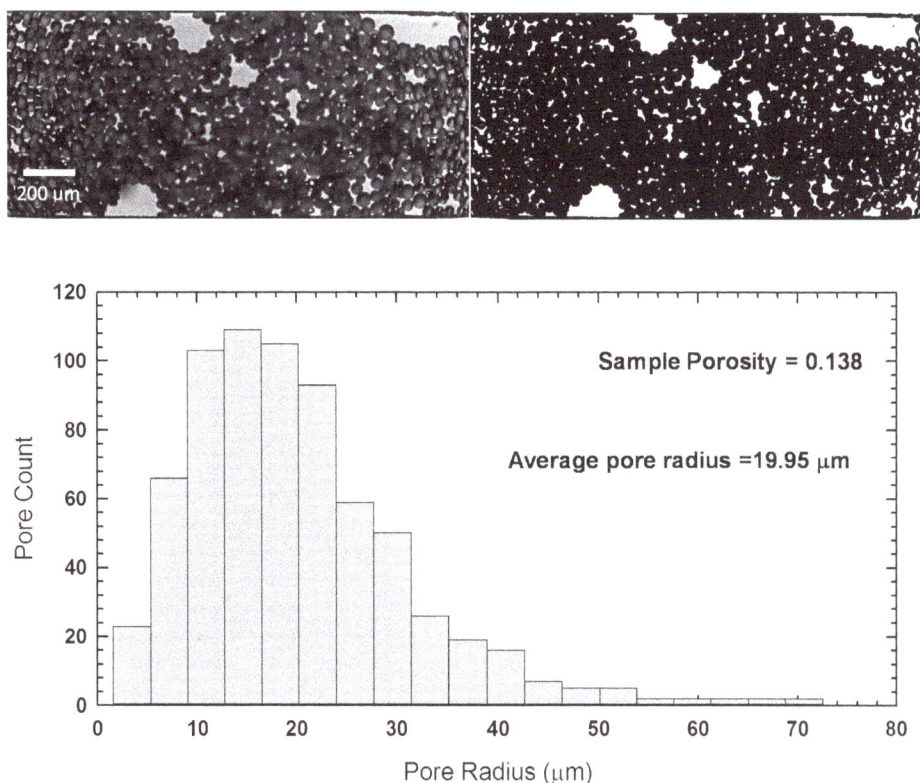

Figure 2. A representative image of our porous structures (**top**, **left**) and a black and white image (**top**, **right**) generated during our measurement of the porosity. (**bottom**) A sample distribution of the pore radius for a representative image.

2.4. Experimental Details

The experiments utilized silicone oil (ν_{oil} = 5 cSt, ρ_{oil} = 0.913 g/mL) as a defending fluid. After porosity measurements were completed, silicone oil (product # 317667, Sigma Aldrich, St. Louis, MO, USA) was injected into the main channel before the main channel was sealed with PDMS to prevent leaks. Deionized water (ν_{water} = 1 cSt, ρ_{water} = 0.99 g/mL), used as the invading fluid, was injected into the microfluidic device through a side channel using a syringe pump (Chemyx Fusion 100, Stafford, TX, USA). The mobility ratio, defined as the ratio of dynamic viscosities (M = ν_{oil} ρ_{oil}/ν_{water} ρ_{water}), is a measure of the ease with which an invading fluid flows in the presence of a defending fluid [45], with lower mobility allowing the invading fluid to flow through the porous media and recover more oil than a mobility is higher.

The capillary number [20] is Ca = ν_{oil} ρ_{oil} ν_{inj}/γ_{wo}, where ν_{inj} is the average velocity of the invading fluid during injection and γ_{wo} is the surface tension between the two fluid phases. The characteristic injection velocity is ν_{inj} = Q/b d, where Q is the injection rate, b is the gap thickness of the device (~54 μm) and d is the median pore-throat size (~20 μm). The injection rate for all experiments in this article was fixed at 0.1 mL/h using a syringe pump. Initial experiments were conducted using deionized water containing 0.1875 mM of rhodamine B dye sourced from Acros Organics (Morris Plains, NJ, USA). The surface tensions for the different aqueous fluids used in our device were evaluated using a surface tension apparatus (Fisher Scientific, Product #14-818).

The surface tension of 5 cSt silicone oil on a solid interface is $\gamma_{os} = 21.3 \pm 0.3$ mN/m based on published results [46]. The contact angle for the oil–water interface was evaluated by injecting both phases into a microfluidic channel and using image analysis [46] to evaluate the resulting angle. From these parameters, we were able to estimate the characteristic capillary number for each of our fluid systems.

Experiments conducted with water injection into the oil-saturated porous medium were compared to experiments in which a surfactant was introduced to reduce surface tension. Two chemical modifiers, sodium dodecyl sulfate (SDS, Sigma Aldrich, Product#75746) and polyvinylpyrrolidone (PVP, Alfa Aesar, MW 40,000, Product #J62417, Ward Hill, MA, USA), were selected for comparison based on their use in the petroleum industry for enhancing oil recovery [47,48]. The fluorescent dye solution was prepared as previously described, and a surfactant or polymer with a concentration of 0.1 wt % was added to this solution. Based on the interpolation of published data, the viscosities of our solutions were approximately 0.99 mPa·s for SDS in water [49] and 0.9 mPa·s for PVP in water [50]. Our injection flow rate was slow, and each experiment took only approximately 30 min to complete, during which time we did not observe PDMS swelling. The injection experiments were conducted in a manner similar to those described for water alone.

3. Theory and Simulation

Modeling two-phase fluid flow in porous media requires a coupled system of nonlinear time-dependent partial differential equations [51]. We used an approach known as the implicit pressure explicit saturation (IMPES) model [52] to simulate the transport of the invading water phase in our device. The model relies on an implicit formulation of conservation of mass and momentum to reduce the computational cost of the simulation [53].

The formulation of the model begins with a mass balance for an incompressible, immiscible two-phase flow [54],

$$\frac{\partial(\phi\rho_\alpha S_\alpha)}{\partial t} + \nabla\cdot(\rho_\alpha u_\alpha) = q_\alpha \qquad \alpha = w,o \tag{1}$$

where φ is the medium porosity, ρ is the fluid density, S is the saturation, u is the volumetric velocity, q is the mass flow rate per unit volume, and α is the phase type (water or oil). Darcy's law is used to model the volumetric velocity of the fluid phases as they flow through the porous medium for a thin section of negligible depth,

$$u_\alpha = -\frac{k_{r\alpha}}{v_\alpha\rho_\alpha}K\nabla P_\alpha \qquad \alpha = w,o \tag{2}$$

where $k_{r\alpha}$ is the relative permeability, K is the absolute permeability tensor of the porous medium, and P is pressure.

Substituting Equation (2) into Equation (1) and applying the assumption that porosity does not change with time yields [52]

$$-\nabla\cdot[K\lambda_\alpha(S_\alpha)\nabla P_\alpha] = q_\alpha \qquad \alpha = w,o \tag{3}$$

where the parameter, λ, is referred to as the phase mobility. The saturation of the water phase is defined as

$$\phi\frac{\partial S_w}{\partial t} + \nabla\cdot(f_w(S_w)u) = \frac{q_w}{\rho_w} \tag{4}$$

where $u = u_w + u_o$ is the total velocity of the system and $f_w = \lambda_w/(\lambda_w + \lambda_o)$ measures the fraction of water flowing through the system. Introducing a set of simple analytical expressions allows us to close the model presented by Equations (3) and (4),

$$\lambda_w(S_w) = \frac{(S^*)^2}{v_w\rho_w}, \quad \lambda_o(S_o) = \frac{(1-S^*)^2}{v_o\rho_o}, \quad S^* = \frac{S_w-S_{wc}}{1-S_{or}-S_{wc}}$$

where S_{wc} is the water trapped in the pores during the formation of the porous medium and S_{or} is the lowest oil saturation that can be achieved by water displacement. The oil and water phase saturations are constrained by $S_o + S_w = 1$.

The solution to these systems of equations is based on the approach of Aarnes et al. [52] and was coded in MATLAB. The simulation approach is summarized here. The model simulates a two-dimensional representation of the oil-saturated porous medium. The fluid properties (density and viscosity) mentioned in the Materials and Methods section are used to estimate the initial mobility ratio of the system. The initial oil saturation distribution is assumed to be uniform throughout our porous medium. The absolute permeability tensor, K, is generated using a random distribution with a log-normal profile. As the time step within the simulation advances, the pressure distribution is calculated using a two-point flux approximation (TPFA) scheme to discretize the pressure equation (Equation (3)) along with edge velocities.

An explicit finite-volume formulation of the saturation equation (Equation (4)) of the form

$$S_i^{n+1} = S_i^n + (\delta_x^t)_i \left(\max(q_i, 0) - \sum_j f(S^m)_{ij} v_{ij} + f(S_i^m) \min(q_i, 0) \right)$$

is used to advance the change in phase saturation with a dimensionless time step, $(\delta_x^t)_i$. The parameter γ_{ij} is the total flux over an edge between two adjacent cells Ω_i and Ω_j, and f_{ij} is the fractional flow function at γ_{ij}. The fractional flow function is:

$$f_w(S)_{ij} = \begin{cases} f_w(S_i) \; if \; v \cdot n_{ij} \geq 0, \\ f_w(S_j) \; if \; v \cdot n_{ij} < 0. \end{cases} \tag{5}$$

The model presented here was used to directly compare with experimental results. This was done by taking the ensemble average of the water phase saturation, $\langle S_w \rangle = N^{-1} \sum_{i=1}^{N} S_{w,i}$, across the simulation domain with N elements (i.e., total numbers of grid blocks) as a function of simulation time. The results were converted to experimental time, which allowed for a direct comparison to experimental results. The ensemble average of experimentally observed water saturation was directly measured by identifying the area occupied by the rhodamine B dye. We obtained a coefficient of determination through water phase saturation to determine the goodness of fit for the simulation.

If the coefficient of determination (COD) was below a value of $R^2 = 0.9$, we ran a series of forward simulations to try and improve the fit to our data. Mobility was used as an adjustable parameter as a way to account for mixing [55] between phases. The mobility for the next forward simulation was drawn from a uniform probability distribution,

$$M_{i+1} = M_f[1 + (2\xi - 1)\delta],$$

where M_f is the mobility ratio associated with the best COD, M_{i+1} is the mobility for the next forward simulation, ξ is a random uniform number that ranges from 0 to 1, and δ is a maximum possible range parameter, which we set to a value of 0.1. The new COD value was compared to the old COD value when the simulation was complete. The factor M_f remained unchanged if the new COD was lower than the old COD. Otherwise, we updated $M_f = M_{i+1}$ if the COD value was found to improve.

4. Results and Discussion

4.1. Oil Displacement Efficiency

The experimental displacement of silicone oil by waterflooding is illustrated in Figure 3. The rhodamine-dyed water initially built up sufficient pressure at the inlet port to break through into the porous media micromodel. As the water displaced silicone oil, some silicone oil still remained in voids (Figure 3B,C), and after approximately 20 s, the water reached the exit port on the opposite side of the porous media micromodel (Figure 3D). Observing the porous media micromodel after the initial exit of the water, we saw a reduction in void size over time as additional silicone oil was either displaced by water or shrank due to the applied pressure of the flow.

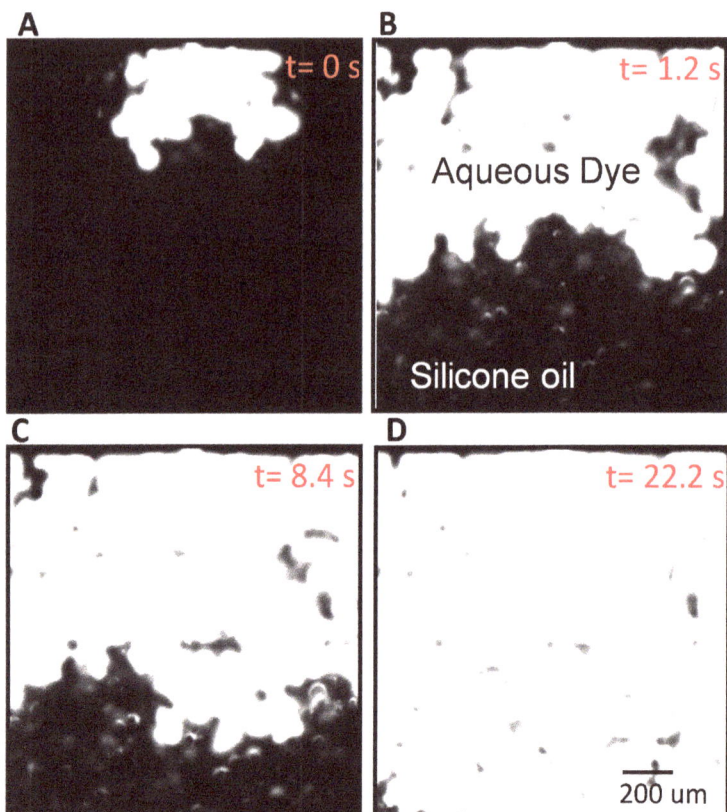

Figure 3. A representative experiment for water with 0.1% SDS at a porosity of 0.143. Initially (**A**); the main channel was full of silicone oil which is black. When the water phase invaded the main channel, rhodamine B was used to track the displacement of silicone oil (**B,C**) until the whole main channel fluoresced (**D**). In this figure, the white represents rhodamine B and black represents silicone oil.

We performed three separate types of waterflooding experiments similar to the one described in Figure 3 (Supplementary Figures S1–S3). The first type of experiment used water mixed only with rhodamine, while the other two types of experiments utilized SDS or PVP. A packed bed of glass beads was prepared before each experiment and characterized using the previously described protocol. MATLAB image analysis code was written to track the area occupied by the rhodamine dye during the experiment, allowing estimation of the average water phase saturation defined by $S_w = A_{dye}/A_{channel}$.

Figure 4 shows the results of our waterflooding experiments, for both the presence and absence of surfactant or polymer. Three separate packed beds of different porosities (0.063, 0.113, and 0.143) were prepared for each fluid type, and Figure 4A shows the results for a media with porosity 0.063. The water-bearing SDS initially invaded the packed bed faster than either the water-bearing PVP or water alone could do. In order to explain these results, we measured contact angle and surface tension for the three water phases examined in this work (Table 1). The measured contact angle for water in the microchannel was similar to contact angle values reported for plasma-treated PDMS surfaces [56–58]. We combined these values with the published value of silicone oil surface tension and Young's equation [59],

$$\cos\theta = \frac{\gamma_{ws} - \gamma_{os}}{\gamma_{wo}},$$

to find the average value of the water–oil surface tension. This calculation, based on our measured values, indicated that SDS reduced the water–oil surface tension to a more significant degree than PVP or water alone, which explains why this solution invaded at a faster rate than the other solutions (Figure 4A).

The experiment shown in Figure 4B was conducted for a packed bed with porosity 0.113. As in Figure 4A, the water-bearing PVP solution achieved a higher saturation than SDS or water alone, and invasion by both surfactant solutions occurred faster than for water alone at this porosity. Water-bearing PVP solution achieved a higher saturation because polymer flooding can change surface wettability to be more water-wet, which will increase the oil recovery [60]. Both the PVP and SDS solutions invaded at comparable rates at the highest porosity, with SDS saturating the porous bed to a slightly greater extent than the PVP solution.

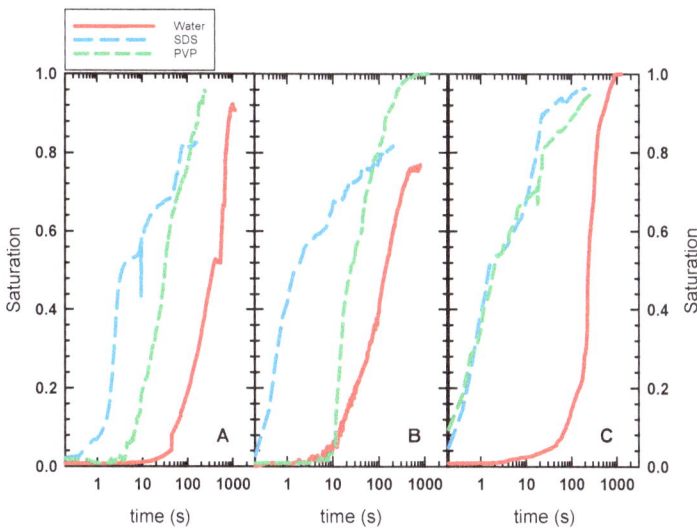

Figure 4. The relationship between saturation of an invading water phase for three water-based fluids (no surfactant, 0.1% sodium dodecyl sulfate (SDS) and 0.1% polyvinylpyrrolidone (PVP)) as they displaced oil from a packed bed of glass microspheres with porosity of 0.063 (**A**); 0.113 (**B**); and 0.143 (**C**).

Table 1. A summary of the measured contact angle and water phase-solid surface tension. The oil–water surface tension and capillary number were calculated based on these parameters.

	Water	0.1% SDS	0.1% PVP
Contact Angle (degrees)	11.2 ± 0.9	42.9 ± 1.4	22.3 ± 0.9
γ_{ws} (mN/m)	72.2 ± 0.1	53.7 ± 0.4	66.8 ± 0.5
γ_{wo} (mN/m)	51.9 ± 0.4	44.2 ± 1.2	49.2 ± 0.7
Average Capillary Number (Ca)/10^{-3}	2.3	2.7	2.4

4.2. Fractal Dimension

We used image analysis to evaluate the fractal dimension of the area occupied by the fluorescent dye in our microfluidic device to better understand the oil displacement kinetics. A MATLAB code was adapted to evaluate the fractal dimension using the Hausdorff technique [61], where a series of boxes were drawn within the region containing the dye. The bright pixels within each box were counted and compared to the total number of boxes drawn by the code. The values obtained by this analysis provide a measure of the fractal dimension for the area occupied by the water. The results of the image analysis (Figure 5A–C) show that water and the PVP solution began with a fractal dimension of D_f ~1.2 or less, representing structures classified as stringy [62]. At the lowest (0.063, Figure 5A) and highest (0.143, Figure 5C) porosity levels, it took about 100 s for the water to occupy an area that morphologically resembled a fractal (D_f ~1.6).

The only exception to this transition in morphology was the channel with porosity 0.11 (Figure 5C), for which the evolution took place over a period of time approximately half that of the other data sets. The SDS solution also exhibited different morphological behavior than the other solutions, achieving a fractal morphology on a short timescale (~1 s or less), indicating that these solutions invaded oil-saturated pore spaces at a far higher rate than the other solutions. All solutions achieved a final fractal dimension of D_f ~1.89, indicating that fluid had percolated [63] through the packed bed in the microchannel.

The fractal dimension is connected to the area saturated by the invading fluid through its radius of gyration [64]. The radius of gyration represents a measure of the extent to which the invading fluid has displaced the oil in the porous medium, and is defined as [65]

$$R_g = \sqrt{M_{2x} + M_{2y}}, \tag{6}$$

where

$$M_{2x} = \frac{1}{A_w}\sum_{i=1}^{N}(x_i - M_{1x})^2, \quad M_{2y} = \frac{1}{A_w}\sum_{i=1}^{N}(y_i - M_{1y})^2 \quad \text{and} \quad M_{1x} = \frac{1}{A_w}\sum_{i=1}^{N}x_i, \quad M_{1y} = \frac{1}{A_w}\sum_{i=1}^{N}y_i$$

where A_w is the measured area occupied by the invading water phase, N is the number of pixels making up the pattern as recorded by the CMOS camera, and x_i and y_i are the coordinates of each pixel in the observed pattern. We found that the relationship between area and radius of gyration was well described by $A_w = C_{rg}R_g^{(D_f-2)}$, where D_f is a function of time, as shown in Figure 5A–C. The relationship between area, radius of gyration, and fractal dimension proposed here is similar to previously proposed models for diffusion-limited viscous fingering in porous media [66]. By comparing data from all nine porosities examined in this work, we found that the data collapsed to a single curve when C_{rg} = 2.5 × 10^6 ± 3.1 × 10^5, where A_w and R_g are in units of microns (see Figure 5D–F).

While the fractal dimension can help us identify morphological differences in the way an invading fluid phase spreads, we would also like to observe differences in rate of invasion (dA_w/dt). Numerical calculation of the rate of invasion from image analysis data is challenging because noise in the data can create artifacts that suggest unrealistically large fluctuations in invasion rate. To minimize the effect of such artifacts, we chose to fit the area data to a model, $A_w = A_{w,max}\left(1 - e^{-t/t_c}\right)$, that is consistent with expected behavior of oil recovered through water injection into porous reservoirs [67]. The parameter, $A_{w,max}$, is the maximum area occupied by the water during the experiment, and t_c is the time constant of the experiment.

Table 2 summarizes the results of fitting an exponential rise to a maximum for our area data as a function of time. The average value of the area parameter was $A_{w,max}$ = 1.208 × 10^6 ± 7.5 × 10^4 μm^2, with the small standard deviation indicating that there was no significant difference between samples based on maximum area occupied by the invading water phase. We did find that there were significant differences in time constants for the samples we examined, with the water samples exhibiting more than a four-fold increase in time constant with increasing porosity. The time constants for SDS and PVP-bearing solutions exhibited an opposite trend (i.e., a decrease in time constant with increasing porosity). This suggests that the polymer and surfactant solutions made it easier to fill larger void volumes when compared to water alone.

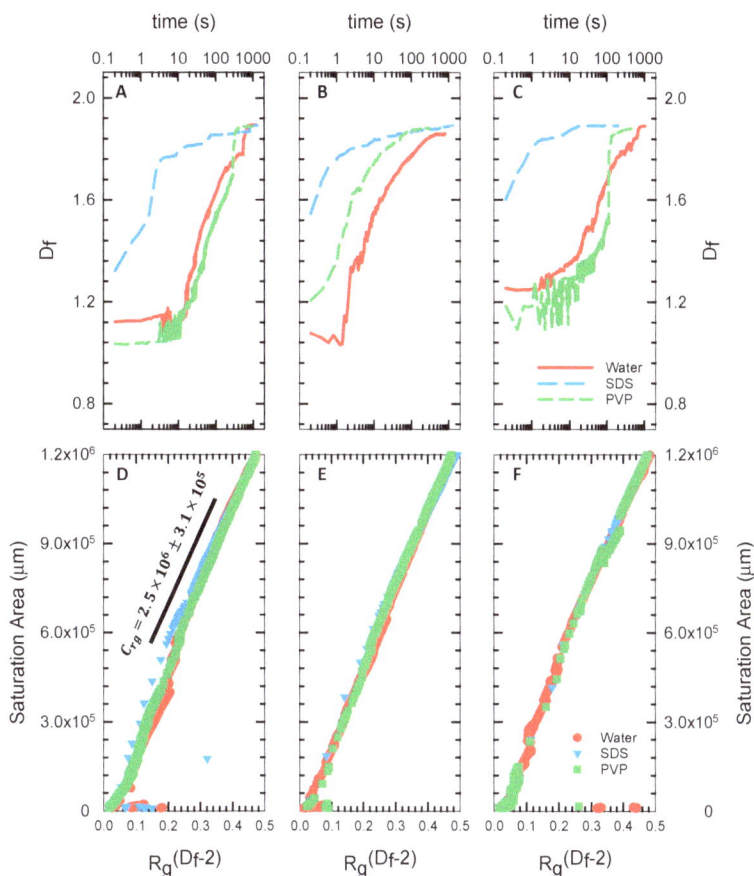

Figure 5. The Hausdorff fractal dimension (**A–C**) as a function of time was evaluated using image analysis. The results in (**A–C**) show that, while all the samples evolved differently with time, they reached a point where they were topologically similar. Combining the fractal dimension with the radius of gyration (**D–F**) reveals that all samples scaled linearly with the saturation area.

Table 2. Fitting parameters for A_{max} and time constant of the three water-based fluid systems examined in this work.

		Water		0.1% SDS		0.1% PVP	
		A_{max} (μm^2)/10^6	t_c (s)	A_{max} (μm^2)/10^6	t_c (s)	A_{max} (μm^2)/10^6	t_c (s)
	0.063	1.266	86.2	1.105	17.6	1.253	285.7
Porosity	0.11	1.059	111.1	1.241	11.0	1.209	25.3
	0.143	1.258	357.1	1.269	2.1	1.209	10.4

The packed glass beads that formed our porous media micromodel formed a 3D network. However, our optical video microscopy measurements were obtained as 2D representations of the flow through the porous media micromodel. This could serve as a source of error in our experiments, although 2D optical video microscopy measurements of 3D fluid flow phenomena have been in good agreement with theoretical predictions based on mass conservation [68]. Finite element 3D modeling of microfluidic channels has shown that the error, assuming 2D flow behavior, is approximately 5% [69]. The error is low because most of the transport occurs in two dimensions, with the top and bottom walls acting to confine fluid flow. In future work, we will further quantify this error using velocity and diffusion data gathered from tracers dispersed within the micromodel and compare the predicted flow field to finite element models.

4.3. IMPES

We implemented an inverse IMPES simulation to be compared to our experimental results. Figure 6 shows a sample simulation result for 0.1% SDS, matching the conditions observed in Figure 3. The key differences between the results of our simulations and the experiment is that the simulation results show the invasion initially occurring at a slightly slower rate that in the actual experiment (Figure 6A,B), and near the midpoint of the simulation (Figure 6C), the invading phase broke through and occupied a wider region than that at the comparable experimental time (Figure 3C). We attribute these differences to the fact that our IMPES simulation models do not count interfacial tension and porous structure effects. To mimic the structure of the randomly packed spheres, we modeled the permeability using values drawn from a log-normal distribution [52], although the procedure was designed to model porous rock rather than a packed bed of spheres.

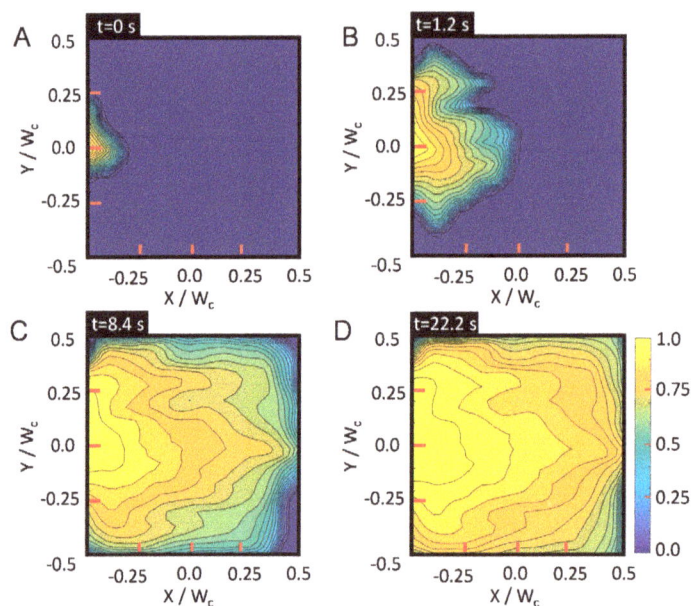

Figure 6. An implicit pressure explicit saturation (IMPES) simulation for the case with 0.1% SDS in water. Snapshots were taken at intervals that correspond to Figure 3 (**A**: t = 0 s, **B**: t = 1.2 s, **C**: t = 8.4 s, and **D**: t = 22.2 s). The color bar shows the value for the local saturation.

Despite these differences, the average saturation curves obtained by simulation closely matched our experimental curves (Figure 7), with high coefficients of determination for most of these curves (Table 3). The mobility ratios that best fit the data exhibited decreases with increasing porosity both for water and 0.1% PVP, indicating that these phases moved through the porous packed bed more easily at higher porosities. The mobility ratios for 0.1% SDS deviated from this trend. At a porosity of 0.113, the 0.1% SDS appeared to have a mobility of 33.3. While the coefficient of determination was low for this this case, the simulation did capture the general saturation trend as a function of time. Overall, our simulations did show that the addition of SDS or PVP improved the mobility of the water phase as it invaded an oil-saturated medium, and except for the 0.113 case, SDS exhibited lower mobility ratios than PVP for the experiments conducted here.

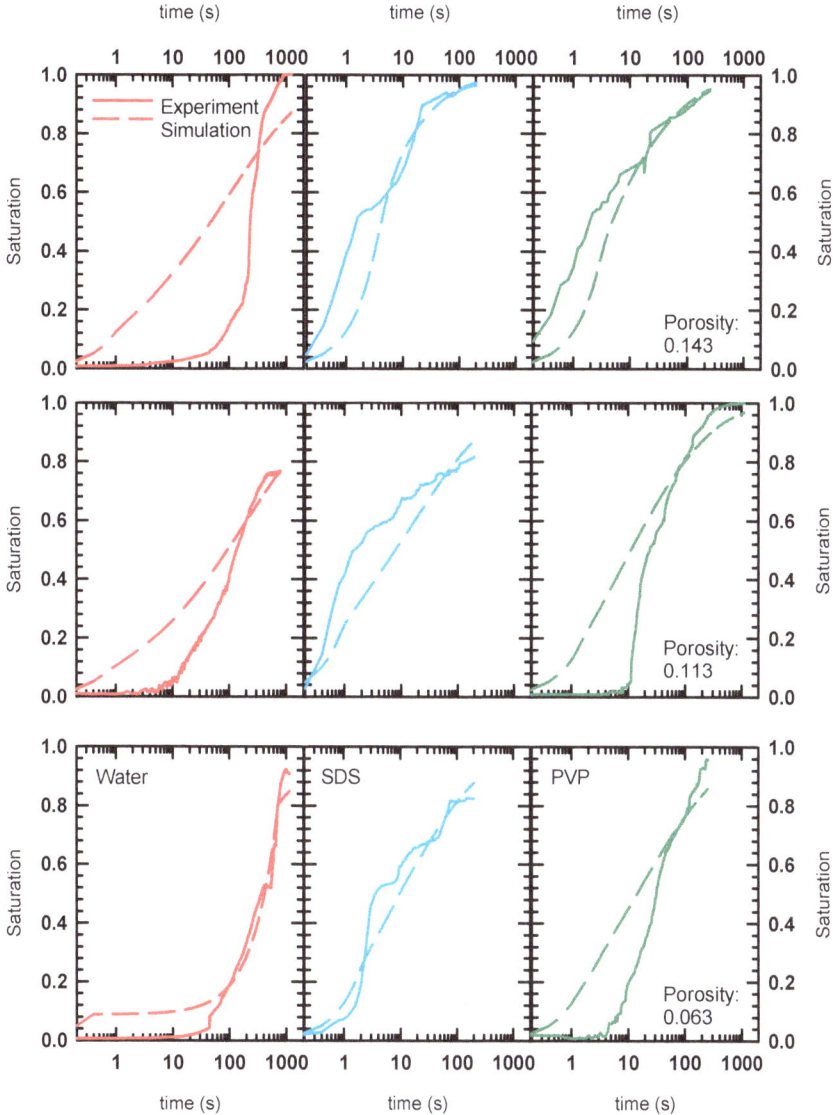

Figure 7. A comparison of our experimental results to an IMPES simulation of oil displacement by an invading water phase in a medium with porosity of 0.063 (**bottom row**), 0.113 (**middle row**), and 0.143 (**top row**).

Table 3. The mobility ratio that best fit the experimental data, and the coefficient of determination (R^2) associated with each fit.

	Water			SDS			PVP		
Porosity	0.063	0.113	0.143	0.063	0.113	0.143	0.063	0.113	0.143
Mobility	255.8	114.7	87.2	11.2	33.3	1.9	18.2	12.5	4.6
R^2	0.95	0.87	0.54	0.87	0.29	0.88	0.8	0.79	0.87

5. Conclusions

We developed a microfluidic platform by which the waterflooding of an oil-saturated porous structure was investigated through optical fluorescence microscopy. Where the characterization of bulk porous media is rate limited [13,14], our microfluidic platform was able to characterize the invasion of a water phase into an oil-saturated medium within a relatively short amount of time. The flexibility of this platform allows us to rapidly analyze chemical modification techniques that are commonly used in the petroleum industry [47,48] to enhance the recovery of oil. Observations of our device using optical fluorescence microscopy enable us to track the invasion of a water phase into an oil-saturated phase and evaluate parameters such as saturation [70–72], fractal morphology [62,64,73], temporal evolution of flow patterns [74,75], and invasion rate [76,77] that are important for evaluating the effectiveness of different types of surfactants used for waterflooding. A simulation based on the IMPES approach [51,52,54] for modeling two-phase flow in porous media was implemented in MATLAB [52] to perform an inverse analysis of our experimental data to estimate the mobility of the invading phase. The results of our inverse analysis demonstrate a significant decrease in mobility for water bearing a polymer or surfactant.

The analysis performed in this article will form the basis of future work aimed at characterizing the effect surface tension and pore structure have in our microfluidic platform. Our IMPES model will be adjusted by using closure expressions [78] for saturation curves that incorporate surface tension as it influences capillary pressure. We also aim to control the pore structure of our system through photolithography techniques [16,26] and perform microparticle image velocimetry measurements [29] to map streamline profiles that can be compared to expected flow distributions as calculated by finite element analysis [79]. The approach we have used here can be used to evaluate other enhanced oil recovery systems, including other types of polymers or surfactants [80], nanoparticles [81,82], and foams [83,84]. Our platform can also be applied to other porous media situations that involve diffusion and transport in biomedical systems [40,85], carbon sequestration [86,87], and the additive manufacturing of complex fluid networks [41,88,89].

Supplementary Materials: Supplementary materials are available online at http://www.mdpi.com/2504-5377/2/2/23/s1.

Author Contributions: H.-L.Y. fabricated devices, conducted the experiments, analyzed data, and performed simulations. J.J.J. designed the experiments and wrote image analysis algorithms to evaluate data. Both H.-L.Y. and J.J.J wrote the manuscript.

Funding: We acknowledge the Department of Mechanical Engineering at Iowa State University for funding this research.

Acknowledgments: J.J.J. acknowledges financial support from the Iowa State University Department of Mechanical Engineering. H.-L.Y. and J.J.J. thank Min Lin and the Keck Microfabrication Facility at Iowa State University for providing photolithography support for this project. We also thank Ashton Archer and Ted Heindel for providing us access to a sieve machine for separating our glass microparticles based on size.

Conflicts of Interest: The authors declare no conflicts of interest.

References

1. Karadimitriou, N.K.; Hassanizadeh, S.M. A review of micromodels and their use in two-phase flow studies. *Vadose Zone J.* **2012**, *11*. [CrossRef]

2.   Muggeridge, A.; Cockin, A.; Webb, K.; Frampton, H.; Collins, I.; Moulds, T.; Salino, P. Recovery rates, enhanced oil recovery and technological limits. *Philos. Trans. R. Soc. A* **2014**, *372*, 20120320. [CrossRef] [PubMed]

3.   Blunt, M.; Fayers, F.J.; Orr, F.M. Carbon dioxide in enhanced oil recovery. *Energy Convers. Manag.* **1993**, *34*, 1197–1204. [CrossRef]

4.   Levitt, D.; Jackson, A.; Heinson, C.; Britton, L.N.; Malik, T.; Dwarakanath, V.; Pope, G.A. Identification and evaluation of high-performance EOR surfactants. *SPE Reserv. Eval. Eng.* **2009**, *12*, 243–253. [CrossRef]

5.   Raffa, P.; Broekhuis, A.A.; Picchioni, F. Polymeric surfactants for enhanced oil recovery: A review. *J. Pet. Sci. Eng.* **2016**, *145*, 723–733. [CrossRef]

6.   Sagi, A.R. Surfactant Enhanced Oil Recovery. Ph.D. Thesis, Rice University, Houston, TX, USA, 2015.

7.   Hirasaki, G.; Miller, C.A.; Puerto, M. Recent advances in surfactant EOR. *SPE J.* **2011**, *16*. [CrossRef]

8.   Wei, B.; Romero-Zerón, L.; Rodrigue, D. Oil displacement mechanisms of viscoelastic polymers in enhanced oil recovery (EOR): A review. *J. Pet. Explor. Prod. Technol.* **2014**, *4*, 113–121. [CrossRef]

9.   Sheng, J.J.; Leonhardt, B.; Azri, N. Status of polymer-flooding technology. *J. Can. Pet. Technol.* **2015**, *54*, 116–126. [CrossRef]

10.   Ezell, R.G.; McCormick, C.L. Electrolyte- and pH-responsive polyampholytes with potential as viscosity-control agents in enhanced petroleum recovery. *J. Appl. Polym. Sci.* **2007**, *104*, 2812–2821. [CrossRef]

11.   Rashidi, M.; Blokhus, A.M.; Skauge, A. Viscosity study of salt tolerant polymers. *J. Appl. Polym. Sci.* **2010**, *117*, 1551–1557. [CrossRef]

12.   Elraies, K.A.; Tan, I.M.; Fathaddin, M.T.; Abo-Jabal, A. Development of a new polymeric surfactant for chemical enhanced oil recovery. *Pet. Sci. Technol.* **2011**, *29*, 1521–1528. [CrossRef]

13.   Morrow, N.R.; Mason, G. Recovery of oil by spontaneous imbibition. *Curr. Opin. Colloid Interface Sci.* **2001**, *6*, 321–337. [CrossRef]

14.   Standnes, D.C.; Austad, T. Wettability alteration in chalk: 2. Mechanism for wettability alteration from oil-wet to water-wet using surfactants. *J. Pet. Sci. Eng.* **2000**, *28*, 123–143. [CrossRef]

15.   Blunt, M.; King, M.J.; Scher, H. Simulation and theory of two-phase flow in porous media. *Phys. Rev. A* **1992**, *46*, 7680–7699. [CrossRef] [PubMed]

16.   Berejnov, V.; Djilali, N.; Sinton, D. Lab-on-chip methodologies for the study of transport in porous media: Energy applications. *Lab Chip* **2008**, *8*, 689–693. [CrossRef] [PubMed]

17.   Chatenever, A.; Calhoun, J.C. Visual examinations of fluid behavior in porous media—Part I. *J. Pet. Technol.* **1952**, *4*, 149–156. [CrossRef]

18.   Krummel, A.T.; Datta, S.S.; Münster, S.; Weitz, D.A. Visualizing multiphase flow and trapped fluid configurations in a model three-dimensional porous medium. *AIChE J.* **2013**, *59*, 1022–1029. [CrossRef]

19.   Anbari, A.; Chien, H.-T.; Datta, S.S.; Deng, W.; Weitz, D.A.; Fan, J. Microfluidic model porous media: Fabrication and applications. *Small* **2018**, *14*, 1703575. [CrossRef] [PubMed]

20.   Lenormand, R.; Touboul, E.; Zarcone, C. Numerical models and experiments on immiscible displacements in porous media. *J. Fluid Mech.* **1988**, *189*, 165–187. [CrossRef]

21.   Ferer, M.; Ji, C.; Bromhal, G.S.; Cook, J.; Ahmadi, G.; Smith, D.H. Crossover from capillary fingering to viscous fingering for immiscible unstable flow: Experiment and modeling. *Phys. Rev. E* **2004**, *70*. [CrossRef] [PubMed]

22.   Cottin, C.; Bodiguel, H.; Colin, A. Drainage in two-dimensional porous media: From capillary fingering to viscous flow. *Phys. Rev. E* **2010**, *82*. [CrossRef] [PubMed]

23.   Lenormand, R.; Zarcone, C.; Sarr, A. Mechanisms of the displacement of one fluid by another in a network of capillary ducts. *J. Fluid Mech.* **1983**, *135*, 337–353. [CrossRef]

24.   Chang, L.-C.; Tsai, J.-P.; Shan, H.-Y.; Chen, H.-H. Experimental study on imbibition displacement mechanisms of two-phase fluid using micro model. *Environ. Earth Sci.* **2009**, *59*, 901. [CrossRef]

25.   Theodoropoulou, M.A.; Sygouni, V.; Karoutsos, V.; Tsakiroglou, C.D. Relative permeability and capillary pressure functions of porous media as related to the displacement growth pattern. *Int. J. Multiph. Flow* **2005**, *31*, 1155–1180. [CrossRef]

26.   Wu, M.; Xiao, F.; Johnson-Paben, R.M.; Retterer, S.T.; Yin, X.; Neeves, K.B. Single- and two-phase flow in microfluidic porous media analogs based on Voronoi tessellation. *Lab Chip* **2011**, *12*, 253–261. [CrossRef] [PubMed]

27. Xiao, F.; Yin, X. Geometry models of porous media based on Voronoi tessellations and their porosity–permeability relations. *Comput. Math. Appl.* **2016**, *72*, 328–348. [CrossRef]

28. Heshmati, M.; Piri, M. Interfacial boundary conditions and residual trapping: A pore-scale investigation of the effects of wetting phase flow rate and viscosity using micro-particle image velocimetry. *Fuel* **2018**, *224*, 560–578. [CrossRef]

29. Roman, S.; Soulaine, C.; AlSaud, M.A.; Kovscek, A.; Tchelepi, H. Particle velocimetry analysis of immiscible two-phase flow in micromodels. *Adv. Water Resour.* **2016**, *95*, 199–211. [CrossRef]

30. Bera, B.; Mitra, S.K.; Vick, D. Understanding the micro structure of Berea Sandstone by the simultaneous use of micro-computed tomography (micro-CT) and focused ion beam-scanning electron microscopy (FIB-SEM). *Micron* **2011**, *42*, 412–418. [CrossRef] [PubMed]

31. Bakke, S.; Øren, P.-E. 3-D Pore-scale modelling of sandstones and flow simulations in the pore networks. *SPE J.* **1997**, *2*, 136–149. [CrossRef]

32. Blunt, M.J. Flow in porous media pore-network models and multiphase flow. *Interface Sci.* **2001**, *6*, 197–207. [CrossRef]

33. Gunda, N.S.K.; Bera, B.; Karadimitriou, N.K.; Mitra, S.K.; Hassanizadeh, S.M. Reservoir-on-a-Chip (ROC): A new paradigm in reservoir engineering. *Lab Chip* **2011**, *11*, 3785–3792. [CrossRef] [PubMed]

34. Xu, K.; Liang, T.; Zhu, P.; Qi, P.; Lu, J.; Huh, C.; Balhoff, M. A 2.5-D glass micromodel for investigation of multi-phase flow in porous media. *Lab Chip* **2017**, *17*, 640–646. [CrossRef] [PubMed]

35. Marchand, S.; Bondino, I.; Ktari, A.; Santanach-Carreras, E. Consideration on data dispersion for two-phase flow micromodel experiments. *Transp. Porous Media* **2017**, *117*, 169–187. [CrossRef]

36. Kim, M.; Sell, A.; Sinton, D. Aquifer-on-a-Chip: Understanding pore-scale salt precipitation dynamics during CO_2 sequestration. *Lab Chip* **2013**, *13*, 2508–2518. [CrossRef] [PubMed]

37. Song, W.; de Haas, T.W.; Fadaei, H.; Sinton, D. Chip-off-the-old-rock: The study of reservoir-relevant geological processes with real-rock micromodels. *Lab Chip* **2014**, *14*, 4382–4390. [CrossRef] [PubMed]

38. Kersting, A.B.; Efurd, D.W.; Finnegan, D.L.; Rokop, D.J.; Smith, D.K.; Thompson, J.L. Migration of plutonium in ground water at the Nevada Test Site. *Nature* **1999**, *397*, 56–59. [CrossRef]

39. Jensen, K.H.; Valente, A.X.C.N.; Stone, H.A. Flow rate through microfilters: Influence of the pore size distribution, hydrodynamic interactions, wall slip, and inertia. *Phys. Fluids* **2014**, *26*, 052004. [CrossRef]

40. Miller, J.S.; Stevens, K.R.; Yang, M.T.; Baker, B.M.; Nguyen, D.-H.T.; Cohen, D.M.; Toro, E.; Chen, A.A.; Galie, P.A.; Yu, X.; et al. Rapid casting of patterned vascular networks for perfusable engineered three-dimensional tissues. *Nat. Mater.* **2012**, *11*, 768–774. [CrossRef] [PubMed]

41. Marschewski, J.; Brenner, L.; Ebejer, N.; Ruch, P.; Michel, B.; Poulikakos, D. 3D-printed fluidic networks for high-power-density heat-managing miniaturized redox flow batteries. *Energy Environ. Sci.* **2017**, *10*, 780–787. [CrossRef]

42. Rabbani, A.; Jamshidi, S.; Salehi, S. Determination of specific surface of rock grains by 2D imaging. *J. Geol. Res.* **2014**, *2014*, 1–7. [CrossRef]

43. Rabbani, A.; Jamshidi, S. Specific surface and porosity relationship for sandstones for prediction of permeability. *Int. J. Rock Mech. Min. Sci.* **2014**, *71*, 25–32. [CrossRef]

44. Van der Marck, S.C. Network approach to void percolation in a pack of unequal spheres. *Phys. Rev. Lett.* **1996**, *77*, 1785–1788. [CrossRef] [PubMed]

45. Kumar, M.; Hoang, V.T.; Satik, C.; Rojas, D.H. High-mobility-ratio waterflood performance prediction: Challenges and new insights. *SPE Reserv. Eval. Eng.* **2008**, *11*, 186–196. [CrossRef]

46. Matlab Central Measuring Angle of Intersection—MATLAB & Simulink Example. Available online: https://www.mathworks.com/help/images/examples/measuring-angle-of-intersection.html (accessed on 6 April 2018).

47. Stahl, G.A.; Schulz, D.N. *Water-Soluble Polymers for Petroleum Recovery*; Springer: New York, NY, USA, 1988; ISBN 978-0-306-42915-6.

48. Taylor, K.C.; Nasr-El-Din, H.A. Water-soluble hydrophobically associating polymers for improved oil recovery: A literature review. *J. Pet. Sci. Eng.* **1998**, *19*, 265–280. [CrossRef]

49. Chari, K.; Antalek, B.; Lin, M.Y.; Sinha, S.K. The viscosity of polymer–surfactant mixtures in water. *J. Chem. Phys.* **1994**, *100*, 5294–5300. [CrossRef]

50. Sadeghi, R.; Azizpour, S. Volumetric, compressibility, and viscometric measurements of binary mixtures of Poly(vinylpyrrolidone) + Water, + Methanol, + Ethanol, + Acetonitrile, + 1-Propanol, + 2-Propanol, and + 1-Butanol. *J. Chem. Eng. Data* **2011**, *56*, 240–250. [CrossRef]

51. Kou, J.; Sun, S. On iterative IMPES formulation for two phase flow with capillarity in heterogeneous porous media. *Int. J. Numer. Anal. Model. Ser. B* **2010**, *1*, 20–40.

52. Aarnes, J.E.; Gimse, T.; Lie, K.-A. An introduction to the numerics of flow in porous media using matlab. In *Geometric Modelling, Numerical Simulation, and Optimization*; Applied Mathematics at SINTEF; Springer: Berlin/Heidelberg, Germany, 2007.

53. Li, B.; Chen, Z.; Huan, G. Comparison of solution schemes for black oil reservoir simulations with unstructured grids. *Comput. Methods Appl. Mech. Eng.* **2004**, *193*, 319–355. [CrossRef]

54. Chen, Z.; Huan, G.; Li, B. An improved IMPES method for two-phase flow in porous media. *Transp. Porous Media* **2004**, *54*, 361–376. [CrossRef]

55. Habermann, B. The efficiency of miscible displacement as a function of mobility ratio. *Trans. Am. Inst. Min. Metall. Eng.* **1960**, *219*, 264–272.

56. Fritz, J.L.; Owen, M.J. Hydrophobic recovery of plasma-treated polydimethylsiloxane. *J. Adhes.* **1995**, *54*, 33–45. [CrossRef]

57. Bhattacharya, S.; Datta, A.; Berg, J.M.; Gangopadhyay, S. Studies on surface wettability of poly(dimethyl) siloxane (PDMS) and glass under oxygen-plasma treatment and correlation with bond strength. *J. Microelectromech. Syst.* **2005**, *14*, 590–597. [CrossRef]

58. Bodas, D.; Khan-Malek, C. Formation of more stable hydrophilic surfaces of PDMS by plasma and chemical treatments. *Microelectron. Eng.* **2006**, *83*, 1277–1279. [CrossRef]

59. Bertrand, E.; Blake, T.D.; De Coninck, J. Dynamics of dewetting. *Colloids Surf. Physicochem. Eng. Asp.* **2010**, *369*, 141–147. [CrossRef]

60. Wang, D.; Cheng, J.; Yang, Q.; Wenchao, G.; Qun, L.; Chen, F. Viscous-elastic polymer can increase microscale displacement efficiency in cores. *Soc. Pet. Eng.* **2000**. [CrossRef]

61. Matlab Central Hausdorff (Box-Counting) Fractal Dimension—File Exchange—MATLAB Central. Available online: http://www.mathworks.com/matlabcentral/fileexchange/30329-hausdorff--box-counting--fractal-dimension (accessed on 21 November 2017).

62. Zhong-Can, O.Y.; Gang, Y.; Bai-Lin, H. From fractal to dendritic: Competition between diffusion and field. *Phys. Rev. Lett.* **1986**, *57*, 3203–3205. [CrossRef] [PubMed]

63. Witten, T.A.; Sander, L.M. Diffusion-limited aggregation. *Phys. Rev. B* **1983**, *27*, 5686–5697. [CrossRef]

64. Gharbi, R.B.C.; Qasem, F.; Peters, E.J. A relationship between the fractal dimension and scaling groups of unstable miscible displacements. *Exp. Fluids* **2001**, *31*, 357–366. [CrossRef]

65. Pons, M.-N.; Weisser, E.M.; Vivier, H.; Boger, D.V. Characterization of viscous fingering in a radial Hele-Shaw cell by image analysis. *Exp. Fluids* **1999**, *26*, 153–160. [CrossRef]

66. Maloy, K.J.; Boger, F.; Feder, J.; Jossang, T.; Meakin, P. Dynamics of viscous-fingering fractals in porous media. *Phys. Rev. A* **1987**, *36*, 318–324. [CrossRef]

67. Aronofsky, J.S.; Masse, L.; Natanson, S.G. A model for the mechanism of oil recovery from the porous matrix due to water invasion in fractured reservoirs. *Pet. Trans.* **1958**, *213*, 17–19.

68. Holden, M.A.; Kumar, S.; Castellana, E.T.; Beskok, A.; Cremer, P.S. Generating fixed concentration arrays in a microfluidic device. *Sens. Actuators B Chem.* **2003**, *92*, 199–207. [CrossRef]

69. Kamholz, A.E.; Schilling, E.A.; Yager, P. Optical measurement of transverse molecular diffusion in a microchannel. *Biophys. J.* **2001**, *80*, 1967–1972. [CrossRef]

70. Cheng, J.T.; Pyrak-Nolte, L.J.; Nolte, D.D.; Giordano, N.J. Linking pressure and saturation through interfacial areas in porous media. *Geophys. Res. Lett.* **2004**, *31*, L08502. [CrossRef]

71. Parseval, Y.D.; Pillai, K.M.; Advani, S.G. A simple model for the variation of permeability due to partial saturation in dual scale porous media. *Transp. Porous Media* **1997**, *27*, 243–264. [CrossRef]

72. Niemet, M.R.; Selker, J.S. A new method for quantification of liquid saturation in 2D translucent porous media systems using light transmission. *Adv. Water Resour.* **2001**, *24*, 651–666. [CrossRef]

73. Cai, J.; Yu, B.; Zou, M.; Mei, M. Fractal analysis of invasion depth of extraneous fluids in porous media. *Chem. Eng. Sci.* **2010**, *65*, 5178–5186. [CrossRef]

74. Zhu, T.; Waluga, C.; Wohlmuth, B.; Manhart, M. A study of the time constant in unsteady porous media flow using direct numerical simulation. *Transp. Porous Media* **2014**, *104*, 161–179. [CrossRef]

75. Zheng, Z.; Guo, B.; Christov, I.C.; Celia, M.A.; Stone, H.A. Flow regimes for fluid injection into a confined porous medium. *J. Fluid Mech.* **2015**, *767*, 881–909. [CrossRef]

76. Martys, N.; Cieplak, M.; Robbins, M.O. Critical phenomena in fluid invasion of porous media. *Phys. Rev. Lett.* **1991**, *66*, 1058–1061. [CrossRef] [PubMed]

77. Geistlinger, H.; Ataei-Dadavi, I. Influence of the heterogeneous wettability on capillary trapping in glass-beads monolayers: Comparison between experiments and the invasion percolation theory. *J. Colloid Interface Sci.* **2015**, *459*, 230–240. [CrossRef] [PubMed]

78. Parker, J.C.; Lenhard, R.J. A model for hysteretic constitutive relations governing multiphase flow: 1. Saturation-pressure relations. *Water Resour. Res.* **1987**, *23*, 2187–2196. [CrossRef]

79. Narsilio, G.A.; Buzzi, O.; Fityus, S.; Yun, T.S.; Smith, D.W. Upscaling of Navier–Stokes equations in porous media: Theoretical, numerical and experimental approach. *Comput. Geotech.* **2009**, *36*, 1200–1206. [CrossRef]

80. Wever, D.A.Z.; Picchioni, F.; Broekhuis, A.A. Polymers for enhanced oil recovery: A paradigm for structure–property relationship in aqueous solution. *Prog. Polym. Sci.* **2011**, *36*, 1558–1628. [CrossRef]

81. Sedaghat, M.H.; Mohammadi, H.; Razmi, R. Application of SiO_2 and TiO_2 nano particles to enhance the efficiency of polymer-surfactant floods. *Energy Sources Part Recovery Util. Environ. Eff.* **2016**, *38*, 22–28. [CrossRef]

82. Mohajeri, M.; Hemmati, M.; Shekarabi, A.S. An experimental study on using a nanosurfactant in an EOR process of heavy oil in a fractured micromodel. *J. Pet. Sci. Eng.* **2015**, *126*, 162–173. [CrossRef]

83. Ma, K.; Liontas, R.; Conn, C.A.; Hirasaki, G.J.; Biswal, S.L. Visualization of improved sweep with foam in heterogeneous porous media using microfluidics. *Soft Matter* **2012**, *8*, 10669–10675. [CrossRef]

84. Conn, C.A.; Ma, K.; Hirasaki, G.J.; Biswal, S.L. Visualizing oil displacement with foam in a microfluidic device with permeability contrast. *Lab Chip* **2014**, *14*, 3968–3977. [CrossRef] [PubMed]

85. Klinzing, G.R.; Zavaliangos, A. A simplified model of moisture transport in hydrophilic porous media with applications to pharmaceutical tablets. *J. Pharm. Sci.* **2016**, *105*, 2410–2418. [CrossRef] [PubMed]

86. Szulczewski, M.L.; MacMinn, C.W.; Herzog, H.J.; Juanes, R. Lifetime of carbon capture and storage as a climate-change mitigation technology. *Proc. Natl. Acad. Sci. USA* **2012**, *109*, 5185–5189. [CrossRef] [PubMed]

87. Zhao, B.; MacMinn, C.W.; Juanes, R. Wettability control on multiphase flow in patterned microfluidics. *Proc. Natl. Acad. Sci. USA* **2016**, *113*, 10251–10256. [CrossRef] [PubMed]

88. Osei-Bonsu, K.; Grassia, P.; Shokri, N. Investigation of foam flow in a 3D printed porous medium in the presence of oil. *J. Colloid Interface Sci.* **2017**, *490*, 850–858. [CrossRef] [PubMed]

89. Chan, H.N.; Chen, Y.; Shu, Y.; Chen, Y.; Tian, Q.; Wu, H. Direct, one-step molding of 3D-printed structures for convenient fabrication of truly 3D PDMS microfluidic chips. *Microfluid. Nanofluid.* **2015**, *19*, 9–18. [CrossRef]

colloids
and interfaces

MDPI

Article

Static and Dynamic Performance of Wet Foam and Polymer-Enhanced Foam in the Presence of Heavy Oil

Ali Telmadarreie [1,2,*] and Japan J. Trivedi [2]

1 Department of Chemical and Petroleum Engineering, Schulich School of Engineering, University of Calgary, Calgary, AB T2N 1N4, Canada
2 School of Mining and Petroleum, Department of Civil and Environmental Engineering, University of Alberta, Edmonton, AB T6G 2R3, Canada; jtrivedi@ualberta.ca
* Correspondence: ali.telmadarreie@ucalgary.ca

Received: 20 July 2018; Accepted: 28 August 2018; Published: 8 September 2018

Abstract: Inadequate sweep efficiency is one of the main concerns in conventional heavy oil recovery processes. Alternative processes are therefore needed to increase heavy oil sweep efficiency. Foam injection has gained interest in conventional oil recovery in recent times as it can control the mobility ratio and improve the sweep efficiency over chemical or gas flooding. However, most of the studies have focused on light crude oil. This study aims to investigate the static and dynamic performances of foam and polymer-enhanced foam (PEF) in the presence of heavy oil. Static and dynamic experiments were conducted to investigate the potential of foam and PEF for heavy oil recovery. Static analysis included foam/PEF stability, decay profile, and image analysis. A linear visual sand pack was used to visualize the performance of CO_2 foam and CO_2 PEF in porous media (dynamic experiments). Nonionic, anionic, and cationic surfactants were used as the foaming agents. Static stability results showed that the anionic surfactant generated relatively more stable foam, even in the presence of heavy oil. Slower liquid drainage and collapse rates for PEF compared to that of foam were the key observations through foam static analyses. Besides improving heavy oil recovery, the addition of polymer accelerated foam generation and propagation in porous media saturated with heavy oil. Visual analysis demonstrated more stable frontal displacement and higher sweep efficiency of PEF compared to conventional foam flooding. Unlike foam injection, lesser channeling (foam collapse) was observed during PEF injection. The results of this study will open a new insight on the potential of foam, especially polymer-enhanced foam, for oil recovery of those reservoirs with viscous oil.

Keywords: CO_2 foam; EOR; heavy oil; SAG; polymer-enhanced foam

1. Introduction

Inadequate sweep efficiencies resulting from unfavorable mobility ratios are the main challenges during enhanced oil recovery (EOR) methods, especially in heavy oil recovery. Foam and polymer-enhanced foam (PEF) flooding can control the mobility ratio and improve the sweep efficiency, especially in heterogeneous reservoirs. Foam can provide better control of the fluids injected and uniformity of the contact as stronger foams can block the flow channels in high-permeability media and divert it toward the low permeable parts [1,2].

Foam has shown its potential for improving reservoir sweep efficiency over gas injection-enhanced oil recovery projects [1,3]. Besides improving sweep efficiency in gas flooding, foam/PEF can be used for mobility control in chemical EOR where the foam is considered an alternative to polymer mobility control in micellar flooding [4]. Zhang et al. (2000) [5] reported on laboratory and field studies of foam in Daqing oilfield in China, where the foam was successfully applied in a heterogeneous porous media and compared with the performance of chemical flooding.

There are several challenges for widespread application of foam in porous media, such as in situ generation and propagation of foam. Although the principal mechanisms for foam generation have been identified, the precise condition when the strong foam can be generated in the reservoir remains unknown. In a homogeneous porous medium, with steady coinjection of gas and liquid, a minimum pressure gradient is required to create foam [6–9]. One of the proposed ways to enhance foam generation is using small, alternating slugs of liquid and gas, i.e., surfactant alternating gas (SAG) injection [7]. SAG injection has several advantages over coinjection of gas and surfactant as it reduces the contact between water and gas in the surface facilities [10,11], and besides improving injectivity, it can possibly improve foam generation in the near-well-bore region [7].

Another important challenge during foam EOR is the detrimental effect of crude oil. There have been several studies on static and dynamic performance of foam–oil systems [12–17]. These studies have mainly focused on performing foam stability experiments (bulk tests) in the presence of different types of light oils and measuring the foam half-life. Some of these researchers have also measured the spreading, entering, and birding coefficients in the foam–oil system and tried to explain the behavior of the foam–oil system. However, the relation between spreading phenomena and foam stability has been inconclusive for general applications. For instance, Andrianov et al. (2012) [13] concluded that there exists a strong correlation between spreading (S) and entering (E) coefficients and foam stability, while Vikingsad et al. (2005) [12] did not find any direct correlation between spreading coefficient and foam stability. Nikolov et al. (1986) [18] mentioned that as the oil drop approaches the liquid–gas interface, the thin liquid film forms between the oil drop and the gas phase called "pseudoemulsion film". According to this study, oil drop cannot enter the interface when the pseudoemulsion film is stable, even if the values of E or S coefficients are different.

Chemical methods are among the most common nonthermal EOR process for heavy oil recovery after water flooding. One of the issues with the chemical method for heavy oil reservoirs is the low injectivity and inadequate sweep efficiency, especially in heterogeneous reservoirs. Foam and PEF can improve sweep efficiency over gas and chemical injections. Foam can reduce viscous fingering and gravity override during gas injection as the effective viscosity of foam is much higher than that of gas [19]. Shallow heavy oil reservoirs in Western Canadian Sedimentary Basin (WCSB), which can also be recovered by nonthermal processes, can be the potential target for foam/PEF EOR application. The high viscosity of the oil and high heterogeneity of these reservoirs can result in inadequate sweep efficiency of conventional nonthermal methods.

Most of the previous studies mentioned above have been performed on light crude oils. In this study, the effect of heavy crude oil is studied on the static and dynamic performance of foam. Moreover, PEF is also introduced for improving the performance of conventional foam in heavy oil reservoirs. For this aim, first, static experiments—including bulk foam stability—and surface tension studies were designed to study the behavior of bulk foam in the presence and absence of heavy oil as well as the effect of polymer addition on its behavior. In the second part, dynamic experiments of foam/PEF propagation through visual sand pack were performed to investigate the dynamic stability with and without the presence of heavy oil.

2. Materials and Methods

2.1. Materials

For the foam/PEF studies, various types of surfactant—nonionic, anionic, and cationic—were selected as foaming agents and one polymer was selected as a foam stabilizer, as follows:

Nonionic surfactant: Surfonic N85 (Huntsman Corporation), which is a nonylphenol-ethoxylated nonionic surfactant with the chemical formula $C_{15}H_{23}(OCH_2CH_2)n\,OH$ (Mw = 594 g/mol), was used.

Anionic Surfactant: Two anionic surfactants—sodium dodecylbenzenesulfonate (DDBS; $C_{12}H_{25}C_6H_4SO_3Na$, Mw = 348.5 g/mol)) and C_{14}–C_{16} alpha olefin sulfonate (AOS; $R\text{-}SO_3^-Na^+$, Mw = 348.5 g/mol)—were used for static foam analysis.

Cationic Surfactant: Cetyltrimethylammonium bromide (CTAB) (Sigma-Aldrich, 99% purity) was also used as a cationic foaming agent ($C_{19}H_{42}BrN$, Mw = 364.5 g/mol).

Polymer: Polymer was used to increase the viscosity of the liquid phase. The anionic polyacrylamide polymer FLOPAAM 3330S (supplied by SNF SAS) was used in the preparation of polymer solutions. It had hydrolysis degree of 25–30% and average molecular weight of 8×10^6. It should be mentioned that all foaming solutions were prepared with tap water (Ca^{2+} = 34 ppm, Mg^{2+} = 10 ppm, and Na^{2+} = 35 ppm) without the addition of any salt. The different salt concentration may have had an effect on the foam stability.

Hydrocarbons: For both static and dynamic experiments, heavy crude oil (sampled from the Canadian oilfields) with dead oil viscosity of 1320 cp (at 22 °C) and dead oil density of 933 kg/m³ was used. In addition, a mineral oil with a viscosity of 27 cp and density of 850 kg/m³ (22 °C) was used for only static analysis of the foam–oil system.

2.2. Foam Bulk (Static) Experiments

Foam and PEF generation: For the preparation of the foaming solution, surfactant (0.29 wt %)—and in some tests, polymer (0.15 wt %)—were mixed in water. A magnetic stirrer (400 rpm for 20 min) was used for mixing to avoid foam generation. Thereafter, foam or PEF was generated using a digital homogenizer (Kinematica Inc., Bohemia, NY, USA). It should be mentioned that all four surfactants were used for foam static analysis and among them, N85 was selected for the static study of PEF.

For foam generation, 100 cc of foaming solution was mixed in a glass cylinder at high speed for two minutes. The shearing speed and shearing time were kept constant for uniformity of foam created throughout all the experiments. In some experiments, 5 cc of oil was added to the foaming solution before high-speed mixing to study the effect of oil on foam/PEF stability.

Static stability: The glass cylinder was closed with a plastic seal after foam generation to avoid evaporation. In each experiment, immediately after mixing, the total height and the height of liquid were measured as a function of time. Foam stability (foam half-life) was recorded based on the time required to drain half of the liquid from the foam. Moreover, the initial foam height value was also recorded as the foamability of foaming solutions. A high-definition camera was used to analyze the foam behavior with and without the presence of oil. For some detailed analysis of foam–oil interaction, a Leica DM 6000M microscope was used to capture high-quality images. All experiments were repeated to assure the reproducibility of results. The reproducibility of foam high and foam half-life was ±1 cm and ± 0.5 min, respectively.

Surface tension and interfacial tension: The surface tension of the surfactant solutions and their mixtures with polymer was measured by the Du Noüy tensiometer (K6, KRÜSS Canada) using a platinum–iridium ring. The ring method directly measures the maximum pull on the interface to find surface tension value. After each measurement, the ring was carefully rinsed with deionized water and then a solvent (usually acetone) to remove impurities. Thereafter, the ring was cleaned with a flame to remove any impurities. The interfacial tension (IFT) measurements were performed using a spinning drop method (SITE100, KRÜSS Canada). The lowest measurement range for this instrument is as low as 10^{-6} mN/m, with rotational speed up to 15,000 rpm (with a capillary diameter of 2.5 mm).

2.3. Foam Dynamic Experiments

To analyze the dynamic performance of foam and PEF in porous media, a linear visual sand pack was used (1-ft in length with an inner diameter of 1-inch). The visual cell was packed with glass beads (40–70 meshes), and a special expandable rubber was used to seal both ends. A metal screen (80 meshes) was used to avoid sand production. After vacuuming the sand pack (at least 3 h), water saturation and permeability measurements were conducted. The measured porosity and permeability of the sand pack were 37 ± 0.5% and 38 ± 0.5 Darcy, respectively. The porous medium was then saturated with heavy crude oil (1320 cp) until no water was produced. A syringe pump (ISCO, Model

500D) and a pressure transducer (OMEGADYNE, Model PX409) were used for liquid injection and pressure record, respectively. A schematic of the dynamic experiments set up is shown in Figure 1.

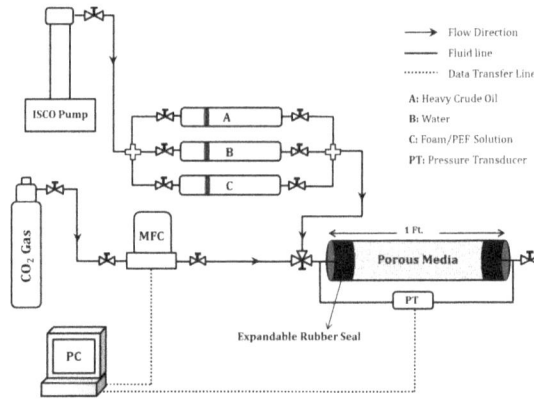

Figure 1. Schematic of CO_2 foam/polymer-enhanced foam (PEF) flooding system.

In this study, the foam was generated in situ by surfactant alternating gas injection. During all the experiments, both liquid and gas rates were kept equal (20 ft/D). The injection rate was selected based on the high permeability of the sand pack and the shear rate of fluid within the porous media (between 1 to 100 s^{-1}). In addition, the high rate ensured that the critical pressure gradient to generate strong foam was exceeded. A gas mass flow controller (EL-flow, Hoskin Scientific Ltd., Saint-Laurent, QC, Canada) was used for the accurate injection of CO_2 gas at the constant volumetric flow rate. The slug volume of both liquid and gas was selected as 0.1 fractions of the total pore volume (0.1 PV). Pressure profile and oil recovery were measured, and sand pack images were captured during all the experiments to compare the performance of different foaming solutions in heavy oil recovery. It should be mentioned that all flooding tests were performed in tertiary recovery mode after reaching a constant water cut of 98% during water flooding. Table 1 summarizes details of the dynamic experiments on visual sand pack. Among the studied surfactants, nonionic N85 and anionic AOS were selected for foam flooding and N85 surfactant was selected for PEF flooding. All experiments were performed at ambient condition (22 °C) without using any backpressure. It should be mentioned that after oil saturation (initial oil saturation 92 ± 0.5%), the sand pack was aged overnight at room temperature.

Table 1. Summary of dynamic experiments performed on visual sand pack with and without the presence of heavy oil.

Experiment	Porous Media Length (cm)	Ø (%)	K (D)	Soi (%)	WF-RF (%)	Total RF (%)
AOS Foam	24.5	36.28	37.75	NA	NA	NA
	24.3	36.85	37.42	92.5	33	91.6
N85 Foam	24.5	36.67	37.88	NA	NA	NA
	24.4	37.22	37.72	93.4	33.1	57
N85 PEF	24.4	37.22	38.19	NA	NA	NA
	24.4	36.82	37.91	92.3	33	98

3. Results and Discussion

3.1. Static Performance of Foam and PEF in the Absence of Heavy Oil

Effect of surfactant type: The changes of normalized foam height (H/H_0) versus time, foamability, and foam stability (half-life) for all studied surfactants are shown in Figure 2. Among all studied surfactants, DDBS (and also AOS) showed better stability (half-life), while CTAB had highest foamability. Foamability is the ability of the surfactant to generate foam. The numbers in Figure 2a represents quality (gas content) of the foam. Several references have reported the good foamability of anionic surfactants [20–22], whereas the nonionic surfactants generally produce less foam. The stability of foams generated with ionic or nonionic surfactants is achieved by repulsive forces between the surfactant monolayers [23,24]. Therefore, the ionic surfactants—DDBS, AOS, and CTAB—generated more foam with relatively higher stability compared to that of nonionic N85 surfactant. According to Figure 2a, initial foam height (foamability) and half-life (foam stability) of anionic surfactants were slightly higher than that of the nonionic surfactant. The presence of ionic surfactants at the interface in the foam film will stabilize the film and induce a repulsive force that opposes the film thinning process. This is called the electric double-layer repulsion [25,26], which depends on the charge density and the film thickness.

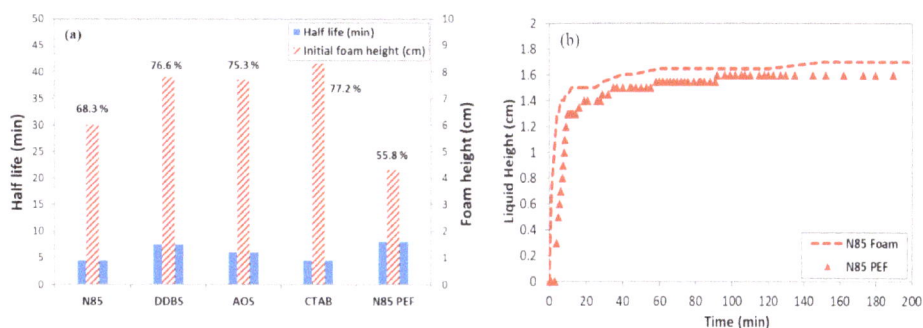

Figure 2. (a) Foamability and half-life values for studied surfactants: anionic dodecylbenzenesulfonate (DDBS) and C_{14}–C_{16} alpha olefin sulfonate (AOS), nonionic N85, and cationic cetyltrimethylammonium bromide (CTAB). Polymer increased foam stability but decreased the foamability of surfactants (numbers represent the quality of foam/PEF). (b) Liquid drainage profiles of the foam and PEF generated with the N85 surfactant.

Effect of polymer addition: One of the major advantages of polymer addition in foaming solution is the viscosity enhancement, which may improve the foam stability by lowering the liquid drainage rate in the foam. The N85 surfactant was selected for the PEF study. The foamability and half-life values for polymer-enhanced foams are shown in Figure 2a. Although the polymer increased the half-life value, it drastically decreased the foamability of surfactant. Polymer increased the viscosity of the liquid solution within the foam lamella, reduced the rate of liquid drainage, and consequently increased the stability of foam, as seen in Figure 2a,b. However, this enhancement was not significant for N85 surfactant, which is a relatively poor foaming agent. Therefore, proper selection of surfactant and polymer is essential to have optimum stability for PEF.

3.2. Static Performance of Foam and PEF in the Presence of Heavy Oil and Mineral Oil

Oils generally destabilize and can also stabilize a foam system [14]. In this study, the effect of heavy crude oil and mineral oil was examined on foam stability. The results are shown in Figure 3. The presence of heavy oil decreased the foamability and stability of foams and PEFs generated with all types of surfactants, especially N85. By contrast, the presence of mineral oil stabilized or destabilized the studied foams depending on the surfactant type. In terms of foamability, the addition of mineral oil had no drastic effect on foamability and foam stability of foams compared to that of heavy oil. Considering foam stability, in some cases, the addition of mineral oil resulted in an increase in the foam stability (N85, AOS, and CTAB foams). The addition of polymer did not improve the stability of N85 foam in the presence of oils. However, the decay profile (Figure 3c) showed that N85 foam completely collapsed after 2 h, whereas this value was more than 4 h for N85 PEF.

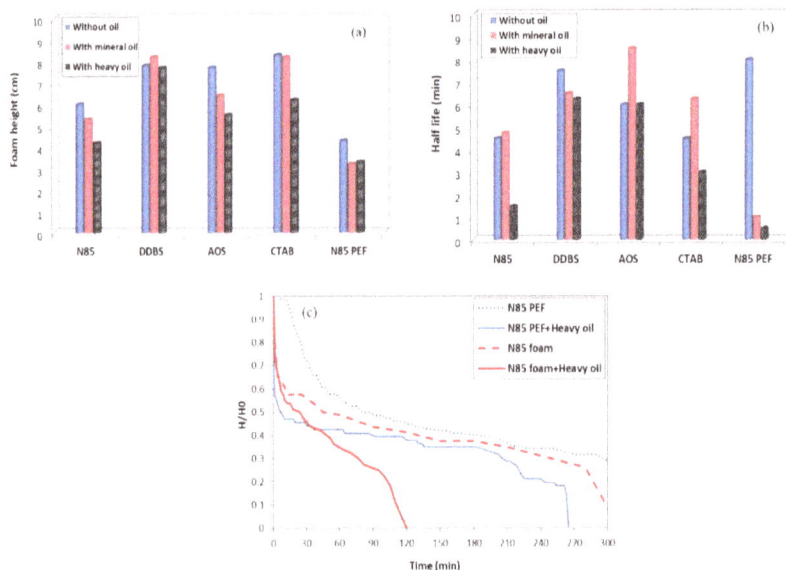

Figure 3. Effect of heavy oil and mineral oil on (**a**) foamability and (**b**) foam stability of studied foams and N85 PEF; (**c**) decay profile of N85 foam and PEF in the presence of heavy oil.

The decay profiles of DDBS, which was similar to AOS (the results are not presented here) and N85 surfactants, were selected for detailed discussion of foam–oil interaction. The foam decay profile was divided into four stages (according to Denkov, 2004) [27] as shown in Figure 4, where the normalized foam height is plotted versus time for different foam–oil systems. Both types of oil increased the foam destruction rate; however, mineral oil had relatively less drastic influence in both foam systems. This was more evident in the N85 foam system. By contrast, mineral oil, interestingly, reduced the liquid drainage rate (stage I) and had no drastic effect on foam decay life. The behavior of each foam–oil system at each stage can be explained as follows:

Stage I: During Stage I, the main process affecting the foam decay profile was liquid drainage. It should be mentioned that some minor bubble coarsening may have occurred during this stage, but the dominant phenomenon was the liquid drainage. Images of the foam bubble texture at the foam top and liquid–foam interface during this stage are shown in Figure 4.

Stage II: There was no significant liquid drainage in Stage II; however, the optical observation of the foam column demonstrated a significant change in bubbles structure during this period. The small bubbles disappeared and bubble coarsening occurred because of gas diffusion through the films (Figure 5). Stage II was too short in the case of N85 surfactant foam system. This is due to N85 (nonionic) surfactant having less potential to generate stable foam, with the foam starting to collapse almost immediately after the drainage stage. Besides surfactant weakness, the antifoaming effect of oil increased the rate of foam collapse. Stage II was about 40 min for the DDBS foam system, which showed the better stability of DDBS surfactant (anionic) to generate the most stable foam. A similar trend was seen in foam–mineral oil system; however, the length of stage II in DDBS–mineral oil system increased to more than 50 min, as seen in Figure 4b.

Stage III: The onset of stage III was identified by rupture of bubbles in the upper layer of the foam column. According to Denkov (2004), [27] when a certain critical value of the compressing capillary pressure (which is higher at the foam top) is reached, the foam starts to collapse. Stage III is called the "antifoaming" stage [27] because the shape of the profile strongly depends on the antifoaming behavior of the oil. In a foam–oil system, after foam generation, oil droplet is immediately drained from the foam film and rests in foam lamella and Plateau borders (PBs) because the droplets are smaller than the foam film thickness. During this stage, the lamella eventually will break by thinning the foam film/lamella and increasing oil droplet size (due to flocculation/coalescence of droplets). Figure 6 shows the collapse of foam lamella at the antifoaming stage, which is most probably due to the drastic effect of oil.

Comparing N85 and DDBS foams in the presence of heavy oil, the N85 foam started to collapse immediately after liquid drainage (stage I) and foam column collapsed in less than 2 h. Heavy oil also significantly reduced the length of stage III (increased rate of foam destruction) in DDBS foam. In mineral oil system, the rate of DDBS foam destruction (slope of the curve at stage III) was less than that of heavy oil, representing the less detrimental effect of mineral oil on foam stability.

Stage IV: Over time in Stage III, the rate of foam destruction gradually decreased in magnitude and stage IV was reached when the foam volume remained almost constant (residual foam height) before the foam died. Unlike N85 foam–heavy oil system, stage IV interval was evident in the DDBS foam system. In N85 foam–heavy oil system, foam collapsed right after the liquid drainage stage and died without any residual foam height, i.e., the height of the foam at stage IV. As noticed earlier, mineral oil showed less drastic effect; therefore, unlike the heavy oil system, stage IV could be seen in N85 foam–mineral oil system.

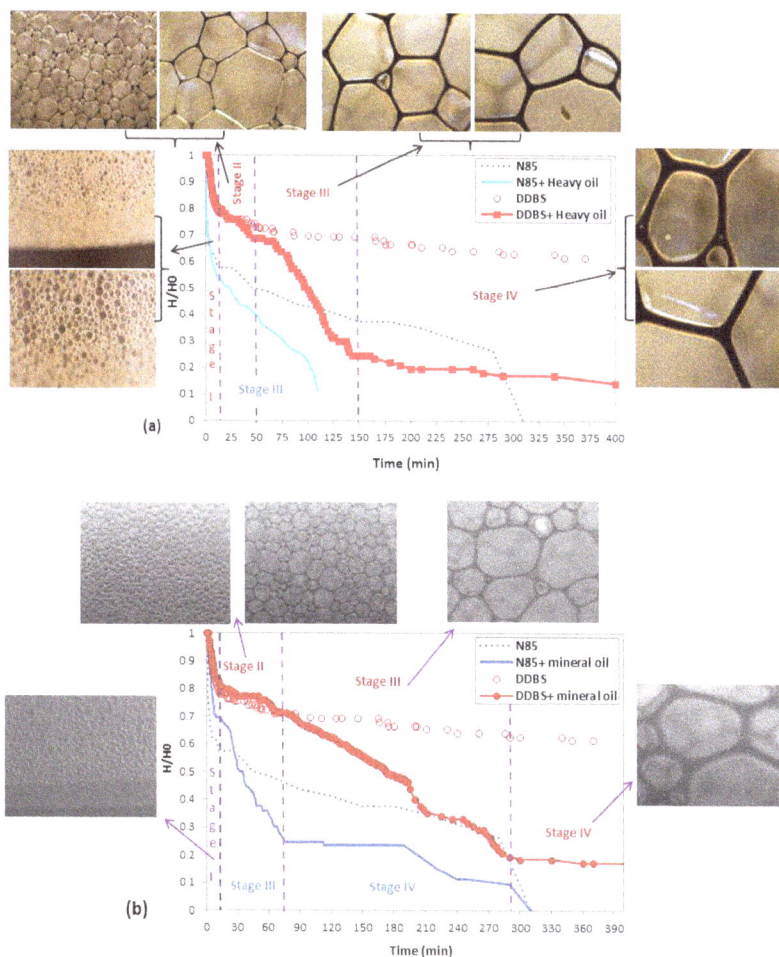

Figure 4. Foam decay profile for N85 (nonionic) and DDBS (anionic) surfactants considering the effect of (**a**) heavy oil and (**b**) mineral oil. Images correspond to DDBS foam (5× magnification).

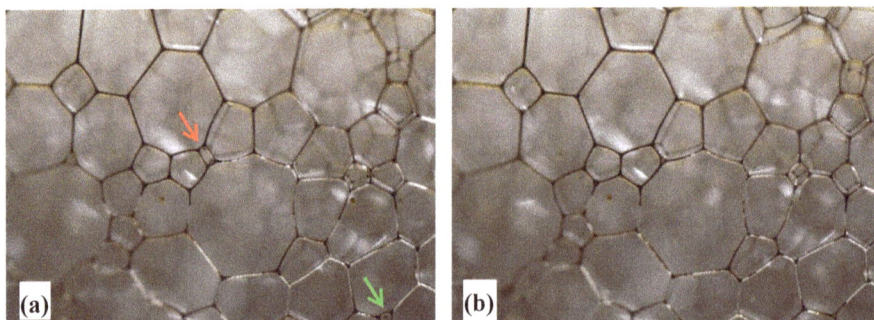

Figure 5. Air diffusion from the small bubbles toward the larger ones leads to the disappearance of the small bubbles and to the gradual accumulation of oil drops in the nodes and the Plateau borders during stage II (DDBS foam with heavy oil). Images (**b**) were captured seconds after images (**a**).

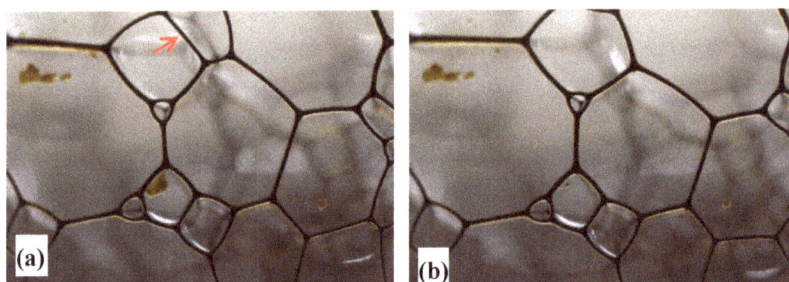

Figure 6. Oil droplets overcome the electrostatic interactions and break a foam lamella during stage III; DDBS foam with heavy oil (2×magnification). Lamella highlighted in image (**a**) breaks in image (**b**). Images (**b**) were captured seconds after images (**a**).

More Insight into Impact of Oil on the Foam Stability

A mixture of gas bubbles and oil droplets are called foamulsions or foamed emulsions [15]. In this structure, emulsion droplets are trapped and jammed between the gas bubbles, which may result in stable foamulsions [28]. The presence of emulsion droplets within the foam structure can slow down the drainage and coalescence [29]. During liquid drainage, the aqueous phase and emulsion drops flow together through the foam structure (foam film and plateau borders). However, oil drops drain slower than that of the aqueous phase. This is the reason for the increase in oil concentration within the lamella over the time, as seen in Figure 7. In this study, most of the oil stayed in the foam structure even after the collapse of the majority of foam lamellas. Oil droplet moves within foam lamella in the form of an emulsion (Figures 8 and 9) and destabilizes a foam system by entering and spreading in the water–gas interface. Oil drop must first overcome the repulsive forces (electrostatic or steric interactions) in the aqueous pseudoemulsion film to destabilize a foam lamella [18,30]. Koczo et al. (1992) [15] studied the effect of emulsion on foam stability. Their study on solubilized oil and emulsified oil systems showed that the latter may improve the stability of foam system, while solubilized oil decreases the foam stability. The packed emulsion droplets (in emulsified oil systems) prevent the liquid drainage through the foam structure due to the increased hydrodynamic resistance.

The S and E coefficients can be calculated by surface tension and interfacial measurement by the following formula [14]:

$$S = \sigma_{wg} - \sigma_{wo} - \sigma_{og}$$

$$E = \sigma_{wg} + \sigma_{wo} - \sigma_{og}$$

where, σ_{wg}, σ_{og}, and σ_{ow} are surface tension of the foaming solution, the surface tension of oil, and interfacial tension of water–oil interface, respectively. It is noteworthy that values of entering (E) and spreading (S) coefficients may give insight into the potential of the oil to destabilize a foam system; however, these coefficients cannot explain the rate of foam destabilization [30,31]. Spreading and entering coefficients values are presented in Table 2. Oil can spread as a lens over the gas–liquid interface when the spreading coefficient S is positive [32]. Similarly, an oil droplet is predicted to enter the aqueous–gas interface when the entering coefficient, E, is positive.

As shown in Table 2, all the surfactants studied here showed a positive entry coefficient, which indicates that oil entry is feasible in all systems. DDBS recorded the lowest spreading and entering coefficients, which is consistent with its highest stability in the presence of oil (Figure 3a). Although all surfactants had positive S and E coefficients, they showed decent stability in the presence of oil in the static test. Despite the positive values of E and S for AOS surfactant, it showed acceptable dynamic stability during heavy oil recovery. Therefore, the overall foam stability cannot be solely explained by these coefficients, and it may relate to the interfacial and bulk properties of the surfactant as well [31]. If the pseudoemulsion film is stable, oil droplet cannot destabilize the foam by entering the water–gas interface [18]. If the surfactants in the aqueous phase can stabilize foam films, then it can be expected that the same surfactants can, but not necessarily will, stabilize the pseudoemulsion film as well [18].

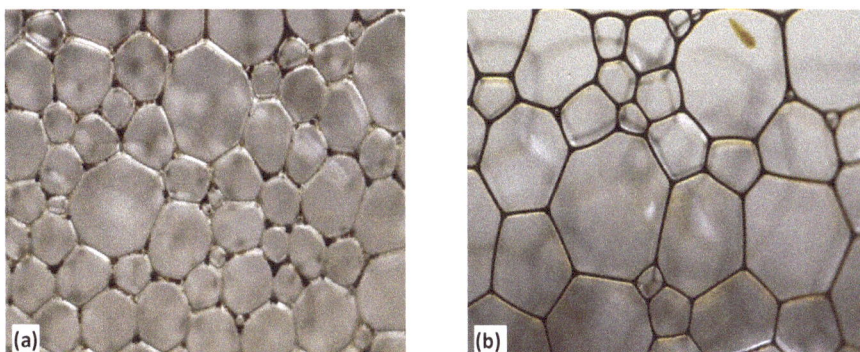

Figure 7. Images of foam generated with DDBS surfactant and heavy oil after (**a**) 3 min and (**b**) one hour; oil saturation increased within the lamella over time (foam top, 2× magnification).

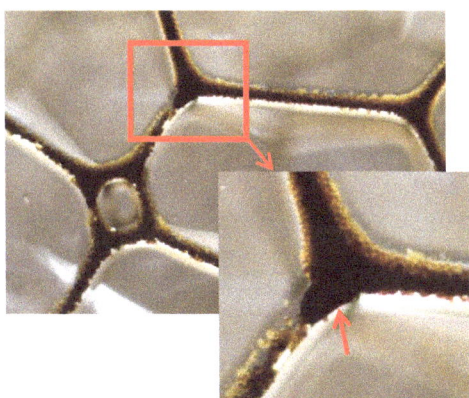

Figure 8. Flocculation and coalescence of oil droplets within the foam lamella resulting in the generation of oil lenses that eventually destabilize the foam (5× magnification).

Table 2. Interfacial tension (IFT) value (at 25 °C), entering and spreading coefficients of studied surfactants in the presence of oils.

Surfactant Solution (0.29 wt%)	Mineral Oil			Heavy Oil		
	IFT (mN/m)	E	S	IFT (mN/m)	E	S
N85	0.51	7.21	6.19	0.58	10.08	8.92
DDBS	0.33	3.03	2.37	0.54	6.04	4.96
AOS	0.50	4.7	3.7	1.10	7.7	5.7
CTAB	0.16	9.56	9.24	0.90	12.9	11.1

As seen in Figure 9, emulsion droplets within the foam lamella destabilized and created a bigger droplet. However, as long as the pseudoemulsion was stable the oil could not enter into the interface, create a lens, and destabilize the foam lamella.

In addition, the presence of emulsion droplets within the foam lamella affected the stability of the foam. Figure 9 shows the emulsion within the foam lamella created by mineral and heavy oil. These images can possibly show the reason for more stable foaming solutions in the presence of mineral oil than heavy oil. The presence of the dense assembly of droplets trapped and jammed in between the bubbles increased the local viscosity and reduced the rate of both films thinning and Plateau borders shrinking [29], resulting in a slowing down of the coarsening phenomena [33]. Note that in a higher fraction of oil, there should be enough free surfactant present in water to improve the foam stability [29]. However, this is not the case for the foam system with the presence of heavy oil. The microscopic images of heavy oil emulsion within the foam lamella demonstrated flocculation of several oil droplets within the lamella. The flocculation eventually resulted in droplet coalescence and formed bigger oil droplet or oil lens, which was detrimental to foam stability. As shown in Figure 4b, mineral oil slightly reduced the liquid drainage rate (stage I) and had no drastic effect on foam total life, i.e., N85 foam died after about 5 h with/without mineral oil.

Polymer increased the liquid viscosity within the foam lamella and significantly decreased the rate of liquid drainage (Figure 3). As a result, a longer time was required for bubble coalescence, and the bubble eventually collapsed due to the thin foam lamella. Consequently, the polymer-enhanced foam lasted much longer than conventional foam even in the presence of heavy oil.

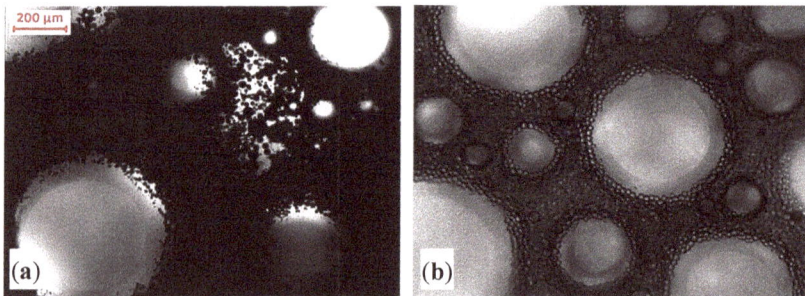

Figure 9. Microscopic image of foam–oil systems showing oil emulsion within the lamella (10× magnification); (a) heavy oil + N85 foam, (b) mineral oil + N85 foam. Images were taken immediately after foam generation.

3.3. Dynamic Performance of Foam and PEF in the Absence of Heavy Oil

The pressure profile during SAG injection in water-saturated sand pack can be divided into three distinct stages, as shown in Figure 10. The images of sand pack at each of the stages are also shown. Because the porous media was not presaturated with surfactant and the injection method was alternative, a relatively long time was required for the foam generation. During this time, surfactant and gas acted as separate slugs (two phases) and the foam was not generated. This was characteristic of Stage I. It can be said that the shorter length of Stage I demonstrated the better performance of the solution in terms of generating the foam faster within the porous media. The onset of foam generation coincided with the beginning of Stage II. An abrupt increase in pressure profile during SAG injection represented the generation of foam [34]. During Stage II, the foam was generated and propagated through porous media and resulted in an increase in the pressure drop. The slope of this stage shows how fast the foam can propagate within the porous media, and it can be used as criteria for comparing the performance of the foaming solutions. At the end of Stage II, foam occupied the whole length of sand pack and the pressure drop remained constant; this was the onset of Stage III. This stage can be termed as steady state foam injection. Pressure drop remained constant during Stage III; however, there might have been some fluctuations in pressure drop. Small pressure fluctuations during this stage demonstrated the temporary channeling or collapse of foam, which would be recovered at the end of same slug or the next successive slug injection. The sand pack images (Figure 10) show the foam-channeling phenomenon during stage III. Less channeling with lower pressure drop fluctuation can be used as criteria to compare the dynamic stability of the generated foam within the porous media.

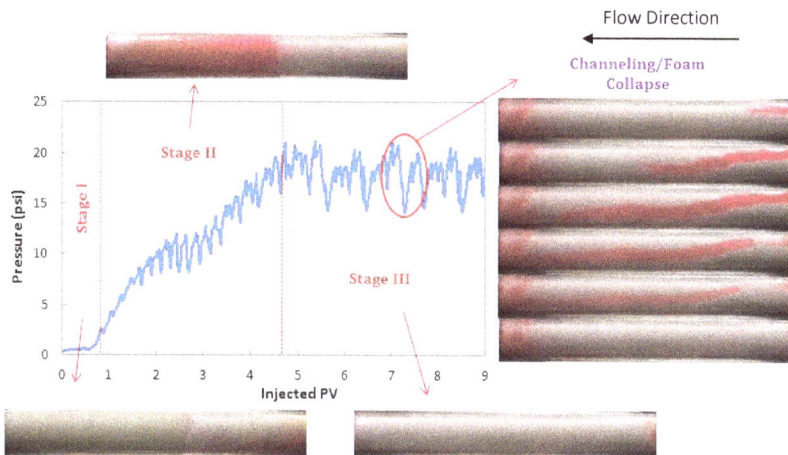

Figure 10. The typical profile of pressure during surfactant alternating gas (SAG) injection in water-saturated porous media. Red and white colors show the existence of liquid and foam, respectively. Images on the right show the foam flow within a pore volume as highlighted in the figure.

The onset of foam generation—Stage II—and abrupt increase in pressure profile was different for all studied foaming solutions studied here (Figure 11). N85 foam had the weakest performance during flow through water-saturated sand pack. Stage III was not observed for N85 foam even after the injection of 9 PV. In addition, the large local pressure drop could be seen during its propagation, demonstrating the low stability of the foam. Although the addition of polymer could not accelerate the N85 foam generation/propagation, it increased its dynamic stability during N85 PEF injection. Pressure drop fluctuations were very low during N85 PEF propagation. Stage III was observed after ~7 PV of injection (Figure 11). These features of N85 PEF compared to that of N85 foam represent its high dynamic stability within the water-saturated porous media.

Mobility reduction factor (MRF) of foam and PEF has a direct relationship to its pressure drop in porous media. MRF here is defined as the ratio of pressure drop across the sand pack when foam is flowing over the pressure drop without foam (i.e., only water is flowing). Higher values of the MRF (or pressure drop) indicate the foam is more finely textured and stronger. If the foam is very strong, the MRF is higher; for weak foams, the MRF values are smaller [35]. Both foam and PEF showed much higher MRF/pressure drop than that for surfactant–polymer (SP) injection. The large pressure difference between SP solution and the corresponding PEF represents the excellent mobility control potential (by increasing apparent viscosity) of PEF for heavy oil recovery, especially in heterogeneous reservoirs by diverting the fluid toward low permeability zones or unswept zones due to unfavorable mobility contrast [36,37].

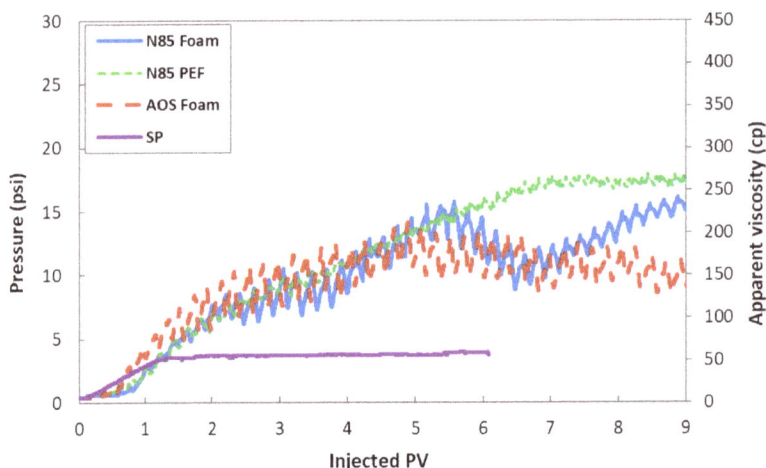

Figure 11. Foam, PEF, and surfactant–polymer (SP) pressure/apparent viscosity profiles during flow through water-saturated sand pack.

Figure 12 shows the sand pack images during N85 foam and N85 PEF injection in water-saturated media. Red color represents the liquid, while the white color is for CO_2 foam or gas. Although injected foam quality was constant, higher saturation of red color during N85 PEF injection represented higher liquid content (i.e., lower foam quality) of PEF compared to that of N85 foam. During dynamic condition, PEF channeling (if any) was smaller and recovered fast. Nevertheless, during N85 foam injection, larger channeling representing the collapse of foam was evident and abundant, representing the lesser dynamic stability of foam compared to that of PEF. It is worth mentioning that, according to the injection rates of liquid and gas in this study, the generated foam in the porous media was wet foam with relatively high liquid content. Therefore, the behavior presented here is representative of wet foam (50% foam quality) performance in porous media.

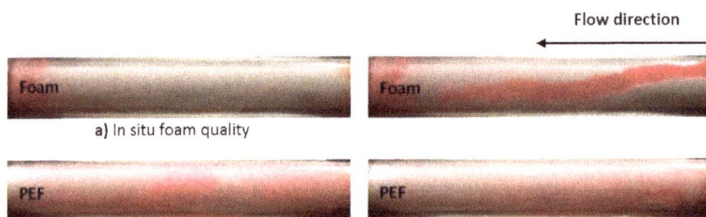

Figure 12. (**a**) In situ foam quality and (**b**) foam dynamic stability of N85 foaming solution in a dynamic condition; PEF showed lower foam quality but higher dynamic stability (less channeling) than that of foam.

3.4. Dynamic Performance of Foam and PEF in the Presence of Heavy Oil

In this section, foam and PEF flooding were conducted by alternate injection of foaming solution and CO_2 gas in the heavy-oil-saturated porous media. Anionic AOS and nonionic N85 surfactants were selected for foam injection in dynamic experiments in the presence of heavy oil. All the flooding experiments were performed after reaching a water cut of ~98% during water flooding (residual oil saturation was 66 ± 1%), i.e., residual oil saturation condition to imitate the tertiary recovery process.

Figure 13 shows the pressure and apparent viscosity profiles of foam and PEF propagation in heavy-oil-saturated sand pack for different foaming solutions. The first peak in pressure profile was due to the development of the oil bank through porous media and then its production. The corresponding oil cut was at its highest value at this point. Heins et al. (2014) also made similar observations [38]. At the same experimental condition of permeability, core dimension, flow rate, and viscosities of oil and foaming solution, the observed pressure drop could have a direct relationship with the quality of the generated foam or PEF at this stage. Higher pressure drop represents higher apparent viscosity of foam or PEF within porous media. At the initial stage of foam injection, bubbles first entered larger pore channels with lower entrance capillary pressure and the foam texture was coarse. Then, the local pressure gradient increased because of the increased foam flow resistance, and therefore foam bubbles entered relatively small pore channels [39]. The foam flow repeated this process until the steady state was reached. The pressure increased, and the foam texture became finer as foam entered smaller pore channels because the porous media shaped the foam [40]. According to Figure 13, the strong foam was generated and started to propagate in the case of N85 PEF. This can be explained by high pressure at the early time of injection. However, this was not the case for N85 foam injection.

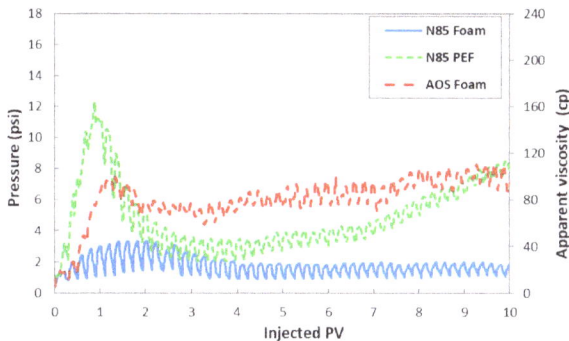

Figure 13. Pressure profiles and apparent viscosity during foam and PEF flow through heavy-oil-saturated sand pack.

For detailed analysis, foam/PEF pressure drop, change of residual oil saturation, oil cut, and images of porous media at different injected pore volume are plotted in Figure 14.

Change in oil saturation is represented as a percentage of the original oil in place. After water flooding, the remaining oil saturation in the sand pack was around 66%. As the water saturation also affects the performance of the foam/PEF, all the tertiary recovery (foam/PEF) was performed with similar water saturation [41]. The weak foaming solutions resulted in a lower apparent viscosity of foam within porous media. For N85 surfactant solution, foam generation was unsuccessful up to 10 PV of SAG injections. After oil bank production, the rate of foam collapse might have been much higher than that of foam generation that resulted in no pressure build up; therefore, stable foam did not propagate through porous media. The addition of polymer into N85 surfactant solution slightly increased the apparent viscosity and resulted in N85 PEF propagation. On the other hand, in the case of AOS foaming solution, foam could generate and propagate through porous media faster than N85 foam. This again shows the importance of selecting a proper foaming agent. The addition of polymer significantly improved the foam propagation through heavy-oil-saturated porous media. Considering the N85 PEF curve, PEF started to propagate (around 4 PV) slightly after oil bank production.

Visual observation of the sand pack (Figure 14) showed that oil appeared to move with the passage of gas bubbles, whereas it had very little movement during liquid injection. These observations are in accordance with Li et al. (2010) [34].

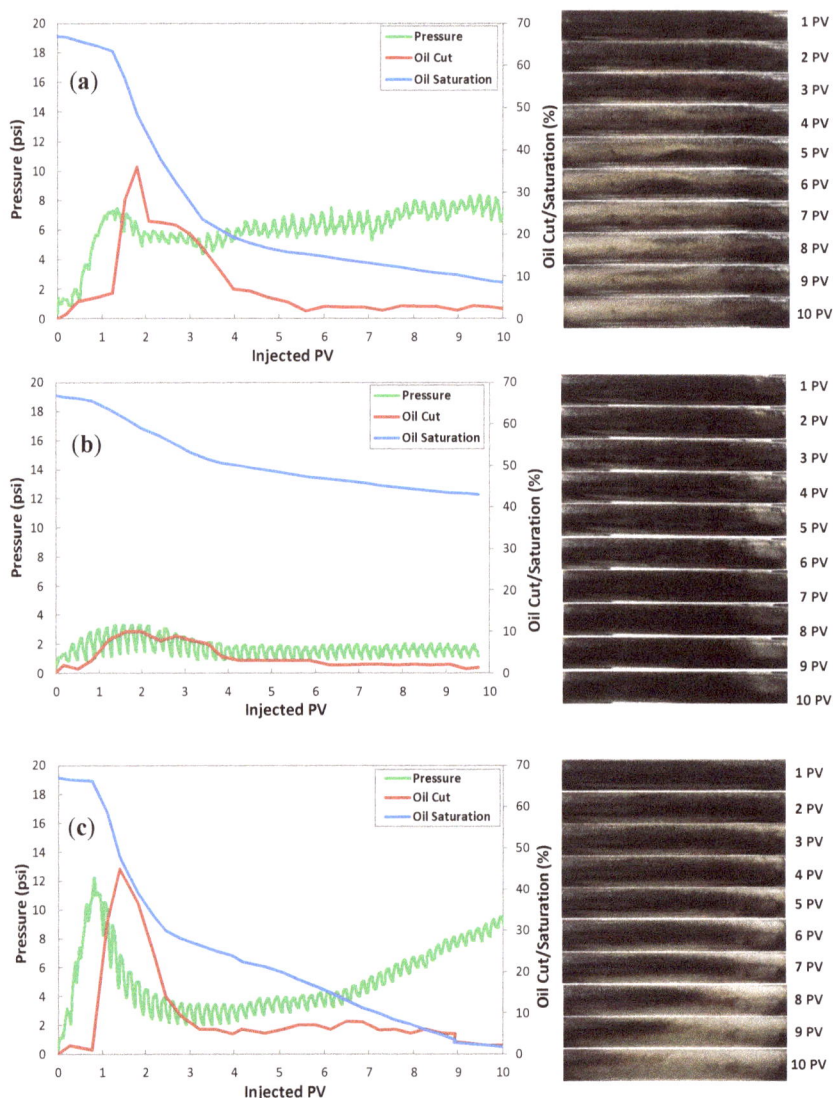

Figure 14. Pressure, oil saturation, and oil cut profiles during heavy oil recovery by foam/PEF; (**a**) AOS foam, (**b**) N85 foam, and (**c**) N85 PEF.

It is worth mentioning that foam generation in the heavy-oil-saturated porous media with residual oil saturation is not similar to that of the water-saturated one. The foam is generated in the areas with lower oil saturation and then propagates to the other area and eventually sweeps the porous media. At this point, the pressure decreases due to the oil bank production and again builds up because of the strong foam propagation. As a result, overall sweep efficiency is improved. Visual images of sand pack confirm this finding. Although ultimate oil recovery of the AOS foam and N85 PEF was similar, piston-like displacement was only observed during N85 PEF injection (Figure 14c). Pressure profile of N85 PEF agrees with the appearance of the porous medium (Figure 14c). After oil bank production around 2 PV, the pressure was almost constant between 2 PV to 5 PV, and there was no significant change in the appearance of the porous medium (change in color). However, after 7 PV, pressure started to increase, and strong foam propagation and piston-like displacement occurred and therefore swept most of the oil. AOS foam also produced a significant amount of residual oil; however, the unstable displacement front and channeling were observed as shown in Figure 14a.

The total heavy oil recovery profiles in water, foam, and PEF flooding are shown in Figure 15. Initial water flood recovery was $33 \pm 1\%$ for all experiments at 98% water cut. The N85 PEF had the best performance during heavy oil recovery (total recovery factor of ~98%,) while N85 foam had the lowest recovery among all cases (~57%). The ultimate oil recovery of AOS foam (~91%) was slightly lower than N85 PEF (98%).

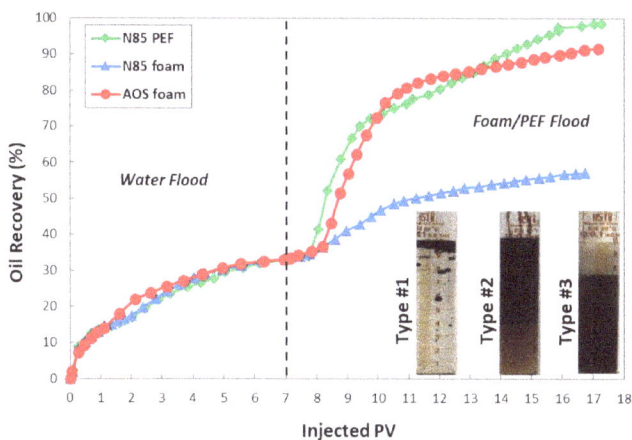

Figure 15. Total heavy oil recovery profiles during water and foam/PEF flooding with different foaming solutions. The images show the foam/PEF flooding effluent samples; Type #1: bulk oil and water, Type #2: oil-in-water emulsion and free gas, and Type #3: oil-in-water emulsion and foam.

Images of typically produced oil samples from foam or PEF flooding are also shown in Figure 15. Emulsified oil was observed from the effluent. The earlier produced samples mainly contained water and some oil as a bulk phase. The produced oil at a later time (after gas breakthrough) was emulsified oil droplet. At end of stage I, when the developed oil bank was produced, the samples consisted of oil-in-water emulsions and free CO_2 gas. After that, the oil was produced in the form of oil-in-water emulsions with gas bubbles. The produced foam during PEF injection demonstrated much higher stability compared to that of foam. In addition, it was observed that the produced foam was denser during PEF flooding. It should be mentioned that the amount of chemical (surfactant and/or polymer) can be further minimized by increasing the foam quality if the foam is stable in the presence of heavy crude oil.

It is worth mentioning that the main goal of this study was to analyze the static and dynamic stability of foam in the presence of heavy oil and also to observe the effect of polymer addition. Oil recovery/sweep efficiency showed an improvement in the 1D sand pack. However, further study needs to be done to understand effectiveness in heterogeneous/layered porous media in the presence of heavy oil [42].

4. Conclusions

The presented work investigated the potential of foam and PEF to increase heavy oil recovery. Static and dynamic experiments were designed and conducted to obtain a new insight into the performance of foam and PEF for heavy oil recovery. Although the physics of foam in bulk and porous media are rather different, the results from the detailed static analysis are essential to predict the performance of foaming solutions for heavy oil recovery. Following is a summary of the main conclusions:

The selection of foaming agent plays a key role in the performance of PEF for heavy oil recovery. However, the addition of polymer to a weak foaming agent (e.g., N85 surfactant in this study) may enhance the performance of foam in the presence of oil in dynamic experiments.

- Entering and spreading coefficient values are required for discussion of oil–foam interaction, but these may not be enough. Pseudoemulsion film stability should also be considered to support the results of foam stability. Some oils (e.g., mineral oil in this study) may increase the stability of a foam system, which is not expected from the entering and spreading coefficients.

- The porous media experiments have shown better performance of PEF in the presence of heavy oil compared to that of foam. The addition of polymer to the N85 foaming solution accelerates the foam generation and increases its stability in heavy-oil-saturated porous media. In our study, N85 PEF produced 98% of residual oil saturation, while this value was only about 57% for N85 foam.

- Overall, both static and dynamic performances of foam and PEF have shown their potential as displacing fluid for enhanced heavy oil recovery.

Author Contributions: Conceptualization, A.T. and J.J.T.; Methodology, A.T.; Formal Analysis, A.T.; Investigation, A.T. and J.J.T.; Resources, J.J.T.; Writing-Original Draft Preparation, A.T.; Supervision, J.J.T.

Funding: This research was funded by Carbon Management Canada (CMC), the Natural Sciences and Engineering Research Council of Canada (NSERC), and Alberta Innovates Technology Futures (AITF).

Acknowledgments: The authors are grateful to Carbon Management Canada (CMC), the Natural Sciences and Engineering Research Council of Canada (NSERC), Alberta Innovates Technology Futures (AITF), and the University of Alberta for supporting this study. As a part of the University of Alberta's Future Energy Systems research initiative, this research was made possible in part thanks to funding from the Canada First Research Excellence Fund.

Conflicts of Interest: The authors declare no conflict of interest.

References

1. Hirasaki, G.J. The Steam-Foam Process. *J. Pet. Technol.* **1989**, *41*, 449–456. [CrossRef]
2. Zhou, Z.H.; Rossen, W.R. Applying Fractional-Flow Theory to Foam Processes at the Limiting Capillary Pressure. *SPE Adv. Technol.* **1995**, *3*, 154–162. [CrossRef]
3. Smith, D.H. (Ed.) *Surfactant-Based Mobility Control: Progress in Miscible-Flood Enhanced Oil Recovery*; American Chemical Society Symposium Seri Number 373; American Chemical Society: Washington, DC, USA, 1988; p. 449.
4. Lake, L.W. *Enhanced Oil Recovery*; Prentice Hall: Upper Saddle River, NJ, USA, 1989; p. 550.
5. Zhang, Y.; Yue, X.; Dong, J.; Yu, L. New and Effective Foam Flooding To Recover Oil in Heterogeneous Reservoir. Presented at the SPE/DOE Improved Oil Recovery Symposium, Tulsa, OK, USA, 3–5 April 2000.
6. Ransohoff, T.C.; Radke, C.J. Mechanisms of Foam Generation in Glass-Bead Packs. *SPE Reserv. Eng.* **1988**, *3*, 573–585. [CrossRef]
7. Rossen, W.R.; Gauglitz, P.A. Percolation Theory of Creation and Mobilization of Foam in Porous Media. *AIChE J.* **1990**, *36*, 1176–1188. [CrossRef]
8. Tanzil, D.; Hirasaki, G.J.; Miller, C.A. Conditions for Foam Generation in Homogeneous Porous Media. Presented at the SPE/DOE Symposium on Improved Oil Recovery, Tulsa, OK, USA, 13–17 April 2002.
9. Gauglitz, P.A.; Friedmann, F.; Kam, S.I.; Rossen, W.R. Foam generation in homogeneous porous media. *J. Chem. Eng. Sci.* **2002**, *57*, 4037–4052. [CrossRef]
10. Mattews, C.S. *Carbon Dioxide Flooding, in Enhanced Oil Recovery II: Processes and Operations*; Donaldson, E.C., Chilingarian, G.V., Yen, T.F., Eds.; Elsevier Scientific Publishing Company: New York, NY, USA, 1989; p. 603.
11. Heller, J.P. *CO₂ Foams in Enhanced Oil Recovery, in Foams: Fundamentals and Applications in the Petroleum Industry*; Schramm, L.L., Ed.; ACS Advances in Chemistry Series, 3, No. 242; American Chemical Society: Washington, DC, USA, 1994; p. 201.
12. Vikingstad, A.K.; Skauge, A.; Hoiland, H.; Aarra, M.G. Foam-oil interactions analyzed by static foam tests. *Colloids Surf. A Physiochem. Eng. Asp.* **2005**, *260*, 189–198. [CrossRef]
13. Andrianov, A.; Farajzadeh, R.; Mahmoodi Nick, M.; Talanana, M.; Zitha, P.L.J. Immiscible foam for enhancing oil recovery: Bulk and porous media experiments. *Ind. Eng. Chem. Res.* **2012**, *51*, 2214–2226. [CrossRef]

14. Schramm, L.L.; Novosad, J.J. The destabilization of foams for improved oil recovery by crude oils: Effect of the nature of the oil. *J. Pet. Sci. Eng.* **1992**, *7*, 77–90. [CrossRef]

15. Koczo, K.; Lobo, L.; Wasan, D.T. Effect of oil on foam stability: Aqueous film stabilized by emulsions. *J. Colloid Interface Sci.* **1992**, *150*, 492–506. [CrossRef]

16. Farajzadeh, R.; Andrianov, A.; Zitha, P.L.J. Investigation of Immiscible and Miscible Foam for Enhancing Oil Recovery. *Ind. Eng. Chem. Res.* **2010**, *49*, 1910–1919. [CrossRef]

17. Simjoo, M.; Rezaei, T.; Andrianov, A.; Zitha, P.L.J. Foam stability in the presence of oil: Effect of surfactant concentration and oil type. *Colloids Surf. A Physiochem. Eng. Asp.* **2013**, *438*, 148–158. [CrossRef]

18. Nikolov, A.D.; Wasan, D.T.; Huang, D.W.; Edwards, D.A. The Effect of Oil on Foam Stability: Mechanisms and Implications for Oil Displacement by Foam in Porous Media. Presented at the SPE Annual Technical Conference and Exhibition, New Orleans, LA, USA, 5–8 October 1986.

19. Yan, W.; Miller, C.A.; Hirasaki, G.J. Foam sweep in fractures for enhanced oil recovery. *Colloids Surf. A Physicochem. Eng. Asp.* **2006**, *282–283*, 348–359. [CrossRef]

20. Flick, E.W. *Industrial Surfactants*, 2nd ed.; Noyes Publications: Park Ridge, NJ, USA, 1993; p. 547.

21. Urban, D.G. *How to Formulate and Compound Industrial Detergents*; Book Surge Publishing: Charleston, CA, USA, 2003; p. 234.

22. Rosen, M.J.; Kunjappu, J.T. *Surfactants and Interfacial Phenomena*, 4th ed.; John Wiley & Sons, Inc.: Hoboken, NJ, USA, 2012; p. 616.

23. Verwey, E.J.W.; Overbeek, J.T.G. *Theory of the Stability of Lyophobic Colloids*; Elsevier: Amsterdam, The Netherlands, 1948; p. 216.

24. Marinova, K.G.; Dimitrova, L.M.; Marinov, R.Y.; Denkov, N.D.; Kingma, A. Impact of the Surfactant Structure on the Foaming/Defoaming Performance of Nonionic Block Copolymers in Na Caseinate Solutions. *Bulg. J. Phys.* **2012**, *39*, 53–64.

25. Israelachvili, J.N. *Intermolecular & Surface Forces*, 3rd ed.; Academic Press: San Diego, CA, USA, 2010; p. 710.

26. Schramm, L.L.; Wassmuth, F. *Foams: Basic Principles in Foams: Fundamentals and Application in the Petroleum Industry*; Schramm, L.L., Ed.; American Chemical Society: Washington, DC, USA, 1994; p. 201.

27. Denkov, N.D. Mechanisms of Foam Destruction by Oil-Based Antifoams. *Langmuir* **2004**, *20*, 9463–9505. [CrossRef] [PubMed]

28. Rio, E.; Drenckhan, W.; Salonen, A.; Langevin, D. Unusually stable liquid foams. *Adv. Colloid Interface Sci.* **2014**, *205*, 74–86. [CrossRef] [PubMed]

29. Salonen, A.; Lhermerout, R.; Rio, E.; Langevin, D.; Saint-Jalmes, A. Dual gas and oil dispersions in water: Production and stability of foamulsion. *Soft Matter* **2012**, *8*, 699–706. [CrossRef]

30. Manlowe, D.J.; Radke, C.J. A Pore-Level Investigation of Foam/Oil Interactions in Porous Media. *SPE Reserv. Eng.* **1990**, *5*, 495–502. [CrossRef]

31. Hadjiiski, A.; Denkov, N.D.; Tcholakova, S.; Ivanov, I.B. Role of entry barriers in the foam destruction by oil drops. In *Adsorption and Aggregation of Surfactants in Solution*; Mittal, K.L., Shah, D.O., Eds.; Marcel Dekker: New York, NY, USA, 2003; pp. 465–498.

32. Ross, S.; Suzin, Y. Measurement of Dynamic Foam Stability. *Langmuir* **1985**, *1*, 145–149. [CrossRef]

33. Martinez, A.C.; Rio, E.; Delon, G.; Saint-Jalmes, A.; Langevin, D.; Binks, B.P. On the origin of the remarkable stability of aqueous foams stabilised by nanoparticles: Link with microscopic surface properties. *Soft Matter* **2008**, *4*, 1531–1535. [CrossRef]

34. Li, R.F.; Yan, W.; Liu, S.; Hirasaki, G.; Miller, C.A. Foam Mobility Control for Surfactant Enhanced Oil Recovery. *SPE J.* **2010**, *15*, 934–948. [CrossRef]

35. Kovscek, A.R. Reservoir Simulation of Foam Displacement Processes. Presented at the 7th UNITAR International Conference on Heavy Crude and Tar Sands, Beijing, China, 27–31 October 1998.

36. Telmadarreie, A.; Trivedi, J.J. Post-Surfactant CO_2 Foam/Polymer-Enhanced Foam Flooding for Heavy Oil Recovery: Pore-Scale Visualization in Fractured Micromodel. *Transp. Porous Media* **2016**, *113*, 717–733. [CrossRef]

37. Telmadarreie, A.; Trivedi, J.J. New Insight on Carbonate-Heavy-Oil Recovery: Pore-Scale Mechanisms of Post-Solvent Carbon Dioxide Foam/Polymer-Enhanced-Foam Flooding. *SPE J.* **2016**, *21*, 1655–1668. [CrossRef]

38. Heins, R.; Simjoo, M.; Zitha, P.L.J.; Rossen, W.R. Oil Relative Permeability during Enhanced Oil Recovery by Foam Flooding. Presented at the SPE Annual Technical Conference and Exhibition, Amsterdam, The Netherlands, 27–29 October 2014. SPE 170810.

39. Farshbaf Zinati, F.; Farajzadeh, R.; Zitha, P.L.J. Modeling and CT scan Study of the Effect of Core Heterogeneity on Foam Flow for Acid Diversion. Presented at the European Formation Damage Conference, Scheveningen, The Netherlands, 30 May–1 June 2007. SPE 107790.

40. Ettinger, R.A.; Radke, C.J. Influence of Texture on Steady Foam Flow in Berea Sandstone. *SPE Reserv. Eng.* **1992**, *7*, 83–90. [CrossRef]

41. Zanganeh, M.N.; Kam, S.I.; LaForce, T.; Rossen, W.R. The Method of Characteristics Applied to Oil Displacement by Foam. *SPE J.* **2011**, *16*, 8–23. [CrossRef]

42. Kovscek, A.R.; Bertin, H.J. Foam Mobility in Heterogeneous Porous Media. *Transp. Porous Media* **2003**, *52*, 17–35. [CrossRef]

colloids and interfaces

MDPI

Review

Interfacial Chemistry in Steam-Based Thermal Recovery of Oil Sands Bitumen with Emphasis on Steam-Assisted Gravity Drainage and the Role of Chemical Additives

Spencer E. Taylor

Department of Chemistry, Centre for Petroleum and Surface Chemistry, University of Surrey, Guildford, Surrey GU2 7XH, UK; s.taylor@surrey.ac.uk; Tel.: +44-1483-681-999

Received: 17 February 2018; Accepted: 26 March 2018; Published: 29 March 2018

Abstract: In this article, the importance of colloids and interfaces in thermal heavy oil or bitumen extraction methods is reviewed, with particular relevance to oil sands. It begins with a brief introduction to the chemical composition and surface chemistry of oil sands, as well as steam-based thermal recovery methods. This is followed by the specific consideration of steam-assisted gravity drainage (SAGD) from the perspective of the interfacial chemistry involved and factors responsible for the displacement of bitumen from reservoir mineral surfaces. Finally, the roles of the different chemical additives proposed to improve thermal recovery are considered in terms of their contributions to recovery mechanisms from interfacial and colloidal perspectives. Where appropriate, unpublished results from the author's laboratory have been used to illustrate the discussions.

Keywords: emulsions; heavy oil and bitumen; interfaces; oil sands; petroleum colloids; SAGD; surfactants; thermal recovery; wettability

1. Introduction

It is estimated that there are about 6 trillion barrels of oil stored in the form of heavy oil and bitumen [1]. One of the key characteristics of these unconventional crude oils is their very high viscosity at reservoir conditions. For example, typical Athabasca reservoir temperatures are about 7–11 °C [2], and a typical Lloydminster reservoir temperature is about 25 °C [3]. At these temperatures, the mobility of the oils is poor, with heavy oil viscosity being in the range of tens of Pa·s, whereas bitumen viscosity is in the range of hundreds of Pa·s [1]. Therefore, the recovery of these petroleum resources requires radically different strategies than those used for the more familiar conventional crude oils.

The approaches for heavy oil and bitumen production can be divided into thermal and non-thermal methods. The most widely used non-thermal methods include waterflooding, gas injection, and cold production, each of which has been thoroughly reviewed elsewhere [4].

Thermal recovery methods used in heavy oil production rely on significant viscosity reduction at increased temperatures [2]. These can also be divided into two classes based on the source of the heat: (i) hot-fluid injection in which the hot fluid (almost exclusively steam) is produced at the surface and injected into the reservoir; (ii) in situ heat-generation, in which the heat is generated in the reservoir, such as in situ combustion and fire-flooding [4].

Thermal (In Situ) Recovery of Bitumen Production from Oil Sands

Currently, there are two main steam-based thermal methods commercially used for recovering heavy oil/bitumen from oil sand reservoirs: cyclic steam stimulation (CSS) and steam-assisted gravity drainage (SAGD) [1].

CSS uses the same well (vertical, deviated, and horizontal) for steam injection and heavy oil production (Figure 1). In the first step, steam is injected with a pressure slightly greater than the fracturing pressure of the formation [1]. A soak period is then followed by production. This process is well-developed, and usually less than 20% of the original oil can be recovered [4]. The major drive mechanisms in CSS include formation re-compaction, solution-gas drive in early cycles and gravity drainage in later cycles [1]. The steam injection/production cycles are repeated until it is uneconomic to continue.

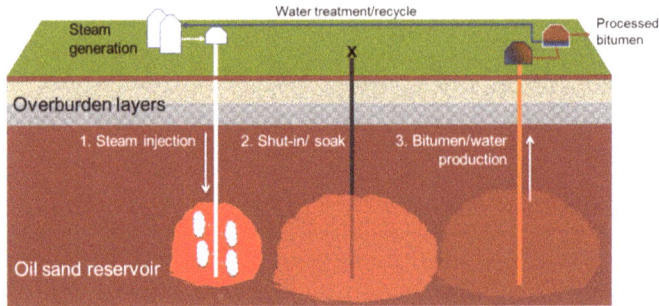

Figure 1. Schematic representation of a CSS operation using a vertical well for the three phases: steam injection, soak and production.

In SAGD, two parallel horizontal wells are used. Normally, steam at a sub-fracture pressure is injected via the injection well which is typically 5–10 m above the production well which is located close to the bottom of the reservoir (Figure 2) [2]. The major drive mechanism of SAGD is gravity drainage [1]. In practice, the SAGD process starts at a high steam pressure, which is reduced as the process evolves [2].

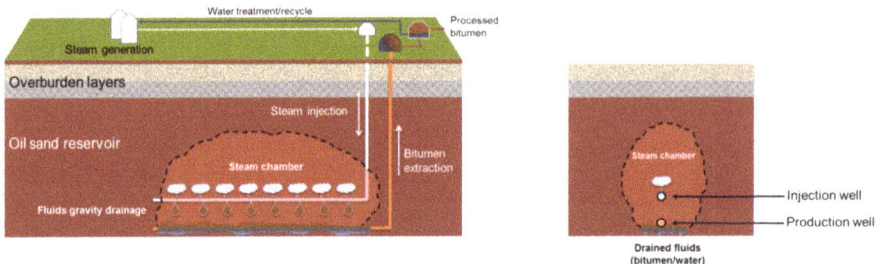

Figure 2. Schematic representation of a SAGD operation. **Left**: an illustration of the main elements of SAGD showing the juxtaposition of injection and production wells. **Right**: a cross-section view of the reservoir highlighting the shape of the steam chamber.

SAGD has been increasingly used for the production of bitumen from Canadian oil sands. Between 2004 and 2014 the production of bitumen using SAGD increased from 72 to 686 thousand barrels/day [5]. However, as reported in summer 2017, approximately 1.4 million barrels/day of bitumen were being produced by thermal in situ methods in Alberta alone. This number is projected to increase to 2.1 million barrels/day by 2030 [6].

Steam generation is an extremely energy-intensive process, which is usually provided by fossil fuel (e.g., natural gas, bitumen) combustion. Therefore, although steam injection offers robust bitumen recovery methods from reservoirs too deep for mining, there are continuing concerns over their high energy requirements and the environmental penalty caused by the associated release of carbon dioxide

(CO_2). For example, the typical SAGD steam/oil ratio (SOR) in commercial operations will be in the range 2–10 [2] and it has been calculated that, for a typical high quality SAGD project operating at a SOR of 2.5, approximately 20% of the bitumen produced is required to raise the steam (based on 80% conversion efficiency in the boiler) [7].

On this basis, it is clear that improvements in the efficiency of thermal recovery methods in terms of lowering energy requirements and CO_2 release require increases in the SOR. To achieve this, an understanding of the interfacial chemistry associated with these recovery processes is needed in order to optimize oil release from reservoir mineral surfaces.

2. Oil Sands Structure and Composition

The presence of water has long been considered to be an integral part of oil sands structure, although there is some conjecture as to its exact nature. Early cryo-transmission electron microscopy (TEM) studies provided evidence that water in natural oil sands exists as a water-in-oil (w/o) emulsion [8]. On the other hand, a model of oil sands with a layer of water (approximately 10 nm thick) separating bitumen from the sand surface was subsequently proposed by Takamura [9]. This latter model was supported by calculations based on composition, pore radius, and surface areas of Athabasca (Canadian) oil sands, which concluded that such water films could exist with a thickness of 5–6 nm [10].

As far as bitumen composition is concerned, analyses based on infrared spectroscopy and elemental analysis of saturates, aromatics, resins, and asphaltenes (SARA) fractions, extracted sequentially from oil sands, suggested that these are not evenly distributed throughout the bitumen layers, with the most polar asphaltene and resin species being concentrated at the sand surfaces [11].

It is also necessary to consider that the oil composition may change during thermal production. When heavy oil or bitumen is produced with steam, for example, a common observation is that the produced oil is lighter than the original, which is believed to be a consequence of in situ upgrading processes [12–14]. It has been suggested [12,13] that minerals in the oil sands or the products of mineral-steam reactions [14] act to catalyze upgrading reactions (e.g., aquathermolysis [15–17]).

It is important to note that, other than the chemical compositions of the oils, to be discussed below, the size distributions of the solids and the wetting phase layer on their surfaces may also play important roles in fluid–fluid and fluid–solid interactions in oil sands. A comprehensive study has been reported very recently based on 42 sets of sandpacks using representative pore and rock configurations of the McMurray Formation, the primary oil sand resource in the world [18].

Compared with lighter, conventional oils, heavy oils and bitumen are enriched with hetero-species, some of which are active at liquid–liquid (oil/water) and/or solid–liquid (sand/oil or sand/water) interfaces. Among the many components present in the oils under consideration here are several classes which are interfacially-active, namely asphaltenes, naphthenic acids, and humic acids. In addition, as alluded to, fine particles can also influence important processes relevant to oilfield recovery and production. These classes are considered in more detail below.

2.1. Asphaltenes

Recognized as the heaviest and most polar components of crude oil, asphaltenes are defined as those oil components that are soluble in aromatic solvents but insoluble in *n*-alkanes. Asphaltene molecules become insoluble in the oil matrix (known as the 'maltenes') due to changes in maltene composition (and hence solvency) as a result of changes in pressure, temperature, etc. The compositions of asphaltenes differ according to the conditions used in their preparation. For example, higher carbon number *n*-alkanes normally lead to asphaltene precipitation at lower volume additions than their lower carbon number counterparts, but the corresponding asphaltene yields are reduced [19].

Asphaltenes have a tendency to form aggregates or clusters that influence their surface and interfacial activity and which, notwithstanding their compositional complexity, has been shown to be due to only a small proportion of the molecular species present [20].

Thus, it has been suggested that asphaltenes isolated using *n*-heptane are mixtures of two main fractions. One (designated A1) has a very low solubility in toluene (~90 mg/L) while the other (designated A2) has a much higher solubility (57 g/L). The strong tendency to aggregate is most likely due to the presence of the less soluble fraction which is present in larger quantities [20]. The adsorption of asphaltenes onto silica is an irreversible second-order adsorption, with the adsorption rate being strongly dependent on the asphaltene concentration: with increasing concentration, the effective adsorption rate decreases as concurrent asphaltene aggregation occurs [20]. Asphaltene adsorption on kaolinite also follows complex behavior [21,22] which has been related to the presence of asphaltene nanoaggregates in solution and agglomeration of the adsorbent particles [23].

Interfacial tension (IFT) studies have also revealed that asphaltene monomers and nanoaggregates have different adsorption kinetics at the oil/water interface [24]. The adsorption of asphaltenes at the toluene–water interface has been suggested to involve two steps, namely asphaltene monomer adsorption, followed by interfacial aggregation [25]. Further flocculation of asphaltene aggregates will lead to the formation of a three-dimensional network which acts as an effective barrier between oil and water. If the acidic fraction soluble in heptane (AFSH) isolated from the same crude oil is added, the interface is found to contain a mixed film containing AFSH and asphaltenes [25]. In this situation, the AFSH reduces the interfacial tension, while the presence of asphaltenes serves as a hydrophobic barrier between the two phases [25].

The status of asphaltenes in the oil is an important factor determining the stability of w/o emulsions. For example, it has been observed that asphaltenes exhibit the maximum capability of stabilizing w/o emulsions when they are close to their stability boundary [26], suggesting that asphaltene nanoaggregates or larger colloidal asphaltenes are more effective as emulsion stabilizers than asphaltene molecules. Increasing the asphaltene solvency or adding resins can reduce asphaltene surface activity and destabilize the emulsion [27,28].

2.2. Organic Acids

In the petroleum industry, organic acids present in crude oils and oil products are termed naphthenic acids. Acids are present in almost all crude oils, with the amounts ranging from 0 to 3 wt % [29]. For Athabasca oil sand, the content of naphthenic acids in the bitumen is approximately 2 wt % [30].

Naphthenic acids are mixtures composed mainly of alkyl-substituted cycloaliphatic carboxylic acids and paraffinic (fatty) acids [29]. The carboxyl group is usually found on the side-chain rather than on the alicyclic ring [30]. As a result of these structures, organic acids have surfactant-like properties and play important roles in the interfacial chemistry of oil production. One study has revealed that the primary naphthenic acids are acidic species with molecular weights in the range 250–425 g/mol and are most effective in reducing IFT [31]. In another study, the self-assembly behavior of the acidic species of bitumen and its distillate fractions was revealed using mass spectrometry [32]. It appeared that the lowest boiling distillate fraction (375–400 °C) showed the strongest tendency toward aggregation, occurring at concentrations as low as 0.05 mg/mL [32]. Compositional characterization revealed that the most abundant acidic species in the 375–400 °C distillation fraction are non-aromatic diprotic acids. On the other hand, the most abundant acidic species forming aggregates in the higher boiling 450–475 °C fraction are aromatic naphthenic acids [32].

Since asphaltenes and organic acids co-exist in crude oils, especially in bitumen, their combined effects on the stabilization of the oil/water interface needs to be considered. It was shown, for example, that naphthenic acids extracted from crude oils show greater IFT reduction in acid-free oil than in the aliphatic-aromatic solvent which is normally used as a model. This suggests that the interfacial properties are influenced by interactions occurring in the bulk oil phase between the naphthenic acids and asphaltenes [33]. A general mechanism is that the basic nitrogen-containing functional groups in asphaltenes interact with the naphthenic acids, leading to the formation of naphthenic acid–asphaltene complexes which have stronger capabilities of stabilizing the oil/water interface [31]. More recent studies have suggested that organic acids could soften the rigid asphaltene film on the oil/water interface and make the emulsion more stable [34].

The polar compounds enriched at the oil/water interface can be separated and characterized using various methods. For example, a "heavy water method" has been developed to isolate the interfacial materials, which contain a mixture of asphaltenes and carboxylic acid salts [35]. Compositional analysis of the interfacial materials using advanced characterization techniques such as high resolution mass spectrometry or 2D GC-MS, for example, has identified the enrichment of oxygen and oxygen/sulfur species at oil/water interfaces. It was also found that the unsaturation level and molecular size of the interfacial materials are important for their interfacial properties [36].

The adsorption of organic acids and asphaltenes has been observed to change the wettability of reservoir minerals, with potential impacts on oil production [37–40].

2.3. Humic Acids

There are significant amounts of toluene-insoluble organic matter (TIOM) in oil sands [41–45]. These materials are able to coat mineral surfaces to produce organic-rich solids (ORS). Generally, there are two types of ORS found in oil sands, a coarse fraction and an ultra-fine fraction. The coarser fraction usually is less than 44 microns and usually comprises aggregates containing humic matter and precipitated minerals. These heavy aggregates can carry associated bitumen into aqueous phases. On the other hand, the ultra-fine fraction contains clay particles with a major dimension of less than 0.3 microns. These clay particles are coated with organic matter, and remain associated with bitumen produced by the flotation method in surface mining processes [42].

Humic materials are able to form complexes with clays in various oil sands, bonded through Fe species [46,47]. Differences between complexes formed from oil sands of different origins is a result of differences in the distribution of the complexes in the oil sand solids. In caustic flooding, NaOH ionizes humic materials, leading to reductions in IFT and increasing oil recovery [47].

To identify the different solid fractions in oil sands, a cold water agitation test was developed by Kotlyar et al. [46]. Solids associated with bitumen were found to be enriched with humic matter, with the insoluble organic carbon content being as high as 36% and the solids containing a higher concentration of Ti, Zr, and Fe. Solids suspended in the aqueous phase had an insoluble organic carbon content of approximately 5%, and were enriched with Al. The remaining solids showed a very low insoluble organic carbon content (<0.3%). ^{13}C nuclear magnetic resonance spectroscopy indicated that the humic matter content of oil sands have a similar maturity as sub-bituminous coal [46]. However, elemental analysis suggested that the organic matter could have a different origin from most of the oil sands bitumen [48].

2.4. Clays and Other Fines

As mentioned in the previous section, fines/clay particles in natural oil sands could be coated with organic species, such as humic matter. During the SAGD process, these can become more oil-wet because of further deposition of asphaltenes. Oil-wet clay particles can stabilize the "rag layer" (comprising mainly w/o emulsions) invariably formed at oil/water interfaces during oil sand production [49]. Upon changing the wettability of these clay particles to water-wet, the emulsion is destabilized [50].

It is also possible that some polar compounds could partition into the water phase during the SAGD process. In turn, polar organic compounds dissolved in the SAGD produced water could play an important role in the interaction between the bitumen and water during SAGD production. One group of the possible polar organics in SAGD water are the humic acids discussed above, which can lead to humic acid–clay complex formation in oil sands [46,47]. Fluorescence excitation–emission matrix spectroscopy (FEEMS [51]) characterization of SAGD-produced water confirmed the existence of humic acids and the related fulvic acids [52]. A systematic characterization revealed that naphthenic acids and larger carboxylic acids are present in SAGD produced water, the major acidic species containing 2–3 rings [53].

3. Nature of Steam-Based Thermal Recovery, Especially SAGD and Its Variants

Conceived and developed by Butler and his coworkers at Imperial Oil in the late 1970s, SAGD is the most widely-used steam-based in situ heavy oil/bitumen recovery method currently available [54]. The original SAGD concept was aimed at improving recovery using gravity as the major driving force involved in separating the oil from the reservoir matrix. Compared with conventional steam drive processes, it was envisaged that this also avoids excessively cooling the produced fluids.

The underlying principle of SAGD involves the continuous introduction of steam into a reservoir via a perforated horizontal well at temperatures of typically 150–250 °C at pressures which depend on the nature of the oil and the formation. As can be seen in Figure 2, the steam permeates upwards and outwards through the formation creating a so-called steam chamber. This reduces the contacted oil viscosity considerably, which is then carried with the steam and condensed water to the cooler growing extremities of the chamber, where the liquids then drain under gravity. These then collect below the steam inlet level where they are extracted via a second horizontal production well a few meters below the injection well. Early work indicated that recovery rates of 79–159 m^3/day can be achieved, with bitumen recoveries of the order of 50% [1,55,56]. SAGD is deemed suitable for unconsolidated reservoirs with high vertical permeability, and is therefore relevant to oil sand deposits for which it is a more environmentally acceptable alternative than mining.

This relatively simple scheme, however, hides many complexities, and the understanding of the mechanics of steam chamber development and heavy oil/bitumen production rates remain at a relatively early stage. For example, Al Bahlani and Babadagli reviewed the status of SAGD based on laboratory experimental studies and highlighted various features of the process that contribute to the process and for which a more detailed understanding is necessary [57]. The areas identified included: steam chamber development, resulting from the low density steam penetrating the formation above it; steam fingering theory, since steam rising through the developing chamber occurs irregularly, and not as a uniform front [58]; co-current and counter-current displacement, reflecting the complexity of the fluid flows occurring during SAGD, one consequence of which can be the formation of a w/o emulsion through condensation of steam in the oil; and heat transfer and distribution through the steam chamber, which is critical to providing a full understanding of the process.

Without recourse to a full appreciation of SAGD theory, several alternative variant strategies have nevertheless been identified to improve factors such as the efficiency of heat transfer, steam chamber reach, and bitumen removal. Al Bahlani and Babadagli classify these variants as chemical and geometrical [57]. Included among those based on chemical additives were: (i) the use of light solvents, either as hybrid processes by introducing the solvent with the steam, or using solvent vapor to completely replace the steam (the Vapex process [59]); and (ii) the introduction of gas and surfactants to generate foams. Thus, expanding solvent-SAGD (ES-SAGD) is one development based on the addition of low levels of solvent (e.g., hexane), which co-condenses at the cooler extremities of the steam chamber, and acts to reduce the viscosity of the mobilized bitumen. Foam-assisted (FA)-SAGD has been proposed by Chen [60] and tested using two-dimensional reservoir simulation models. The use of foam was shown to have an overall positive impact on the energy efficiency of the process, compared with SAGD alone, through controlling steam breakthrough in the inter-well region. One negative effect of the introduction of foaming steam into the simulation, however, was a reduction in the steam mobility in the upper regions of the chamber, which had a consequent negative impact on recovery rates.

SAGD operations are also potentially demanding in terms of water recycling. Treatment of produced water for re-use in steam generation, for example, is an essential part of sustainable water recovery operations [61]. The chemistry associated with SAGD operations can make this challenging, however, owing to the significant dissolution of reservoir minerals—e.g., silica—which increases substantially at temperatures above 100 °C. In turn, this increases the probability of silicate mineral deposition occurring on susceptible surfaces [62]. In the case of silica, solubility is greatly enhanced under high pH conditions, although by taking appropriate measures steam production is not necessarily compromised [63], as discussed in Section 5.

3.1. Factors Affecting SAGD Performance

In addition to the factors discussed above, there are many others that are considered to affect SAGD recovery rates, including the heterogeneity and permeability of the reservoir [64], influencing both propagation of the steam chamber and non-uniform heat transfer through the formation, as well as phenomena such as steam fingering and viscous coupling (related to emulsion formation) involving the steam (water) and oil phases. The latter factor exemplifies one of the most important aspects of any scheme based on gravity drainage, that of the viscosity of the draining fluid. This is evident from the various mathematical treatments based on Butler's original SAGD concept for the rate of oil production, *q*, which have been further developed in the intervening years, but the basic form is [65]

$$q = L\sqrt{\frac{B\phi\Delta S_0 kg\alpha h}{mv_s}} \tag{1}$$

In Equation (1), L is the length of the production well, ϕ is the porosity of the reservoir, ΔS_0 is the difference between initial and residual oil saturation levels, k is the effective oil permeability, α is the thermal rock diffusivity, h is the height of the steam chamber, v_s is the kinematic oil viscosity under the reservoir conditions, m is a constant relating to the temperature dependence of viscosity, and B is a constant relating to the system configuration.

Some parameters in Equation (1) are characteristics of the reservoir and are hence very difficult to influence. It is also evident that the drainage (hence recovery) rate is proportional to (viscosity)$^{-0.5}$, and therefore minimizing oil viscosity is an obvious but key factor in optimizing the SAGD process. Positive effects can also be envisaged, however, both by minimizing residual oil saturation and increasing the porosity; these properties are related to the effectiveness of oil removal from the porous structure. Associated with these properties will be the size of the steam chamber, since more efficient displacement of oil will lead to improved steam chamber propagation.

As mentioned above, heavy oils and bitumen exhibit a natural tendency to form w/o emulsions owing to the presence of natural surfactants present in the oil phase, possibly also favored by the viscosity ratio between oil and water. Therefore, flow rate restrictions could be expected to arise from the in situ formation of w/o emulsions, since these will be more viscous than the heavy oil or bitumen at any given temperature. The viscosity of emulsions is highly dependent on the dispersed water phase content (volume fraction, ϕ), according to empirical relationships, such as that given by Krieger and Dougherty [66,67], given in Equation (2).

$$\eta = \eta_0 \left(1 - \frac{\phi}{\phi_{max}}\right)^{-2.5\phi_{max}} \tag{2}$$

In Equation (2), η_0 and ϕ_{max} are the (shear) viscosity of the oil phase and the maximum close-packed phase volume, respectively, and the factor 2.5 is the intrinsic viscosity for spheres. This equation indicates, for example, that the incorporation of 10% dispersed water phase (by volume) will result in approximately 30% increase in viscosity, and for 30% water incorporation, the viscosity is expected to increase by approximately 160% (calculations assuming that $\phi_{max} = 0.74$, the maximum packing fraction for monodisperse spherical droplets, although it is often found that this value may be significantly different, which will affect the quoted increases in viscosity).

However, from the point of view of gravity drainage, intermixing of steam, condensed water, and bitumen under conditions of high temperature, pressure and shear, as well as a water/bitumen ratio of at least 3 at the extremities of the steam chamber provides extremely favorable conditions for bitumen emulsification in water to take place. In fact, produced fluids are found to comprise mixed water-in-oil and oil-in-water (w/o + o/w) and multiple (principally water-in-oil-in-water (w/o/w) [61]) emulsions. From the standpoint of production efficiency, formation of o/w emulsions may be considered to be more beneficial, since these will invariably have lower viscosities and potentially drain more efficiently than the corresponding w/o emulsions. As an example, Figure 3 shows the main produced emulsions (o/w and w/o) from an Athabasca oil sands SAGD operation in which both bitumen and water droplet sizes are seen to be less than 20 microns, as described previously [68]; many are seen to be considerably less than 5 microns for both emulsion types. Higher magnification o/w emulsion images than are shown here indicate that the bitumen droplets contain inclusions of finely dispersed water droplets (i.e., w/o/w), consistent with operational experience [61]. In practical terms, however, these examples can only be considered as being a snapshot of the fluids produced at a moment in the operational history, since the predominant emulsion types in SAGD will be influenced by the water/bitumen ratio, which will vary over the production lifetime of the operation.

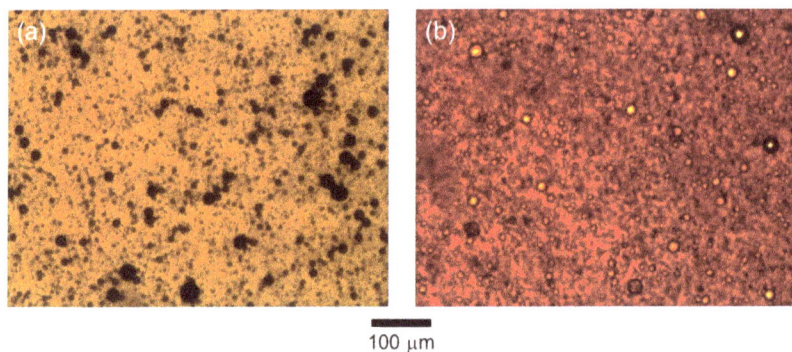

Figure 3. Examples of photomicrographs of emulsions produced from a SAGD process. (**a**) Oil-in-water emulsion; (**b**) Water-in-oil emulsion comprising the bulk bitumen phase. Unpublished work from the author's laboratory.

Emulsions of this type have also been encountered in produced fluids from steamfloods incorporating chemical additives. Thus, Sarbar et al. investigated the effect of surfactant (sodium dodecyl benzene sulfonate) and alkaline (NaOH) solutions on the flow of o/w emulsions through porous media [69]. Some instability of the emulsions was apparent, as it was found that 10% o/w emulsions were transported through a 100 m length of porous media (unconsolidated sandpack), with only around 70% of the emulsion remaining intact. Increasing the oil content of the emulsion increased the pressure drop across the porous media, in accordance with the increasing viscosity (Equation (2)), but also possibly reflecting a reduction in stability (through inversion) of the o/w emulsion.

Arguably, o/w emulsions are favored during pre-production in the reservoir; however, in post-production they can potentially pose some operational problems requiring specific chemical treatments [70].

3.2. Oil Mobilization Mechanisms in SAGD Processes

As described above, the basis of SAGD is the formation of a hot, oil-depleted zone in the reservoir by the horizontal introduction of superheated steam. The resulting steam chamber produced grows as the steam gives up its latent heat to the reservoir rocks and the oil, which drains under gravity together with condensed water. During the growth of the steam chamber, various mechanisms play a part in mobilizing the oil and affecting recovery rates, including:

- Solvency/dissolution (e.g., using oil-miscible solvents)
- Emulsification
- Displacement/detergency
- Foam formation

The various interactions between steam (and any additional components), oil and the reservoir mineralogy determine the efficiency and extent of steam chamber growth, oil displacement, and ultimately potential recovery rates (some aspects of which are broadly expressed by Equation (1)). In the absence of additional steam components, significant residual oil is left behind in the steam chamber (up to 50% recoveries have been quoted, e.g., [1,55,56]), and water is incorporated into the recovered oil.

This suggests that there should be sufficient scope for improvements to be made in respect of reducing residual oil saturation and improving overall oil recovery. For example, the inclusion of suitable solvents to steam would lead to improved oil mobilization through improved drainage arising from lower oil viscosities, but perhaps would not restrict the formation of w/o emulsions. On the other hand, inclusion of low concentrations of chemical demulsifiers would alleviate the formation of w/o emulsions, without necessarily enhancing overall recovery. Displacement mechanisms, for example involving surface-active agents, would address overall recoveries, reducing residual oil saturations, and the nature of the agents involved would also reduce the tendency for w/o emulsion formation, preferring instead to stabilize o/w emulsions.

Not all of the above mechanisms would necessarily apply to all SAGD situations, but their consideration is relevant in attempts to improve the efficiency of SAGD processes.

3.2.1. What Happens at the Liquid/Liquid Interface? Water-in-Oil Emulsions

When steam is injected into oil sands, water-in-bitumen emulsions have a natural tendency to form from the condensed water [68,71]. By co-injecting water and bitumen into sandpacks, Chen et al. [71] demonstrated the importance of the wettability of the porous medium and the water flow rate; oil-wet media favored w/o emulsification and the extent of emulsification increased as the water flow rate was increased [72].

As alluded to above, the formation of w/o emulsions is commonly observed during steam-based thermal recovery. Although some benefit of w/o emulsion formation may accrue under certain circumstances, such as from plugging unproductive high permeability zones [73], there is a greater concern of the increased viscosity of the w/o emulsion on production. However, based on an analysis of production data from a CSS operation in Cold Lake (AB, Canada), Vittoratos proposed that improved permeability arises from the formation of water-in-bitumen emulsions through greater effective bitumen occupancy of the pore spaces [74]. It was assumed that the increased emulsion viscosity is "more than compensated for by the increased relative permeability to oil resulting from increased bitumen saturation due to swelling" [74]. Currently, most operational SAGD projects are at relatively early stages where the bitumen saturation is still high. However, when the bitumen saturation decreases below a critical value, the SAGD process may become uneconomic, making alternative recovery methods necessary in future.

Related to the foregoing, the involvement of asphaltenes in the respective stabilities of heavy and conventional oil emulsions under high temperature conditions has been the subject of a recent study [75] in which it was shown that water emulsification in a heavy oil increases with increasing temperature (studied up to 200 °C), accompanied by increased stability and higher viscosity of the resultant emulsions. By contrast, the opposite trend was found for a conventional oil [75]. The authors considered a mechanism for heavy oils involving asphaltene disaggregation leading to a greater number of asphaltene molecules available to stabilize the oil/water interfaces [75]. However, it has also to be considered that the viscosity increase can be counteracted to some extent by a dilution effect of higher concentrations of dissolved water at the high temperatures [76,77], leading to significant viscosity reductions at SAGD temperatures, e.g., approximately 20% reduction seen at 200 °C [77].

Chung and Butler concluded that the primary mechanism for the in situ formation of w/o emulsions during SAGD is condensation of steam in contact with bitumen, with the condensing water effectively being engulfed by spreading oil films [71]. "Precursor" oil films that preferentially spread from bitumen over a water surface were studied by Drelich et al. more than 20 years ago [78] and have recently been shown [79] to contain a range of polar surface-active components, which could therefore explain water uptake by the oil by this mechanism. However, a second mechanism involves counter-current flow of steam and oil within the steam chamber. The result is that water droplets with diameters in the range 10–20 μm are uniformly distributed throughout the oil phase (see also Figure 3b). In arriving at these conclusions, these authors also determined that emulsification (measured in terms of the emulsified water/oil ratio) was not significantly affected by the quality of steam used in their experiments, the porosity of the reservoir, or the injection steam pressure. On the other hand, less water was emulsified (approximately 10% versus 20%) if the formation initially contained 12.5% connate water. More recently, emulsification of water at the steam/oil interface was proposed by Azom and Srinivasan to facilitate convective thermal transfer, thereby leading to improved recovery, and suggesting that such a mechanism could explain the underestimation of recovery rates given by some SAGD models [80].

Visualization of the process of emulsion formation at the steam chamber interface using scaled two-dimensional reservoir models containing porous packing materials has been reported by Sasaki et al. [81]. Fine water droplets of the order of 10 μm were observed to form at the interface between steam and heavy oil phases. The droplets were then incorporated into the flowing oil phase. The recovered fluids from the model comprised condensed water and w/o emulsion phases [81], as shown schematically in Figure 4 with the proposed change in the wettability of the reservoir either side of the steam front being indicated. Incomplete removal of oil from the mineral surfaces will modify the original oil-wet characteristics, most likely resulting in a mixed-wet condition behind the steam/condensed water front. Ahead of the front, mobilization of the oil together with water entrainment are shown.

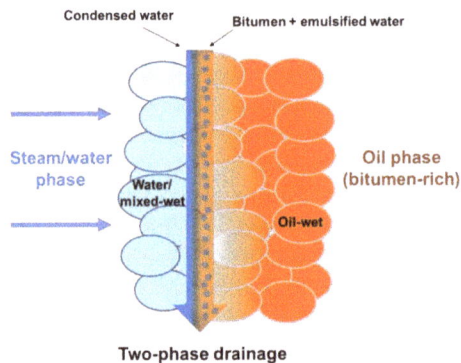

Figure 4. Schematic of drainage and the formation of w/o emulsions at the steam chamber walls (modified from Sasaki et al. [81]).

3.2.2. What Happens at the Solid/Liquid Interface? Displacement/Detergency/Wetting

As mentioned earlier, Equation (1) indicates that a reduction in oil saturation will lead to an improvement in recovery rate. Although not necessarily a primary consideration for improving SAGD efficiency, more effective oil separation from reservoir surfaces could be brought about by the classic roll-up detergency mechanism (see below). This occurs through bitumen de-wetting the solid surface and is caused by a reduction of water/solid and water/bitumen interfacial tensions, with a consequent increase in the contact angle θ between the bitumen and solid. This is depicted in Figure 5 for the equilibrium contact between bitumen and a solid surface surrounded by water. Interfacial tensions (energies) are indicated between the different pairs of phases. By considering the balance of forces at the three-phase contact point, this criterion is met if the water/solid and water/bitumen interfacial tensions are reduced—usually arising upon the addition of a suitable surfactant.

Young's equation for the situation shown in Figure 5 is given by

$$\gamma_{\text{solid/water}} = \gamma_{\text{solid/bitumen}} + \gamma_{\text{water/bitumen}} \cos\theta \tag{3}$$

so that if the surfactant reduces both $\gamma_{\text{solid/water}}$ and $\gamma_{\text{water/bitumen}}$, then $\cos\theta$ will change according to

$$\cos\theta = \frac{\gamma_{\text{solid/water}} - \gamma_{\text{solid/bitumen}}}{\gamma_{\text{water/bitumen}}} \tag{4}$$

Since $\gamma_{\text{water/bitumen}}$ will generally be affected to a greater extent than $\gamma_{\text{solid/water}}$, θ will show a tendency to increase, resulting in release of the oil from the surface, consistent with classical detergency theory. Using the same considerations, interfacial displacement by thin film spreading agents (TFSAs, see later) would lead to similar recovery improvements.

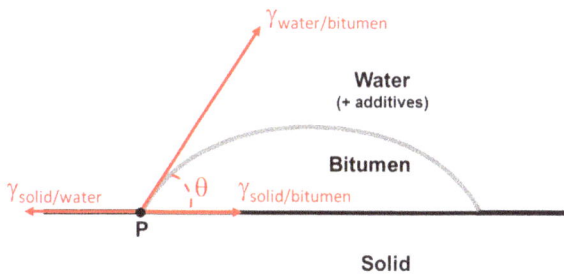

Figure 5. Schematic depiction of the three-phase contact point P involving a bitumen drop on a solid surface surrounded by an aqueous additive solution.

The effect of steam alone has also been demonstrated to improve permeability. A limestone core with permeability impaired as a result of asphaltene deposition was treated using steam (at 263 °C), which improved the core permeability by an average approximately 95% [82]. A similar result was reported for steam treatment in an Oman field [83].

Surface modification has also been shown to give promising results for oil recovery in laboratory and field tests [84]. Contact angle measurements showed that deposition of fine calcium carbonate ($CaCO_3$) particles on sand substantially increased water-wettability, and subsequent laboratory corefloods demonstrated that $CaCO_3$ deposition improved oil recovery. This approach was then trialed in a CSS field test at Elk Point in Alberta, Canada. Oil production (both oil cut and oil rate) had declined over four cycles (see Figure 2) from 87 vol % oil at 30.7 m^3/day to 17 vol % oil at 3.6 m^3/day, which was believed to be a result of decreasing water wettability [84]. Therefore, prior to the fifth cycle, the near-wellbore region was treated by deposition of a thin layer of $CaCO_3$ particles. This resulted in an increase in oil cut and oil rate (23 vol % and 8.2 m^3/day, respectively), and a corresponding reduction in steam/oil ratio (from 22.9 to 8.5) [84].

On the other hand, in a SAGD process, changing the wettability around the production well from water-wet to oil-wet was shown to enhance bitumen recovery [85]. Laboratory experiments and numerical simulation suggested that such a wettability change can reduce the fluid communication time (between injection and production wells), and increase the bitumen production rate, at least during the early stages of production [85]. An oil-wet production well zone can also partially prevent water production. Additionally, the simulation study also suggested that it may not be advantageous to maintain an oil-wet production zone during the full production period [86].

On the basis of these examples, it is evident that different impacts of wettability on steam-based thermal recovery seem to be dependent on, e.g., the thermal production method (SAGD versus CSS), stage of production (early stage, plateau stage or late stage), etc. In CSS, the steam injection and oil/water production is via the same well. For SAGD, the steam is injected via an injection well while the oil is produced from a separate production well.

Increasing the pH can reduce bitumen–rock interactions. Thus, for example, a certain pH can be reached where bitumen still adheres to the rock surface, but the low contact angle of the original bitumen on the surface increases to form drops, as exemplified in Figure 6 which shows the roll-up of bitumen on sand. The sequence shown in Figure 6 shows the bitumen gradually (over a period of several minutes) dewetting the sand surface (a → d) and eventually being liberated as droplets. This has been the main principle underlying flotation treatment used in bitumen recovery from mined oil sands. In the case of alkali treatment, it has been found that the type of alkali used during this process is important. Thus, Flury et al. [87] found that ammonium hydroxide (NH_4OH), although a weak base, is remarkably more effective than sodium hydroxide (caustic soda; NaOH) in separating bitumen from the rock. This was ascribed to a more hydrophobic surface being generated with NH_4OH, as well as a lower (negative) zeta potential of bitumen due to the release of lower concentrations of natural surfactants [87].

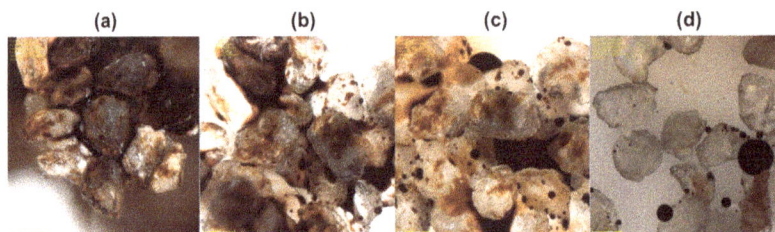

Figure 6. Sequence (**a–d**) of images showing roll-up and detachment of bitumen coating on sand (bars represent 200 microns). This effect can be brought about in a number of different ways, including chemical addition and pH change. Unpublished images from the author's laboratory.

4. Chemical Additives in Steam-Based Thermal Recovery

Over the years, several types of additive have been evaluated in conjunction with steam to try to improve oil recovery. Different mechanisms are applicable to each of these which are summarized in Table 1.

Table 1. Summary of additive classes and relevant recovery mechanisms.

Additive Type	Mechanisms
Solvents	Dissolution of oil; viscosity reduction
Alkaline agents	Interfacial tension reduction by activating indigenous surface active oil components; displacement and emulsification of oil droplets
Surfactants	Foaming—steam (and non-condensable gas) mobility control Interfacial tension reduction—oil displacement by surface dewetting, followed by emulsification of oil droplets Oil displacement by microemulsion formation Oil displacement by thin-film spreading agents Demulsification of w/o emulsions
Nanoparticles	Enhanced spreading and wettability modification Interfacial tension reduction Oil upgrading (catalytic nanoparticles) leading to viscosity reduction

The earliest uses of additives in conjunction with steam, ranging from the late 1960s through to the early 1990s, were concerned with reducing the propensity for w/o emulsions to be formed; the application of ppm concentrations of chemical demulsifiers was shown to be beneficial in this respect and led to increased production and reduced post-production treatment. Castanier and Brigham [88] summarized data from 16 field tests involving the use of additives to improve steam-based recovery during the 1980s. There was a reasonable amount of activity during this period, and overall, the inclusion of additives was positive. CSS operations used alkali and TFSAs, with surfactants being used in steam drives and cyclic applications and injected continuously, in slugs or cyclically. The mechanisms considered to be applicable in these cases include:

- Detergency, including dissolution of asphaltenes, reducing oil/water interfacial tension, and modification of the wetting characteristics of the rock.
- Foam-assisted diversion of the steam to unswept areas of the reservoir.

Castanier and Brigham also made the point that both types of mechanism could be combined to optimize recovery [88], although no tests had been conducted at that time (or hitherto, to the present author's knowledge) to test that suggestion. Almost exclusively, sulfonate surfactants from various sources were used. The use of alkali also appeared to be successful.

Shedid and Abbas considered the recovery of Egyptian heavy oil using vertically [89] and horizontally [90] configured steamflooding in laboratory sandpacks. The effects of alkali and nonionic surfactant (Triton X-100, an ethoxylated octyl phenol) were found to be very similar under the experimental conditions, although the combination of alkali and surfactant was found to be most effective in enhancing the overall recovery [89].

Yamazaki et al. used a laboratory rig to evaluate the use of steam in conjunction with solvents or alkali to enhance recovery from Athabasca oil sand at temperatures up to 200 °C [91]. The injection of solvents with steam was found to improve bitumen recovery, with benzene producing the best effect. The addition of sodium silicate (an alkaline agent, used at 2000 ppm) improved recovery over steam alone, and it also lowered the extraction temperature (from 90 °C in the case of pure steam to 70 °C). These effects might be a consequence of a reduction in interfacial tension. In more recent studies using alkaline flooding alone, Madhaven and Mamora obtained improved recovery of Duri heavy oil compared with San Ardo [92]. Significantly, the interfacial tensions of the respective oils measured against 5 wt % sodium hydroxide solution were 9.5 and 26.5 mN/m, to be compared with approximately 0.04 mN/m reported by Abbas and Shedid [90]; the incremental oil recoveries found by the former workers therefore appear to indicate that a significant reduction in interfacial tension is necessary to improve recovery.

Simple nitrogen-containing compounds were previously utilized by Brown et al., for which they were considered to act as interfacial tension reducers [93]. Introduction of aqueous solutions of a number of these compounds into the steam flow-line was shown to have a beneficial effect on recovery when used in conjunction with alkaline solutions. Thus, compounds such as nitroanisole and various pyridines and quinolines were shown to lead to improvements. However, in the only specific example given, the use of quinoline in aqueous 0.02 mol/L sodium hydroxide was compared with steam only, thereby not unequivocally accounting for the possibility of additional effects arising from the alkaline conditions. No explanation or mechanism for the behavior was proposed, nor comments made regarding the additive stability under the test conditions. In other studies, however, it has been shown that pyridine and quinoline are stable in pure water under aquathermolysis conditions [94], but it is possible that they will be more reactive under alkaline conditions, thereby generating other species that could be involved mechanistically.

Many steam-based enhanced recovery approaches suffer from conformance problems resulting from the oil-bearing regions being bypassed as a result of gravity segregation and viscous fingering. Steam sweep efficiency can be improved by the introduction of more viscous foams [95].

Recognizing that the oil can affect foam stability (e.g., see [60]), Dilgren and Owens evaluated various olefin sulfonates for their ability to sustain steam-containing foams and thereby aid oil recovery by flowing steam and non-condensable gas vertically upward through a sandpack [96]. Pressure drops recorded in the sandpack increased substantially, indicative of the influence of foam, with preferred steam foam-forming compositions comprising specific olefin sulfonates and electrolyte.

Maini and Ma conducted some of the most comprehensive evaluations of the stability of foams at high temperatures and pressures, as well as determining the corresponding thermal stability of the surfactants [97]. Novosad et al. considered adsorption losses to sand of two of the contemporary candidate foam-producing surfactants available at the time (1980s) that were previously shown to be thermally stable [98]. Results obtained at 50–150 °C indicated that losses through adsorption on sand were relatively low (<1 mg/g).

4.1. The Emerging Role of Nanotechnology

In recent years, the role of nanotechnology in enhancing oil recovery has been given serious consideration. The review articles by Cheraghian and Hendraningrat [99,100], Negin et al. [101] and Hashemi et al. [102] provide appropriate background as far as the present discussion is concerned. Most specifically, however, in the context of heavy oil and bitumen recovery, the application of a relatively new group of additives based on colloidally-dispersed inorganic oxide nanoparticles, known as nanofluids, is currently being explored. The dispersed nanoparticles exhibit solid-like ordering at the edge of the spreading fluid [103] as well as affecting the wettability of solids [104,105] and reducing oil/water interfacial tension [105], making them potential candidates for the displacement of oil from surfaces [106]. This occurs as a result of a disjoining pressure being created by the build-up of a nanoparticle wedge between the oil and the solid surface [104], as shown in Figure 7. Shrinkage of the oil/solid contact line on a microscopic level effectively increases the macroscopic contact angle, thereby displacing the oil from the surface.

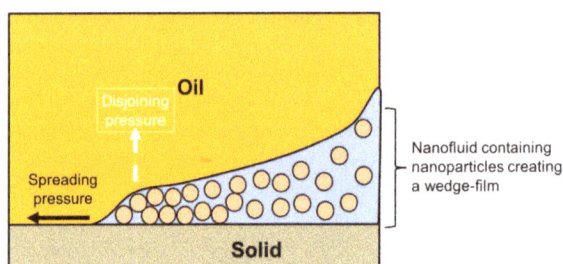

Figure 7. Sketch showing nanoparticle structuring into a wedge-film at the oil/solid/nanofluid spreading front (adapted from Wasan et al. [104]).

Other types of nanoparticles, including organic polymers and metals, have also been studied, the latter because of their associated catalytic properties [102], but oxides such as SiO_2, ZrO_2, Al_2O_3, and TiO_2 offer suitable surface chemistry and thermal stability properties that are potentially more appropriate for displacing heavy oil. Titanium dioxide (in both amorphous form and as crystalline anatase), for instance, has been shown to improve heavy oil recovery from sandstone cores (from 49% recovery using water/brine to 80% using 0.01% anatase under equivalent conditions) [107]. Silica nanoparticles have been shown to be very effective in altering the wettability of calcite [108] and sandstone [109] from oil-wet to water-wet, the change being generally deemed to be beneficial for oil recovery under certain conditions, as discussed above. Zirconium oxide nanoparticles have also been shown to be promising for enhancing oil recovery from carbonate [110].

In addition to the mechanisms indicated above (i.e., production of a structural disjoining pressure [111], wettability alteration, and interfacial tension reduction), the comprehensive review by Sun et al. emphasizes that asphaltene precipitation can also be reduced or prevented in the presence of nanoparticles [112].

As alluded to above, heavy oil and bitumen can be upgraded as a result of reactions, such as aquathermolysis, leading to a reduction in viscosity. A further example of this is the demonstration that nickel nanoparticles applied in a CSS process promoted aquathermolysis and improved oil recovery [113], with a variety of other metal particles providing evidence for heavy oil/bitumen viscosity reduction [114]. More detailed discussions of the application of nanoparticles in in situ upgrading and heavy oil recovery are given elsewhere [115–117].

4.2. Additive Requirements for SAGD

The majority of the above discussions relate primarily to the use of additives in traditional steam stimulation processes, which differ significantly from SAGD; in either situation, however, selection of a particular additive or combination of additives very much depends on the functions required for the specific application. As has been described above, various scenarios are possible—from aiding displacement of the oil from reservoir matrix surfaces to emulsifying a proportion of it, or even upgrading it.

Equation (1) provides some insight into the practical factors associated with additive selection. All the additives and mechanisms described so far can conceivably lead to improved SAGD performance and oil production rates. Thus, for example, surfactant additives favoring oil dispersion in water (rather than water dispersion in oil) will ensure that the effective viscosity of the draining fluids is minimized. This can be achieved using water-soluble surfactants or w/o demulsification chemicals.

Additives that act to maximize the displacement of entrained oil will thereby open up the reservoir porosity. Thin film spreading agents (TFSAs) offer one approach in which surfactants facilitate oil displacement by a combination of displacement/detergent action and emulsification. These additives are also considered to reduce the residual oil saturation, hence increasing the ΔS_0 term in Equation (1). The basis of TFSAs, as proposed and championed by Charles Blair (Magna Corp., Santa Fe, CA, USA) in the late 1970s [118], is the removal of surface films from rock surfaces. Most reservoir minerals in their natural state exhibit hydrophilic characteristics, yet in contact with crude oil the wettability of their surfaces is modified through adsorption of polar species, including asphaltenes and naphthenic acids. TFSAs comprise highly surface-active polymeric species which are believed to act by spreading readily over the surface and displacing the organic contaminants responsible for oil wetting. The molecular compositions of TFSAs are similar to demulsifiers used to resolve crude oil emulsions [119], and it is therefore unsurprising that the two modes of action are similar [120].

If additives are to be deployed on a continuous basis, they should ideally be introduced with the steam, either co-vaporized or possibly dispersed directly into the steam flow. Low volatility surfactants, for example, would fall into the latter category, whereas more volatile, lower molecular weight candidates will be able to form a homogeneous admixture with the steam, and will also co-condense within the cooler parts of the formation. In addition, although it is not necessarily the most efficient approach, squeezing low volatility additives into the near-wellbore region prior to steaming could still allow sufficient carryover into the growing steam chamber.

In the particular case of SAGD, the additives should retain a major proportion of their (surface) activity upon reaching the steam chamber; thereafter, the level of activity retained will depend on the function they are required to fulfill during further propagation of the chamber. Thermal stability is therefore a significant factor governing the selection of suitable additives. In general, most compounds subjected to high temperature and possibly high pH hydrolytic conditions will be susceptible to some chemical transformation. The key, therefore, is to identify kinetic stability in compounds for which thermodynamic stability may be limited.

Zeidani and Gupta [121] described the criteria for surfactants suitable for SAGD applications as

- Effective at reducing IFT
- Changes the reservoir wettability to water-wet
- Capable of being vaporized under SAGD operational conditions

- Thermally stable under SAGD conditions
- Able to stabilize oil-in-water emulsions
- Fully compatible with formation water

These researchers also suggested that certain nonionic surfactants possess good attributes as candidates for SAGD applications. One such class of surfactants advocated are tertiary acetylenic diols. Laboratory tests demonstrated that the surfactants were able to increase the oil recovery factor by 6–16% [121,122].

As suggested, the pH conditions will play an important part in determining the relative stability of certain compounds. Knowledge of the likely reservoir conditions to which the additives will be exposed would be helpful in this respect. Potentially, an experimental ex situ assessment of these conditions could be made in laboratory tests under appropriate temperature and pressure conditions. Generally, neutral-to-high pH conditions are preferable as far as hydrolytic stability is concerned, with acidic conditions largely to be avoided. Recognizing this, Srivastava et al. [123,124] targeted steam-volatile chemical additives to enhance oil recovery under SAGD conditions. These materials are claimed to possess advantages over alkali and surfactants in that they are more easily carried throughout the reservoir with steam [125]. These additives react with, and neutralize, indigenous acid species in the oil, generating surface-active molecules which facilitate the formation of o/w emulsions that are more efficiently co-produced with the condensed water. Laboratory tests typically showed approximately 13% increases in cumulative recovery.

There have also been claims of beneficial effects for oil recovery being obtained using other low molecular weight chemical additives not normally considered in the oil recovery context. Intriguingly, Campos and Hernandez found that the introduction of urea (H_2NCONH_2) in steamflooding reduced oil viscosity [126]. Under hydrothermal conditions, urea is expected to decompose to ammonia and carbon dioxide according to [127]

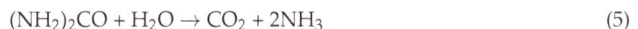

$$(NH_2)_2CO + H_2O \rightarrow CO_2 + 2NH_3 \tag{5}$$

Tests using Hamaca and Cerro Negro bitumen produced a reduction in viscosity of greater than 90% by the addition of 2% urea (relative to steam) at 192 °C. In the field, the steam and urea could be mixed at the surface or downhole in the well, or the urea "can be fed to the boiler feed water for making the steam" [126].

In a patent application, urea has also been proposed as a thermal-trigger agent to accelerate communication between injection and production wells in SAGD [128]. It is likely that a pH increase initiated by release of ammonia under the high temperature aqueous conditions (as in Equation (5)) causes ionization of acid groups on exposed bitumen surfaces in the near-wellbore region, resulting in bitumen de-wetting of the hydrophilic sand surfaces and ultimately release of emulsified bitumen droplets.

In a very recent study [129], various additives were co-injected with steam in order to test their impact on SAGD efficiency. The additives tested included surfactants, ionic liquids, biodiesel, high pH solutions (using sodium metaborate, $NaBO_2$) and silica nanoparticles. The results showed that biodiesel leads to increased oil recovery and reduced the steam/oil ratio during gravity-driven flow, but water consumption was increased. Biodiesel, ionic liquids, and silica nanoparticles showed promising results, but further verification is needed.

One particular approach in the above list which makes specific use of colloid and interfacial chemistry takes advantage of the activation of latent emulsifiers in heavy oils, principally carboxylic (naphthenic and fatty) acids. The effects of alkaline flooding have been described by Liu et al. (using Na_2CO_3) [130], Dong et al. (Na_2CO_3, NaOH) [131] and Pei et al. (Na_2CO_3) [132] under conditions of low interfacial disturbance. These alkaline agents operate by reducing the bitumen/water interfacial tension to very low values through ionization of interfacial carboxylic acid (HA) groups to carboxylate, A^-), as shown schematically in Figure 8. The situation is made more complex due to partitioning as well as ionization of species at the interface. The scheme shown in Figure 8 is an over-simplification as it does not allow for equilibria involving the different sources of hydroxide ions as indicated above, nor the potential role of cations.

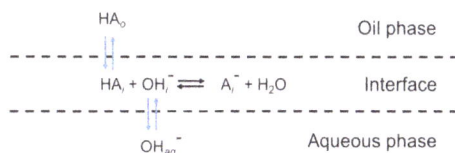

Figure 8. Partitioning of species responsible for interfacial tension reduction at the oil/water interface.

The creation of the negative surface charge (located in the interfacial region in Figure 8) also helps to stabilize bitumen droplets in aqueous media. As indicated above, some recent studies have also used $NaBO_2$ as the alkaline agent [133] which led to 27% improvement in recovery of a Chinese heavy oil in sandpack tests. Our own work has also considered this alkaline agent as it is more tolerant to divalent metal ions than the corresponding carbonates. Figure 9 shows the effects of varying the $NaBO_2$ concentration on Athabasca bitumen emulsions formed under low intensity agitation in our laboratory. Using higher $NaBO_2$ concentrations produced inferior emulsification, which was identified as being due to the associated higher ionic strength. This is demonstrated in Figure 10 where it can be seen that changing the ionic strength (using NaCl) for a constant $NaBO_2$ concentration results in reduced emulsification, indicating the likely role of the electric double layer in stabilizing the emulsions.

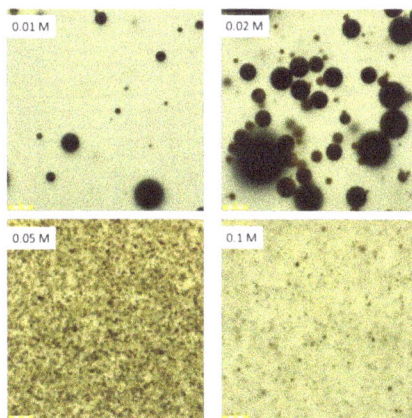

Figure 9. Effect of $NaBO_2$ concentration on low intensity Athabasca bitumen emulsification. The bars represent 50 microns.

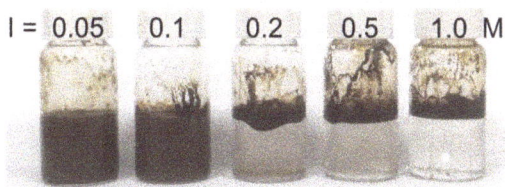

Figure 10. Demonstrating the effect of ionic strength (NaCl) on 0.05 M $NaBO_2$-stabilized Athabasca bitumen-in-water emulsions.

The most common method used for studying steam-based bitumen recovery involves coreflood tests using real core or artificial sandpacks. However, a novel microfluidic approach has recently been applied to the examination of the SAGD process, where it was used to evaluate the role of an alkaline steam-based additive in SAGD, which enabled emulsions, both w/o and o/w, formed at the pore-scale to be visualized and quantified [134]. The o/w droplet sizes are seen to be significantly reduced in the presence of the additive compared to the baseline pure steam case and the (Cold Lake, AB, Canada) bitumen recovery is increased by 50%. Also interesting was the observation of tighter w/o emulsions using additized steam. The formation of w/o emulsions of this type has been suggested as being an additional recovery mechanism, perhaps applicable to SAGD, since the resultant expansion of the oil phase could increase the effective contact angle between bitumen and the solid surface [135].

The use of foams has recently been explored by Chen et al. in a process they term foam-assisted SAGD (FA-SAGD) [136]. The proposed process involves co-injection of surfactant solution with steam (continuously or intermittently) to improve the uniform development of steam chambers in heterogeneous reservoirs. As determined by numerical simulation, the results indicated opposing effects of steam foam: on the one hand, the presence of foam impairs gravity drainage and reduces the oil production rate during SAGD plateau production, whereas on the other, improved energy efficiency leads to overall improved performance.

Foams have been used for many years in steam-based recovery processes, and therefore significant experience has been built up in their deployment and selection of foaming agent. To date, however, no experimental FA-SAGD results have been reported to the author's knowledge, but Chen acknowledged that foam stability in the presence of oil is "a potential unknown quantity" [136].

5. Composition of Produced Water

From an environmental standpoint, produced water from thermal recovery operations has long been one of the most challenging aspects of heavy oil and bitumen production in which colloidal and interfacial factors play important roles. Not only are high concentrations of dissolved inorganic species (such as silica and other dissolved solids) present as a result of the high steam temperatures, but a considerable amount of small oil droplets (see Figure 3a) and dissolved organic matter can also accumulate. For different reasons, these species cause problems for recycling as boiler feed water (resulting in silicate scale formation) [137] and water disposal (owing to the toxicity of the organic species present) [52,138,139]. Control of silicate boiler scale formation is therefore a necessity, which currently employs different softening strategies [140].

Additionally, the existence of mineral fines in the bitumen produced by SAGD has been found to reduce the performance of surfactant-based demulsifiers [141]. Surface-active species from the bitumen adsorb onto the high surface area fines (comprising clays and quartz), making the solids more oil-wet. As a result, adhesion of the solids to the oil droplets is enhanced and the separation of bitumen from water becomes more difficult [141].

Two-dimensional GC-ToFMS analysis has indicated that there is a broad range of dissolved organic compounds in the SAGD produced water, and that the most prevalent constituents among hetero-species (which might possess significant surface activity) include methyl- and ethyl-phenols [142]. Characterization based on FT-ICR high resolution mass spectrometry also showed that the SAGD produced water contains acidic as well as basic species [143]. Razi et al. studied the dynamic interfacial tension between diluted bitumen and SAGD produced water and identified interactions between the polar components in the bitumen and the humic acids in SAGD produced water at basic pH (9–10) relevant to SAGD operations [144].

Under these process conditions, the effective separation of the different components is essential, and requires appropriate colloid chemical treatment strategies. As detailed in the patent and scientific literature, the use of cationic flocculating agents provides the basis for precipitating solid and bitumen contaminants, which can be removed, for example, using flotation [145] or filtration [146]. A more general review is given by Gupta et al. [147].

6. General Discussion and Conclusions

The scope of this review has, by necessity, been wide-ranging. It has attempted to apply colloid and interface science concepts to high temperature (steam-based) environments. In so doing, various examples of the use of steam-based additives in thermal recovery processes for heavy oil or bitumen recovery have been identified. When discussing the use of additives in SAGD, for example, it is necessary to consider how they will be introduced into the reservoir, as well as their chemical stability under the relatively harsh hydrothermal conditions.

For the present purposes, the additives have been divided into different classes based on their contributions to different proposed recovery mechanisms. Much of the reported work has been extrapolated from conventional crude oil enhanced oil recovery. Thus, surfactant-based processes utilizing products developed for tertiary recovery of lighter oils have also been directed at heavier oils. Anionic surfactants, usually sulfonates, have been the most common commercial products evaluated. Sulfonates represent the most thermally stable class of anionic surfactants, owing to the stability of the C–S bond; in contrast, C–O bonds are susceptible to hydrolysis under high temperature aqueous conditions, more typical of steamflooding. Sulfonates are particularly stable at high temperatures under alkaline pH conditions. Therefore, when used in alkaline surfactant steam floods, for example, sulfonate surfactants should be chemically stable.

As discussed at some length, based on the interfacial chemistry of the reservoir, the principal requirements of additives to enhance hydrocarbon recovery are to increase water wettability and reduce oil/water interfacial tension. To date, transport of additives into the reservoir has been directly via the steam flow, which is particularly applicable to water-soluble species (e.g., [148,149]), although, remarkably, some higher molecular weight surfactants have been advocated as being volatile under SAGD conditions [121,122]. More attractive, perhaps, is the potential use of volatile additives that can be co-vaporized with the steam, which would then result in improved uniformity of distribution throughout the formation, e.g., within the steam chamber in SAGD [134,149]. Suitable species for this purpose are generally low molecular weight, with amines, including ammonia being described [125]. In principle, these additives satisfy the two effects mentioned above owing to their ability as bases to activate interfacial surface-active species, thereby reducing interfacial tension and increasing the water wettability of the system [149]. In addition, these additives would direct the formation of the preferred lower viscosity o/w emulsions. In another example [150], co-functionalizing amines with glycol ethers combines basicity with water-wetting properties, although at the expense of volatility.

Only relatively minor attention has been given to the application of nonionic surfactants in heavy oil thermal recovery. However, in a very recent investigation, Sunnatov [151] co-injected the nonionic surfactant Triton X-100 with steam (200 °C, 100 psig) in horizontal and vertical laboratory test cells. Although the stability of the surfactant was not specifically considered, its addition was shown to produce an average incremental (20 °API) oil recovery of 7% of the original oil in place compared with steam alone. This suggests that this surfactant survived sufficiently well under the test conditions to produce a positive effect.

As will be apparent from the foregoing, much of the literature that has focused on heavy oil and bitumen thermal recovery has been based on experience derived from conventional enhanced oil recovery. The key differences have been considered throughout the above discussions, but the underlying colloid and interface chemistry principles involved are similar. In the case of the heavier hydrocarbons, however, there is an overriding environmental driver behind attempts to improve recovery. The efficiency of steam-based thermal recovery methods is measured in terms of the steam/oil ratio (SOR) as described in Section 1. The use of additives, whether solvent-based or relying on colloid/interfacial principles, is seen as an important approach to reducing the SOR, thereby improving thermal recovery technology, and in particular SAGD.

Acknowledgments: This review builds on a heavy oil project initiated in 2012 within the Centre for Petroleum and Surface Chemistry under the original auspices of BP America. Present and past colleagues from that project are thanked for their insight into a technological area that is rich in colloids and interfaces. A special acknowledgement must be given to Huang Zeng, who provided extensive scientific support and stimulating discussions throughout the project. Support from the Department of Chemistry, University of Surrey, enabling continuation of the work, is also acknowledged.

Conflicts of Interest: The author declares no conflicts of interest.

References

1. Gates, I.D.; Wang, J.Y.J. Evolution of In Situ Oil Sands Recovery Technology in the Field: What Happened and what's New? In *SPE Heavy Oil Conference and Exhibition*; Society of Petroleum Engineers: Kuwait City, Kuwait, 2011.
2. Gates, I.D.; Chakrabarty, N. Optimization of Steam Assisted Gravity Drainage in McMurray Reservoir. *J. Can. Pet. Technol.* **2006**, *45*, 54–62. [CrossRef]
3. Freitag, N.P.; Kristoff, B.J. Comparison of Carbon Dioxide and Methane as Additives at Steamflood Conditions. *SPE J.* **1998**, *3*, 14–18. [CrossRef]
4. Speight, J.G. *Enhanced Recovery Methods for Heavy Oil and Tar Sands*, 1st ed.; Gulf Publishing Company: Houston, TX, USA, 2009; ISBN 978-1-933762-25-8.
5. Holly, C.; Mader, M.; Soni, S.; Toor, J. *Alberta Energy Technical Services Alberta Energy: Oil Sands Production Profile 2004–2014*; Alberta Provincial Government: Edmonton, AB, Canada, 2016.
6. Alberta Provincial Government Alberta Oil Sands Industry Quarterly Update: Summer 2017. Available online: http://www.albertacanada.com/business/statistics/oil-sands-quarterly.aspx (accessed on 25 March 2018).
7. McCormack, M.E. Design of Steam-Hexane Injection Wells for Gravity Drainage Systems. *J. Can. Pet. Technol.* **2009**, *48*, 22–28. [CrossRef]
8. Zajic, J.E.; Cooper, D.G.; Marshall, J.A.; Gerson, D.F. Microstructure of Athabasca Bituminous Sand by Freeze-Fracture Preparation and Transmission Electron Microscopy. *Fuel* **1981**, *60*, 619–623. [CrossRef]
9. Takamura, K. Microscopic Structure of Athabasca Oil Sand. *Can. J. Chem. Eng.* **1982**, *60*, 538–545. [CrossRef]
10. Hall, A.C.; Collins, S.H.; Melrose, J.C. Stability of Aqueous Wetting Films in Athabasca Tar Sands. *Soc. Pet. Eng. J.* **1983**, *23*, 249–258. [CrossRef]
11. He, L.; Li, X.; Wu, G.; Lin, F.; Sui, H. Distribution of Saturates, Aromatics, Resins, and Asphaltenes Fractions in the Bituminous Layer of Athabasca Oil Sands. *Energy Fuels* **2013**, *27*, 4677–4683. [CrossRef]
12. Montgomery, W.; Sephton, M.; Watson, J.; Zeng, H. The Effects of Minerals on Heavy Oil and Bitumen Chemistry When Recovered Using Steam-assisted Methods. In Proceedings of the SPE Heavy Oil Conference-Canada, Calgary, AB, Canada, 10–12 June 2014.
13. Hongfu, F.; Yongjian, L.; Liying, Z.; Xiaofei, Z. The Study on Composition Changes of Heavy Oils during Steam Stimulation Processes. *Fuel* **2002**, *81*, 1733–1738. [CrossRef]
14. Fan, H.-F.; Liu, Y.-J.; Zhong, L.-G. Studies on the Synergetic Effects of Mineral and Steam on the Composition Changes of Heavy Oils. *Energy Fuels* **2001**, *15*, 1475–1479. [CrossRef]
15. Clark, P.; Hyne, J. Steam-Oil Chemical Reactions; Mechanisms for the Aquathermolysis of Heavy Oils. *AOSTRA J. Res.* **1984**, *1*, 15–20.
16. Montgomery, W.; Court, R.W.; Rees, A.C.; Sephton, M.A. High Temperature Reactions of Water with Heavy Oil and Bitumen: Insights into Aquathermolysis Chemistry during Steam-Assisted Recovery. *Fuel* **2013**, *113*, 426–434. [CrossRef]
17. Jia, N.; Zhao, H.; Yang, T.; Ibatullin, T.; Gao, J. Experimental Measurements of Bitumen−Water Aquathermolysis during a Steam-Injection Process. *Energy Fuels* **2016**, *30*, 5291–5299. [CrossRef]
18. Mohammadmoradi, P.; Taheri, S.; Kantzas, A. Interfacial Areas in Athabasca Oil Sands. *Energy Fuels* **2017**, *31*, 8131–8145. [CrossRef]
19. Wiehe, I.A.; Yarranton, H.W.; Akbarzadeh, K.; Rahimi, P.M.; Teclemariam, A. The Paradox of Asphaltene Precipitation with Normal Paraffins. *Energy Fuels* **2005**, *19*, 1261–1267. [CrossRef]
20. Acevedo, S.; Ranaudo, M.A.; García, C.; Castillo, J.; Fernández, A. Adsorption of Asphaltenes at the Toluene−Silica Interface: A Kinetic Study. *Energy Fuels* **2003**, *17*, 257–261. [CrossRef]

21. Montoya, T.; Coral, D.; Franco, C.A.; Nassar, N.N.; Cortés, F.B. A Novel Solid−Liquid Equilibrium Model for Describing the Adsorption of Associating Asphaltene Molecules onto Solid Surfaces Based on the "Chemical Theory". *Energy Fuels* **2014**, *28*, 4963–4975. [CrossRef]

22. Wang, S.; Liu, Q.; Tan, X.; Xu, C.; Gray, M.R. Adsorption of Asphaltenes on Kaolinite as an Irreversible Process. Colloids Surfaces A: Physicochem. *Eng. Aspects* **2016**, *504*, 280–286. [CrossRef]

23. Tsiamis, A.; Taylor, S.E. Adsorption Behavior of Asphaltenes and Resins on Kaolinite. *Energy Fuels* **2017**, *31*, 10576–10587. [CrossRef]

24. Rane, J.P.; Harbottle, D.; Pauchard, V.; Couzis, A.; Banerjee, S. Adsorption Kinetics of Asphaltenes at the Oil–Water Interface and Nanoaggregation in the Bulk. *Langmuir* **2012**, *28*, 9986–9995. [CrossRef] [PubMed]

25. Acevedo, S.; Borges, B.; Quintero, F.; Piscitelly, V.; Gutierrez, L.B. Asphaltenes and Other Natural Surfactants from Cerro Negro Crude Oil. Stepwise Adsorption at the Water/Toluene Interface: Film Formation and Hydrophobic Effects. *Energy Fuels* **2005**, *19*, 1948–1953. [CrossRef]

26. Kilpatrick, P.K. Water-in-Crude Oil Emulsion Stabilization: Review and Unanswered Questions. *Energy Fuels* **2012**, *26*, 4017–4026. [CrossRef]

27. McLean, J.D.; Kilpatrick, P.K. Effects of Asphaltene Aggregation in Model Heptane-Toluene Mixtures on Stability of Water-in-Oil Emulsions. *J. Colloid Interface Sci.* **1997**, *196*, 23–34. [CrossRef] [PubMed]

28. McLean, J.D.; Kilpatrick, P.K. Effects of Asphaltene Solvency on Stability of Water-in-Crude-Oil Emulsions. *J. Colloid Interface Sci.* **1997**, *189*, 242–253. [CrossRef]

29. Brient, J.A.; Wessner, P.J.; Doyle, M.N. Naphthenic Acids. *Kirk-Othmer Encycl. Chem. Technol.* **1995**. [CrossRef]

30. Headley, J.V.; McMartin, D.W. A Review of the Occurrence and Fate of Naphthenic Acids in Aquatic Environments. *J. Environ. Sci. Health Part A* **2004**, *39*, 1989–2010. [CrossRef]

31. Varadaraj, R.; Brons, C. Molecular Origins of Heavy Crude Oil Interfacial Activity. Part 2: Fundamental Interfacial Properties of Model Naphthenic Acids and Naphthenic Acids Separated from Heavy Crude Oils. *Energy Fuels* **2007**, *21*, 199–204. [CrossRef]

32. Smith, D.F.; Schaub, T.M.; Rahimi, P.; Teclemariam, A.; Rodgers, R.P.; Marshall, A.G. Self-Association of Organic Acids in Petroleum and Canadian Bitumen Characterized by Low- and High-Resolution Mass Spectrometry. *Energy Fuels* **2007**, *21*, 1309–1316. [CrossRef]

33. Varadaraj, R.; Brons, C. Molecular Origins of Heavy Oil Interfacial Activity. Part 1: Fundamental Interfacial Properties of Asphaltenes Derived from Heavy Crude Oils and Their Correlation to Chemical Composition. *Energy Fuels* **2007**, *21*, 195–198. [CrossRef]

34. Wang, X.; Pensini, E.; Liang, Y.; Xu, Z.; Chandra, M.S.; Andersen, S.I.; Abdallah, W.; Buiting, J.J. Fatty Acid-Asphaltene Interactions at Oil/Water Interface. *Colloids Surfaces A: Physicochem. Eng. Aspects* **2017**, *513*, 168–177. [CrossRef]

35. Wu, X. Investigating the Stability Mechanism of Water-in-Diluted Bitumen Emulsions through Isolation and Characterization of the Stabilizing Materials at the Interface. *Energy Fuels* **2003**, *17*, 179–190. [CrossRef]

36. Stanford, L.A.; Rodgers, R.P.; Marshall, A.G.; Czarnecki, J.; Wu, X.A.; Taylor, S. Detailed Elemental Compositions of Emulsion Interfacial Material versus Parent Oil for Nine Geographically Distinct Light, Medium, and Heavy Crude Oils, Detected by Negative- and Positive-Ion Electrospray Ionization Fourier Transform Ion Cyclotron Resonance Mass Spectrometry. *Energy Fuels* **2007**, *21*, 973–981. [CrossRef]

37. Collins, S.H.; Melrose, J.C. Adsorption of Asphaltenes and Water on Reservoir Rock Minerals. In Proceedings of the SPE Oilfield and Geothermal Chemistry Symposium, Denver, CO, USA, 1–3 June 1983; Society of Petroleum Engineers: Denver, CO, USA.

38. Gonzalez, G.; Travalloni-Louvisse, A.M. Adsorption of Asphaltenes and Its Effect on Oil Production. *SPE Prod. Facil.* **1993**, *8*, 91–96. [CrossRef]

39. Keleşoğlu, S.; Volden, S.; Kes, M.; Sjöblom, J. Adsorption of Naphthenic Acids onto Mineral Surfaces Studied by Quartz Crystal Microbalance with Dissipation Monitoring (QCM-D). *Energy Fuels* **2012**, *26*, 5060–5068. [CrossRef]

40. Madsen, L.; Ida, L. Adsorption of Carboxylic Acids on Reservoir Minerals from Organic and Aqueous Phase. *SPE Res. Eval. Eng.* **1998**, *1*, 47–51. [CrossRef]

41. Osacky, M.; Geramian, M.; Ivey, D.G.; Liu, Q.; Etsell, T.H. Mineralogical and Chemical Composition of Petrologic End Members of Alberta Oil Sands. *Fuel* **2013**, *113*, 148–157. [CrossRef]

42. Sparks, B.D.; Kotlyar, L.S.; O'Carroll, J.B.; Chung, K.H. Athabasca Oil Sands: Effect of Organic Coated Solids on Bitumen Recovery and Quality. *J. Pet. Sci. Eng.* **2003**, *39*, 417–430. [CrossRef]

43. Kotlyar, L.S.; Sparks, B.D.; Woods, J.R.; Chung, K.H. Solids Associated with the Asphaltene Fraction of Oil Sands Bitumen. *Energy Fuels* **1999**, *13*, 346–350. [CrossRef]

44. Guttierez, L.; Pawlik, M. Influence of Humic Acids on Oil Sand Processing. Part I: Detection and Quantification of Humic Acids in Oil Sand Ores and Their Effect on Bitumen Wettability. *Int. J. Min. Proc.* **2014**, *126*, 117–125. [CrossRef]

45. Guttierez, L.; Pawlik, M. Influence of Humic Acids on Oil Sand Processing. Part II: Relationship between Bitumen Extraction, Humic Acids Concentration and Power Draw Measurements on Oil Sand Slurries. *Int. J. Min. Proc.* **2014**, *126*, 126–135. [CrossRef]

46. Kotlyar, L.S.; Ripmeester, J.A.; Sparks, B.D.; Montgomery, D.S. Characterization of Organic-Rich Solids Fractions Isolated from Athabasca Oil Sand Using a Cold Water Agitation Test. *Fuel* **1988**, *67*, 221–226. [CrossRef]

47. Margeson, J.; Hornof, V. Characterization of the Humic-Clay Complex and Its Influence on Bitumen Displacement from Athabasca Oil Sand. *J. Can. Pet. Technol.* **1989**, *28*, 57–62. [CrossRef]

48. Majid, A.; Ripmeester, J.A. Isolation and Characterization of Humic Acids from Alberta Oil Sands and Related Materials. *Fuel* **1990**, *69*, 1527–1536. [CrossRef]

49. Kupai, M.M.; Yang, F.; Harbottle, D.; Moran, K.; Masliyah, J.; Xu, Z. Characterising Rag-Forming Solids. *Can. J. Chem. Eng.* **2013**, *91*, 1395–1401. [CrossRef]

50. Jiang, T.; Hirasaki, G.J.; Miller, C.A.; Ng, S. Effects of Clay Wettability and Process Variables on Separation of Diluted Bitumen Emulsion. *Energy Fuels* **2011**, *25*, 545–554. [CrossRef]

51. Marhaba, T. Fluorescence Technique for Rapid Identification of DOM Fractions. *J. Environ. Eng.* **2000**, *126*, 145–152. [CrossRef]

52. Thakurta, S.G.; Maiti, A.; Pernitsky, D.J.; Bhattacharjee, S. Dissolved Organic Matter in Steam Assisted Gravity Drainage Boiler Blow-Down Water. *Energy Fuels* **2013**, *27*, 3883–3890. [CrossRef]

53. Pillai, R.G.; Yang, N.; Thi, S.; Fatema, J.; Sadrzadeh, M.; Pernitsky, D. Characterization and Comparison of Dissolved Organic Matter Signatures in Steam-Assisted Gravity Drainage Process Water Samples from Athabasca Oil Sands. *Energy Fuels* **2017**, *31*, 8363–8373. [CrossRef]

54. Chung, T.; Bae, W.; Lee, J.; Lee, W.; Jung, B. A Review of Practical Experience and Management of the SAGD Process for Oil Sands Development. *Energy Sources Part Recovery Util. Environ. Eff.* **2011**, *34*, 219–226. [CrossRef]

55. Butler, R.M.; Yee, C.T. Progress in the In Situ Recovery of Heavy Oils and Bitumen. *J. Can. Pet. Technol.* **2002**, *41*. [CrossRef]

56. Jiang, Q.; Thornton, B.; Houston, J.R.; Spence, S. Review of Thermal Recovery Technologies for the Clearwater and Lower Grand Rapids Formations in the Cold Lake Area in Alberta. *J. Can. Pet. Technol.* **2010**, *49*, 2–13. [CrossRef]

57. Al Bahlani, A.M.; Babadagli, T. A Critical Review of the Status of SAGD: Where Are We and What Is Next? In Proceedings of the SPE Western Regional and Pacific Section AAPG Joint Meeting, Bakersfield, CA, USA, 29 March–4 April 2008; Society of Petroleum Engineers: Bakersfield, CA, USA.

58. Gotawala, D.R.; Gates, I.D. Steam Fingering at the Edge of a Steam Chamber in a Heavy Oil Reservoir. *Can. J. Chem. Eng.* **2008**, *86*, 1011–1022. [CrossRef]

59. Upreti, S.R.; Lohi, A.; Kapadia, R.A.; El-Haj, R. Vapor Extraction of Heavy Oil and Bitumen: A Review. *Energy Fuels* **2007**, *21*, 1562–1574. [CrossRef]

60. Chen, Q. Assessing and Improving Steam-Assisted Gravity Drainage: Reservoir Heterogeneities, Hydraulic Fractures, and Mobility Control Foams. Ph.D. Thesis, Stanford University, Stanford, CA, USA, 2009.

61. Acosta, E. Achieving Sustainable, Optimal SAGD Operations. *J. Pet. Technol.* **2010**, *62*, 24–28. [CrossRef]

62. Thimm, H.F. Understanding the Generation of Dissolved Silica in Thermal Projects: Theoretical Progress. *J. Can. Pet. Technol.* **2008**, *47*, 22–25. [CrossRef]

63. Bowman, R.W.; Gramms, L.C.; Craycraft, R.R. High-Silica Waters in Steamflood Operations. *SPE Prod. Facil.* **2000**, *15*, 123–125. [CrossRef]

64. Barillas, J.L.M.; Dutra, T.V.; Mata, W. Reservoir and Operational Parameters Influence in SAGD Process. *J. Pet. Sci. Eng.* **2006**, *54*, 34–42. [CrossRef]

65. Butler, R.M.; McNab, G.S.; Lo, H.Y. Theoretical Studies on the Gravity Drainage of Heavy Oil During In-Situ Steam Heating. *Can. J. Chem. Eng.* **1981**, *59*, 455–460. [CrossRef]

66. Krieger, I.M.; Dougherty, T.J. A Mechanism for Non-Newtonian Flow in Suspensions of Rigid Spheres. *Trans. Soc. Rheol.* **1959**, *3*, 137–152. [CrossRef]
67. Mueller, S.; Llewellin, E.W.; Mader, H.M. The Rheology of Suspensions of Solid Particles. *Proc. R. Soc. Math. Phys. Eng. Sci.* **2010**, *466*, 1201–1228. [CrossRef]
68. Bennion, D.B.; Chan, M.Y.S.; Sarioglu, G.; Courtnage, D.; Wansleeben, J.; Hirata, T. The In-Situ Formation of Bitumen-Water-Stable Emulsions in Porous Media during Thermal Stimulation. In Proceedings of the SPE International Thermal Operations Symposium, Bakersfield, CA, USA, 8–10 February 1993; Society of Petroleum Engineers: Bakersfield, CA, USA.
69. Sarbar, M.; Livesey, D.B.; Wes, W.; Flock, D.L. The Effect of Chemical Additives on the Stability of Oil-in-Water Emulsion Flow through Porous Media. In Proceedings of the 38th Annual CIM Petroleum Society Technical Meeting, Calgary, AB, Canada, 7–10 June 1987.
70. Bosch, R.; Axcell, E.; Little, V.; Cleary, R.; Wang, S.; Gabel, R.; Moreland, B. A Novel Approach for Resolving Reverse Emulsions in SAGD Production Systems. *Can. J. Chem. Eng.* **2004**, *82*, 836–839. [CrossRef]
71. Chung, K.; Butler, A. In Situ Emulsification by the Condensation of Steam in Contact with Bitumen. *J. Can. Pet. Technol.* **1989**, *28*, 48–55. [CrossRef]
72. Chen, T.; Yuan, J.Y.; Serres, A.J.; Rancier, D.G. Emulsification of Water into Bitumen in Co-injection Experiments. *J. Can. Pet. Technol.* **1999**, *38*. [CrossRef]
73. Bryan, J.; Wang, J.; Kantzas, A. Measurement of Emulsion Flow in Porous Media: Improvements in Heavy Oil Recovery. *J. Phys. Conf. Ser.* **2009**, *147*, 012058. [CrossRef]
74. Vittoratos, E. Flow Regimes during Cyclic Steam Stimulation at Cold Lake. *J. Can. Pet. Technol.* **1991**, *30*, 82–86. [CrossRef]
75. Wang, D.; Lin, M.; Dong, Z.; Li, L.; Jin, S.; Pan, D.; Yang, Z. Mechanism of High Stability of Water-in-Oil Emulsions at High Temperature. *Energy Fuels* **2016**, *30*, 1947–1957. [CrossRef]
76. Glandt, C.A.; Chapman, W.G. Effect of Water Dissolution on Oil Viscosity. *SPE Reserv. Eng.* **1995**, *10*, 59–64. [CrossRef]
77. Zirrahi, M.; Hassanzadeh, H.; Abedi, J. Experimental and Modeling Studies of MacKay River Bitumen and Water. *J. Pet. Sci. Eng.* **2017**, *151*, 305–310. [CrossRef]
78. Drelich, J.; Leliński, D.; Miller, J.D. Bitumen Spreading and Formation of Thin Bitumen Films at a Water Surface. Colloids Surfaces A: Physicochem. *Eng. Asp.* **1996**, *116*, 211–223. [CrossRef]
79. Gonzalez, V.; Taylor, S.E. Physical and Chemical Aspects of "Precursor Films" Spreading on Water from Natural Bitumen. *J. Petrol. Sci. Eng.* Submitted.
80. Azom, P.; Srinivasan, S. Mechanistic Modeling of Emulsion Formation and Heat Transfer during the Steam-Assisted Gravity Drainage (SAGD) Process. In Proceedings of the SPE Annual Technical Conference and Exhibition, New Orleans, LA, USA, 4–7 October 2009; Society of Petroleum Engineers: New Orleans, LA, USA.
81. Sasaki, K.; Akibayashi, S.; Yazawa, N.; Kaneko, F. Microscopic Visualization with High Resolution Optical-Fiber Scope at Steam Chamber Interface on Initial Stage of SAGD Process. In Proceedings of the SPE/DOE Improved Oil Recovery Symposium, Tulsa, OK, USA, 13–17 April 2002; Society of Petroleum Engineers: Tulsa, OK, USA.
82. Zekri, A.; Shedid, S.; Hassan, A. A Novel Technique for Treating Asphaltene Deposition using Laser Technology. In Proceedings of the SPE Permian Basin Oil and Gas Recovery Conference, Midland, TX, USA, 15–17 May 2001; Society of Petroleum Engineers: Midland, TX, USA.
83. Al-Hadhrami, H.; Blunt, M. Thermally Induced Wettability Alteration to Improve Oil Recovery in Fractured Reservoirs. *SPE Reservoir Eval. Eng.* **2001**, *4*, 179–186. [CrossRef]
84. Rao, D. Wettability Effects in Thermal Recovery Operations. *SPE Res. Eval. Eng.* **1999**, *2*, 420–430. [CrossRef]
85. Isaacs, E.; Nasr, T.; Babchin, A. Enhanced Oil Recovery by Altering Wettability. U.S. Patent 6,186,232, 13 February 2001.
86. Yuan, J.Y.; Law, D.H.S.; Nasr, T.N. Benefit of Wettability Change near the Production Well in SAGD. In Proceedings of the Canadian International Petroleum Conference, Tulsa, OK, USA, 11–13 June 2002; Petroleum Society of Canada: Tulsa, OK, USA.
87. Flury, C.; Afacan, A.; Tamiz Bakhtiari, M.; Sjoblom, J.; Xu, Z. Effect of Caustic Type on Bitumen Extraction from Canadian Oil Sands. *Energy Fuels* **2014**, *28*, 431–438. [CrossRef]

88. Castanier, L.M.; Brigham, W.E. An Evaluation of Field Projects of Steam with Additives. *SPE Res. Eng.* **1991**, *6*, 62–68. [CrossRef]

89. Shedid, S.A.; El Abbas, A.A. Experimental Study of Surfactant Alkaline Steam Flood through Vertical Wells. In Proceedings of the SPE/AAPG Western Regional Meeting, Long Beach, CA, USA, 19–22 June 2000; Society of Petroleum Engineers: Long Beach, CA, USA.

90. El Abbas, A.; Shedid, S. Experimental Investigation of the Feasibility of Steam/Chemical Steam Flooding Processes through Horizontal Wells. In Proceedings of the SPE Asia Pacific Oil and Gas Conference and Exhibition, Jakarta, Indonesia, 17–19 April 2001; Society of Petroleum Engineers: Jakarta, Indonesia.

91. Yamazaki, T.; Matsuzawa, N.; Abdelkarim, O.; Ono, Y. Recovery of Bitumen from Oil Sand by Steam with Chemicals. *J. Pet. Sci. Eng.* **1989**, *3*, 147–159. [CrossRef]

92. Madhavan, R.M.; Mamora, D.D. Experimental Investigation of Caustic Steam Injection for Heavy Oils. In Proceedings of the SPE Improved Oil Recovery Symposium, Tulsa, OK, USA, 24–28 April 2010; Society of Petroleum Engineers: Tulsa, OK, USA.

93. Brown, A.; Carlin, J.; Fontaine, M.; Haynes, S. Method for Recovery of Hydrocarbons Utilizing Steam Injection. U.S. Patent 3,732,926, 15 May 1973.

94. Katritzky, A.R.; Allin, S.M.; Siskin, M. Aquathermolysis: Reactions of Organic Compounds with Superheated Water. *Acc. Chem. Res.* **1996**, *29*, 399–406. [CrossRef]

95. Gall, J.W. Steam Diversion by Surfactants. In *SPE Annual Technical Conference and Exhibition*; Society of Petroleum Engineers: Las Vegas, NV, USA, 1985.

96. Dilgren, R.E.; Owens, K.B. Olefin Sulfonate-Improved Steam Foam Drive. U.S. Patent 4,393,937, 19 July 1983.

97. Maini, B.B.; Ma, V. Laboratory Evaluation of Foaming Agents for High-Temperature Applications: I. Measurements of Foam Stability at Elevated Temperatures and Pressures. *J. Can. Pet. Technol.* **1986**, *25*. [CrossRef]

98. Novosad, J.; Maini, B.B.; Huang, A. Retention of Foam-Forming Surfactants at Elevated Temperatures. *J. Can. Pet. Technol.* **1986**, *25*. [CrossRef]

99. Cheraghian, G.; Hendraningrat, L. A Review on Applications of Nanotechnology in the Enhanced Oil Recovery. Part A: Effects of Nanoparticles on Interfacial Tension. *Int. Nano Lett.* **2016**, *6*, 129–138. [CrossRef]

100. Cheraghian, G.; Hendraningrat, L. A Review on Applications of Nanotechnology in the Enhanced Oil Recovery. Part B: Effects of Nanoparticles on Flooding. *Int. Nano Lett.* **2016**, *6*, 1–10. [CrossRef]

101. Negin, C.; Ali, S.; Xie, Q. Application of Nanotechnology for Enhancing Oil Recovery—A Review. *Petroleum* **2016**, *2*, 324–333. [CrossRef]

102. Hashemi, R.; Nassar, N.N.; Almao, P.P. Nanoparticle Technology for Heavy Oil In-Situ Upgrading and Recovery Enhancement: Opportunities and Challenges. *Appl. Energy* **2014**, *133*, 374–387. [CrossRef]

103. Wasan, D.T.; Nikolov, A.D. Spreading of Nanofluids on Solids. *Nature* **2003**, *423*, 156–159. [CrossRef] [PubMed]

104. Wasan, D.; Nikolov, A.; Kondiparty, K. The Wetting and Spreading of Nanofluids on Solids: Role of the Structural Disjoining Pressure. *Curr. Opin. Colloid Interface Sci.* **2011**, *16*, 344–349. [CrossRef]

105. Lim, S.; Horiuchi, H.; Nikolov, A.D.; Wasan, D. Nanofluids Alter the Surface Wettability of Solids. *Langmuir* **2015**, *31*, 5827–5835. [CrossRef] [PubMed]

106. Wu, S.; Nikolov, A.; Wasan, D. Cleansing Dynamics of Oily Soil using Nanofluids. *J. Colloid Interface Sci.* **2013**, *396*, 293–306. [CrossRef] [PubMed]

107. Ehtesabi, H.; Ahadian, M.M.; Taghikhani, V.; Ghazanfari, M.H. Enhanced Heavy Oil Recovery in Sandstone Cores Using TiO$_2$ Nanofluids. *Energy Fuels* **2013**, *28*, 423–430. [CrossRef]

108. Al-Anssari, S.; Arif, M.; Wang, S.; Barifcani, A.; Lebedev, M.; Iglauer, S. Wettability of Nanofluid-Modified Oil-Wet Calcite at Reservoir Conditions. *Fuel* **2018**, *211*, 405–414. [CrossRef]

109. Huibers, B.M.J.; Pales, A.R.; Bai, L.; Li, C.; Mu, L.; Ladner, D.; Daigle, H.; Darnault, C.J.G. Wettability Alteration of Sandstones by Silica Nanoparticle Dispersions in Light and Heavy Crude Oil. *J. Nanopart. Res.* **2017**, *19*, 323. [CrossRef]

110. Wei, Y.; Babadagli, T. Selection of Proper Chemicals to Improve the Performance of Steam Based Thermal Applications in Sands and Carbonates. In Proceedings of the SPE Latin America and Caribbean Heavy and Extra Heavy Oil Conference, Lima, Peru, 19–20 October 2016; Society of Petroleum Engineers: Lima, Peru.

111. Zhang, H.; Nikolov, A.; Wasan, D. Enhanced Oil Recovery (EOR) Using Nanoparticle Dispersions: Underlying Mechanism and Imbibition Experiments. *Energy Fuels* **2014**, *28*, 3002–3009. [CrossRef]

112. Sun, X.; Zhang, Y.; Chen, G.; Gai, Z. Application of Nanoparticles in Enhanced Oil Recovery: A Critical Review of Recent Progress. *Energies* **2017**, *10*, 345. [CrossRef]
113. Shokrlu, Y.H.; Babadagli, T. In-Situ Upgrading of Heavy Oil/Bitumen during Steam Injection by Use of Metal Nanoparticles: A Study on In-Situ Catalysis and Catalyst Transportation. *SPE Res. Eval. Eng.* **2013**, *16*, 333–344. [CrossRef]
114. Shokrlu, Y.H.; Babadagli, T. Viscosity Reduction of Heavy Oil/Bitumen using Micro- and Nano-Metal Particles during Aqueous and Non-Aqueous Thermal Applications. *J. Petrol. Sci. Eng.* **2014**, *119*, 210–220. [CrossRef]
115. Guo, K.; Li, H.; Yu, Z. Metallic Nanoparticles for Enhanced Heavy Oil Recovery: Promises and Challenges. *Energy Procedia* **2015**, *75*, 2068–2073. [CrossRef]
116. Hashemi, R.; Nassar, N.N.; Almao, P.P. Enhanced Heavy Oil Recovery by in Situ Prepared Ultradispersed Multimetallic Nanoparticles: A Study of Hot Fluid Flooding for Athabasca Bitumen Recovery. *Energy Fuels* **2013**, *27*, 2194–2201. [CrossRef]
117. Hashemi, R.; Nassar, N.N.; Almao, P.P. In Situ Upgrading of Athabasca Bitumen Using Multimetallic Ultradispersed Nanocatalysts in an Oil Sands Packed-Bed Column: Part 1. Produced Liquid Quality Enhancement. *Energy Fuels* **2014**, *28*, 1338–1350. [CrossRef]
118. Blair, C.M. Method of Recovering Petroleum from a Subterranean Reservoir Incorporating Resinous Polyalkylene Oxide Adducts. U.S. Patent 4,260,019, 7 April 1981.
119. Taylor, S.E. Resolving Crude Oil Emulsions. *Chem. Ind.* **1992**, *20*, 770–773.
120. Blair, C.M.; Scribner, R.E.; Stout, C.A. Persistent Action of Thin-Film Spreading Agents used in Cyclic Steam Stimulation. In Proceedings of the SPE California Regional Meeting, Ventura, CA, USA, 23–25 March 1983; Society of Petroleum Engineers: Ventura, CA, USA.
121. Zeidani, K.; Gupta, S. Hydrocarbon Recovery from Bituminous Sands with Injection of Surfactant Vapour. U.S. Patent Application 2013/0081808, 4 April 2013.
122. Zeidani, K.; Gupta, S.C. Surfactant-Steam Process: An Innovative Enhanced Heavy Oil Recovery Method for Thermal Applications. In Proceedings of the SPE Heavy Oil Conference Canada, Calgary, AB, Canada, 11–13 June 2013; Society of Petroleum Engineers: Calgary, AB, Canada.
123. Srivastava, P.; Debord, J.; Sadetsky, V.; Stefan, B.J.; Orr, B.W. Laboratory Evaluation of a Chemical Additive to Increase Production in Steam Assisted Gravity Drainage (SAGD). In Proceedings of the SPE Improved Oil Recovery Symposium, Tulsa, OK, USA, 24–28 April 2010; Society of Petroleum Engineers: Tulsa, OK, USA.
124. Srivastava, P.; Sadetsky, V.; Debord, J.; Stefan, B.; Orr, B. Development of a Steam-Additive Technology to Enhance Thermal Recovery of Heavy Oil. In Proceedings of the SPE Annual Technical Conference, Florence, Italy, 19–22 September 2010; Society of Petroleum Engineers: Florence, Italy.
125. Hart, P.R.; Stefan, B.J.; Srivastava, P.; Debord, J.D. Method for Enhancing Heavy Hydrocarbon Recovery. U.S. Patent 7,938,183, 11 May 2011.
126. Campos, R.E.; Hernandez, J.A. In-Situ Reduction of Oil Viscosity during Steam Injection Process in EOR. U.S. Patent 5,314,615, 24 May 1994.
127. Koebel, M.; Strutz, E.O. Thermal and Hydrolytic Decomposition of Urea for Automotive Selective Catalytic Reduction Systems: Thermochemical and Practical Aspects. *Ind. Eng. Chem. Res.* **2003**, *42*, 2093–2100. [CrossRef]
128. Rees, A.C.; Coulter, C.; Engelman, R.; Guerrero-Aconcha, U.; Taylor, S.E.; Peats, A.; Zeng, H. Systems and Methods for Accelerating Production of Viscous Hydrocarbons in a Subterranean Reservoir with Thermally Activated Chemical Agents. U.S. Patent Application 2014/0262241, 18 September 2014.
129. Bruns, F.; Babadagli, T. Recovery Improvement of Gravity Driven Steam Applications Using New Generation Chemical Additives. In Proceedings of the SPE Western Regional Meeting, Bakersfield, CA, USA, 23–27 April 2017; Society of Petroleum Engineers: Bakersfield, CA, USA.
130. Liu, Q.; Dong, M.; Yue, X.; Hou, J. Synergy of Alkali and Surfactant in Emulsification of Heavy Oil in Brine. *Colloids Surfaces A Physicochem. Eng. Asp.* **2006**, *273*, 219–228. [CrossRef]
131. Dong, M.; Ma, S.; Liu, Q. Enhanced Heavy Oil Recovery through Interfacial Instability: A Study of Chemical Flooding for Brintnell Heavy Oil. *Fuel* **2009**, *88*, 1049–1056. [CrossRef]
132. Pei, H.; Zhang, G.; Ge, J.; Jin, L.; Ding, L. Study on the Variation of Dynamic Interfacial Tension in the Process of Alkaline Flooding for Heavy Oil. *Fuel* **2013**, *104*, 372–378. [CrossRef]

133. Tang, M.; Zhang, G.; Ge, J.; Jiang, P.; Liu, Q.; Pei, H.; Chen, L. Investigation into the Mechanisms of Heavy Oil Recovery by Novel Alkaline Flooding. *Colloids Surfaces A Physicochem. Eng. Asp.* **2006**, *421*, 91–100. [CrossRef]

134. De Haas, T.W.; Fadaei, H.; Guerrero, U.; Sinton, D. Steam-on-a-Chip for Oil Recovery: The Role of Alkaline Additives in Steam Assisted Gravity Drainage. *Lab. Chip* **2013**, *13*, 3832. [CrossRef] [PubMed]

135. Gonzalez, V.; Jones, M.; Taylor, S.E. Spin–Spin Relaxation Time Investigation of Oil/Brine/Sand Systems. Kinetics, Effects of Salinity, and Implications for Wettability and Bitumen Recovery. *Energy Fuels* **2016**, *30*, 844–853. [CrossRef]

136. Chen, Q.; Gerritsen, M.G.; Kovscek, A.R. Improving Steam-Assisted Gravity Drainage Using Mobility Control Foams: Foam Assisted-SAGD (FA-SAGD). In Proceedings of the SPE Improved Oil Recovery Symposium, Tulsa, OK, USA, 24–28 April 2010; Society of Petroleum Engineers: Tulsa, OK, USA.

137. Maiti, A.; Sadrezadeh, M.; Thakurta, S.G.; Pernitsky, D.J.; Bhattacharjee, S. Characterization of Boiler Blowdown Water from Steam-Assisted Gravity Drainage and Silica–Organic Coprecipitation during Acidification and Ultrafiltration. *Energy Fuels* **2012**, *26*, 5604–5612. [CrossRef]

138. Clemente, J.S.; Fedorak, P.M. A Review of the Occurrence, Analyses, Toxicity, and Biodegradation of Naphthenic Acids. *Chemosphere* **2005**, *60*, 585–600. [CrossRef] [PubMed]

139. Kawaguchi, H.; Li, Z.; Masuda, Y.; Sato, K.; Nakagawa, H. Dissolved Organic Compounds in Reused Process Water for Steam-Assisted Gravity Drainage Oil Sands Extraction. *Water Res.* **2012**, *46*, 5566–5574. [CrossRef] [PubMed]

140. Xu, Y.; Dong, B.; Dai, X. Effect of the Silica-Rich, Oilfield-Produced Water with Different Degrees of Softening on Characteristics of Scales in Steam-Injection Boiler. *Desalination* **2015**, *361*, 38–45. [CrossRef]

141. Angle, C.W.; Dabros, T.; Hamza, H.A. Demulsifier Effectiveness in Treating Heavy Oil Emulsion in the Presence of Fine Sands in the Production Fluids. *Energy Fuels* **2007**, *21*, 912–919. [CrossRef]

142. Petersen, M.A.; Grade, H. Analysis of Steam Assisted Gravity Drainage Produced Water using Two-Dimensional Gas Chromatography with Time-of-Flight Mass Spectrometry. *Ind. Eng. Chem. Res.* **2011**, *50*, 12217–12224. [CrossRef]

143. Lewis, A.T.; Tekavec, T.N.; Jarvis, J.M.; Juyal, P.; McKenna, A.M.; Yen, A.T.; Rodgers, R.P. Evaluation of the Extraction Method and Characterization of Water-Soluble Organics from Produced Water by Fourier Transform Ion Cyclotron Resonance Mass Spectrometry. *Energy Fuels* **2013**, *27*, 1846–1855. [CrossRef]

144. Razi, M.; Sinha, S.; Waghmare, P.R.; Das, S.; Thundat, T. Effect of Steam-Assisted Gravity Drainage Produced Water Properties on Oil/Water Transient Interfacial Tension. *Energy Fuels* **2016**, *30*, 10714–10720. [CrossRef]

145. Sikes, C.S.; Sikes, D.; Hochwalt, M.A. Flotation and Separation of Flocculated Oils and Solids from Waste Waters. U.S. Patent 9,321,663, 26 April 2016.

146. Alpatova, A.; Kim, E.-S.; Dong, S.; Sun, N.; Chelme-Ayala, P.; El-Din, M.G. Treatment of Oil Sands Process-Affected Water with Ceramic Ultrafiltration Membrane: Effects of Operating Conditions on Membrane Performance. *Sep. Purific. Technol.* **2014**, *122*, 170–182. [CrossRef]

147. Gupta, R.K.; Dunderdale, G.J.; England, M.W.; Hozumi, A. Oil/Water Separation Techniques: A Review of Recent Progresses and Future Directions. *J. Mater. Chem. A* **2017**, *5*, 16025–16058. [CrossRef]

148. Cross, K.J. Additives for Improving Hydrocarbon Recovery. U.S. Patent Application 2012/0312532, 12 December 2012.

149. Guerrero, U. In Situ Hydrocarbon Recovery Operations such as SAGD with Injection of Water-Wetting Agents. WO Patent Application 2015/061903, 7 May 2015.

150. Ayika, A.A.; Donate, F.A. Method to Extract Bitumen from Oil Sands. U.S. Patent Application 2015/0307787, 29 October 2015.

151. Sunnatov, D.M. Experimental Study of Steam Surfactant Flood. Master's Thesis, Texas A&M University, College Station, TX, USA, 2010.

colloids
and interfaces

MDPI

Article

Microbial-Enhanced Heavy Oil Recovery under Laboratory Conditions by *Bacillus firmus* BG4 and *Bacillus halodurans* BG5 Isolated from Heavy Oil Fields

Biji Shibulal [1], Saif N. Al-Bahry [1,*], Yahya M. Al-Wahaibi [2], Abdulkadir E. Elshafie [1], Ali S. Al-Bemani [2] and Sanket J. Joshi [1,3]

[1]   Department of Biology, College of Science, Sultan Qaboos University, Muscat 123, Oman;
      bijisd@gmail.com (B.S.); akelshafie@gmail.com (A.E.E.); sanketjj@gmail.com (S.J.J.)
[2]   Department of Petroleum and Chemical Engineering, College of Engineering, Sultan Qaboos University,
      Muscat 123, Oman; ymn@squ.edu.om (Y.M.A.-W.); bemani@squ.edu.om (A.S.A.-B.)
[3]   Central Analytical and Applied Research Unit, College of Science, Sultan Qaboos University,
      Muscat 123, Oman
*   Correspondence: snbahry@squ.edu.om; Tel.: +968-24-146-868

Received: 15 November 2017; Accepted: 4 January 2018; Published: 7 January 2018

Abstract: Microbial Enhanced Oil Recovery (MEOR) is one of the tertiary recovery methods. The high viscosity and low flow characteristics of heavy oil makes it difficult for the extraction from oil reservoirs. Many spore-forming bacteria were isolated from Oman oil fields, which can biotransform heavy crude oil by changing its viscosity by converting heavier components into lighter ones. Two of the isolates, *Bacillus firmus* BG4 and *Bacillus halodurans* BG5, which showed maximum growth in higher concentrations of heavy crude oil were selected for the study. Gas chromatography analysis of the heavy crude oil treated with the isolates for nine days showed 81.4% biotransformation for *B. firmus* and 81.9% for *B. halodurans*. In both cases, it was found that the aromatic components in the heavy crude oil were utilized by the isolates, converting them to aliphatic species. Core flooding experiments conducted at 50 °C, mimicking reservoir conditions to prove the efficiency of the isolates in MEOR, resulted in 10.4% and 7.7% for *B. firmus* and *B. halodurans*, respectively, after the nine-day shut-in period. These investigations demonstrated the potential of *B. firmus* BG4 and *B. halodurans* BG5 as an environmentally attractive approach for heavy oil recovery.

Keywords: spore forming bacteria; *Bacillus firmus*; *Bacillus halodurans*; Microbial Enhanced Oil Recovery; biotransformation; heavy oil recovery

1. Introduction

Global energy requirements demand an increased production of crude oil. During conventional recovery methods, about 30–40% of crude oil is recovered while rest remains trapped in the reservoir [1–4]. Enhanced oil recovery (EOR) targets the trapped crude oil. Crude oil is a fossil fuel which is considered as non-renewable energy source. It is composed of a mixture of different hydrocarbons (including alkanes/paraffins, alkenes/olefins, cycloalkanes/naphthenes, and aromatics), complex hydrocarbons (such as polycyclic aromatic hydrocarbons), resins, asphaltenes, along with certain other hetero-species, containing nitrogen, oxygen and sulfur [5]. Heavy crude oil is characterized by high density or specific gravity, more resistant to flow with an American Petroleum Institute (API) gravity of less than 20°. Extraction of heavy crude oil needs higher energy input. Current methods of extraction include open-pit mining, steam stimulation, the addition of sand to the oil, and the injection of air into well to create subterranean fires that burn heavier hydrocarbons to generate heat. Transportation of these types of crude oil through pipelines poses much difficulty and requires certain diluting agents. Sometimes heavy and light crude oils are mixed to facilitate transport through pipeline. This will result in contamination of the light crude and a reduction in its value [6].

Enhanced oil recovery (EOR) is a tertiary method of extracting residual oil from the reservoirs after the primary and secondary phases of production. EOR methods adopted will either modify the properties of reservoir fluids and/or the reservoir rock characteristics such as reducing the interfacial tension between oil and water, reducing oil viscosity, and displacing oil through porous rocks [7,8].

Microbial enhanced oil recovery (MEOR) has become an important, fast developing tertiary recovery method which uses microorganisms or their metabolites to enhance the recovery of residual oil [7–10]. MEOR is different from conventional EOR methods such as CO_2 injection, steam injection, chemical surfactant and polymer flooding, in that it involves injecting live microorganisms and nutrients into the reservoir so that bacteria and their metabolic products mobilize the residual oil. It is considered to be a more environmentally friendly method since it does not involve any toxic chemicals and it is easy to carry out in fields since it does not need any modifications of existing water-injection amenities [11–13]. MEOR takes place by different mechanisms, such as reduction of oil-water interfacial tension and alteration of wettability by surfactant production, selective plugging by microorganisms and their metabolites, oil viscosity reduction by gas production or degradation or biotransformation of long-chain saturated hydrocarbons, and production of acids which improves absolute permeability by dissolving minerals in the rock [14]. The microbial metabolic products include biosurfactants, biopolymers, acids, solvents, gases, and enzymes. The bacteria used in MEOR are usually hydrocarbon-utilizing, non-pathogenic, and are naturally occurring in petroleum reservoirs [15].

Biological processing of heavy oil is a cost-effective and eco-friendly approach which provides a higher selectivity to specific reactions to upgrade heavy oil. Microbial systems which are capable of biotransforming oil fractions are used in heavy oil reservoirs for increased oil recovery by reducing the oil viscosity [16]. Many microorganisms capable of biotransforming hydrocarbons using crude oil as the sole carbon source have been identified [16–22]. A successful field trial using oil biotransforming bacteria without injection of nutrients has been reported [3,23]. The role of spore-forming bacteria in crude oil biotransformation, and competent *Bacillus* strains existing in many oil-polluted sites have been widely studied [24–28]. The economy of countries, such as Oman, is highly dependent on revenues generated from crude oil production and a cost effective, environmentally friendly alternative method of upgrading and producing heavy crude oil will be a significant benefit. Also, the transportation of heavy oil through pipelines will be facilitated by biotransformation. The goal of this study was therefore to demonstrate the potential of *Bacillus halodurans* and *Bacillus firmus* for the biotransformation of heavy crude oil (4.57° API).

2. Materials and Methods

All chemicals and media were from Sigma-Aldrich Co. (St. Louis, MO, USA), Analytical Reagent (AR) grade.

2.1. Culture Media and Cultivation

Two different media were used for the isolation of bacterial cells for the biotransformation study, Bushnell-Haas (BH) [29] and mineral salt (Medium C) [30]. Medium C (pH = 7 ± 0.2) contained (g L^{-1}): NH_4NO_3 (4.002); KH_2PO_4 (4.083); Na_2HPO_4 (7.119); $MgSO_4 \cdot 7H_2O$ (0.197). To this was added, 1 mL of trace metal solution containing (g L^{-1}): $CaCl_2$ (0.00077); $FeSO_4.7H_2O$ (0.0011); $MnSO_4 \cdot 4H_2O$ (0.00067); Na-EDTA (0.00148). The BH medium (pH = 7 ± 0.2) consisted of (g L^{-1}): $MgSO_4$ (0.2); $CaCl_2$ (0.02); KH_2PO_4 (1.0); K_2HPO_4 (1.0); NH_4NO_3 (1.0); $FeCl_3$ (0.050). All media were sterilized by autoclaving at 121 °C at 15 psi for 15 min.

2.2. Characterization of Soil and Oil Samples

A total of 10 different soil samples were characterized for pH, mineralogy analysis using X-ray diffraction (XRD), extractable total petroleum hydrocarbons (eTPH) and moisture content as described previously [20]. Briefly, the heavy crude oil contaminated soil samples were mixed with anhydrous sodium sulfate to remove moisture in a capped conical flask. The eTPH was estimated by mixing 10 g of sample with 30 mL of dichloromethane (DCM, high pressure liquid chromatography (HPLC) grade, 99% pure), capped tightly, mixed well by inverting the flasks several times and then transferred to a mechanical shaker for 4–5 h and allowed the sediments to settle for 1 h. The solvent with the hydrocarbon was filtered through Whatman® qualitative filter paper, Grade 1 110 mm into a pre-weighted conical flask and allowed to concentrate overnight [31–33]. The moisture content of the soil samples were found to be in the range 0.018 to 0.024 m^3/m^3. The heavy crude oil viscosity was measured using a Rheolab QC rotational viscometer and API gravity with a DSA 5000 M density meter.

2.3. Isolation of Spore Forming Bacterial Strains Using Heavy Crude Oil as Carbon Source

Spore-forming bacteria were isolated from soil samples contaminated with heavy crude oil. The sampling site was a contaminated area near oil wells of one of the oil rigs in Oman. The subsurface soil samples (8 cm below surface) were aseptically collected from seven different regions in random manner around each well and mixed together. The soil samples were collected with pre-sterilized shovels into sterilized bags, properly labelled and transferred to the laboratory and stored at 4 °C until use. Heavy crude oil samples used in the study were collected from the oil field in sterile bottles and stored for further studies.

For the isolation of spore forming isolates, 1 g of soil sample mixed with 10 mL distilled water was vortexed thoroughly and the vegetative cells were killed by boiling the mixture in a water bath at 90 °C for 30 min [20,26]. 5 mL of the supernatant served as an inoculum for the first enrichment in both media in 250 mL conical flasks. 1% (w/v) heavy crude oil was added to the media used for the isolation as the sole carbon source. The flasks inoculated with the supernatant were incubated at 40 °C, 160 rpm for two weeks. A negative control flask without heavy crude oil was set up and incubated at the same conditions. A 1% (w/v) aliquot from the first enrichment served as the inoculum for second enrichment which was incubated at the same conditions for a further one week period. The enrichment technique for the isolation of bacteria has already been reported [34,35]. The dilutions from both the first and second enrichments were spread-plated on corresponding fresh agar plates and incubated at 40 °C for 24 h. Well-isolated single colonies were picked up carefully and by successive streaking in fresh agar plates resulted in pure isolates, which were stored in 60% (v/v) glycerol stock solution at −80 °C.

2.4. Identification of Bacillus firmus and Bacillus halodurans

Among the 40 isolates studied, the ones which showed maximum growth on agar plates were identified using a MALDI Biotyper (Bruker Daltonik GmbH, Bremen, Germany) [36] and 16S rDNA sequencing. For the Biotyper identification, a direct smearing method was used where 24 h-grown pure cultures were smeared on the target plate and layered with 1 μL sinapinic acid. The target plate was inserted in the Matrix Assisted Laser Desorption/Ionization-Time of Flight-Mass Spectrometer (MALDI-TOF/MS) instrument and the protein fingerprints were generated. The integrated software generates an outcome list, by comparing the fingerprint of the reference sample with the reference spectra in the database, in which species with the most similar fingerprints are ordered according to their logarithmic score value (log (score value)) [37].

16S ribosomal DNA (rDNA) sequencing was performed using 27F and 1492R primers of the genomic DNA isolated using PowerSoil DNA isolation kit (Mo Bio Laboratories Inc., Carlsbad, CA, USA), as reported before [20]. The amplification reaction (Polymerase Chain Reaction (PCR)), was performed using T100 thermal cycler. The amplification reaction was performed on a total volume of 25 μL containing: 12.5 μL master mix (Taq polymerase and deoxynucleotide triphosphate (dNTP) mix), 9.5 μL double distilled water, 1 μL extracted DNA and 1 μL of each primer. PCR amplification was performed with an initial denaturation step at 94 °C for 3 min followed by 35 cycles of a 1 min denaturation step at 94 °C, a 2 min annealing step at 53 °C, and a 2 min elongation step at 72 °C, with a final extension step at 72 °C for 7 min using a 2720 thermal cycler. The PCR products were detected in 1.6% agarose gel electrophoresis. The PCR products were purified using QIAquick PCR purification kit (QIAGen, Carlsbad, CA, USA). The BigDye® Terminator v3.1 Cycle Sequencing Kit (Applied Biosystems™, Foster City, CA, USA) was used for de-novo sequencing. The sequencing was done using 3130 XL Genetic Analyzer (Applied Biosystem-Hitachi, Waltham, MA, USA). The sequences of the 16S ribosomal RNA (rRNA) genes identified in this study were submitted to the NCBI GenBank databases under the accession numbers KP119100 and KP119100.

2.5. Growth Characteristics during Biotransformation under Aerobic Conditions

The effect of heavy crude oil concentration on the growth of the isolates in BH medium was studied for a period of 10 days. The BH medium with 1% *w/v*, 3% *w/v*, 5% *w/v* and 7% *w/v* of heavy crude oil was inoculated with the two strains of bacteria, *B. firmus* and *B. halodurans* and incubated at 40 °C and 160 rpm. One-way ANOVA was conducted to determine if the heavy crude oil concentration had an effect on the growth of the isolate. A Kruskal-Wallis test was done to evaluate the effect of crude oil concentration in the pH of the culture medium.

2.6. Biotransformation Studies Using GC-MS

Isolates, *B. firmus* and *B. halodurans* were incubated in BH medium containing 1% heavy crude oil as the sole carbon source for a period of nine days to determine the biotransformation potential of the isolate under aerobic conditions. Seed cultures of the corresponding isolates were prepared from 24 h grown isolates in Luria-Bertani broth at 40 °C and 160 rpm. One percent (*v/v*) of the seed culture served as the inoculum for 100 mL BH medium with 1% heavy crude oil and incubated at the same conditions described. The contents of each flask for each isolate were extracted on the third, sixth, and ninth days of incubation for GC-MS analysis. All experiments were done in triplicate. The cell free extract was analyzed for the production of biosurfactant using Drop Shape Analyzing system-DSA 100 (Krüss GmbH, Hamburg, Germany) by measuring the surface tension (ST) and interfacial tension (IFT). IFT was measured against *n*-hexadecane.

The extraction of the biotransformed heavy crude oil at the third, sixth, and ninth days of incubation by the isolates were done by vigorously mixing the contents with 20 mL DCM in a separating funnel allowing the mixture to separate to different fractions. The DCM fraction with biotransformed heavy crude oil was collected carefully in a glass collection tube. The collected fraction was then purified by passing through silica G-60. The column was sequentially eluted with hexane to obtain the aliphatic fractions and then with hexane:DCM (1:1) to elute the aromatic fractions [38].

The fractions were analyzed by GC MS/MS with DB 5 capillary column (30 m × 0.32 mm internal diameter, 0.1 mm thickness) (Waters, Quattro Micro™ GC MS/MS, Micromass UK Ltd., Wilmslow, UK) following EPA Method 1655 [39]. Helium was used as a carrier gas and a constant flow rate of 2 mL/min was set. Injector and detector temperatures were 350 and 370 °C, respectively. The oven temperature program was: initial temperature 50 °C for 1 min, raised to 350 °C at a rate of 10 °C/min, and a hold at 370 °C for 1 min.

2.7. Core Flooding Experiments

The core flooding experiments were performed to study the ability of the isolates to degrade heavy crude oil under anoxic conditions and to evaluate the potential of the strain in heavy oil recovery. The heavy crude oil sample used in the core flood experiments was degassed and dehydrated. The brine was purged with nitrogen. The Berea sandstone cores (absolute permeability $350–360 \times 10^{-2}$ μm^2) were cleaned in methanol using a Soxhlet apparatus. The cleaned cores after being dried at 80 °C for

24 h were saturated with filtered, sterilized formation water for 12 h in a desiccator under vacuum. The formation water was collected from one of the Oman heavy oil fields. The characteristics of brine was as reported before [20]. The cores were then placed in the core flood apparatus and heated in the oven provided in the system to 50 °C, mimicking the reservoir condition. The pore volume was calculated as the difference in the wet and dry weights of the core and was flooded with four pore volumes (PV) of brine at 0.4 cm^3/min to ensure 100% brine saturation and to degas the core. The cores were then injected with heavy crude oil until no more water was produced until it reached the irreducible water saturation (S$_{wr}$). The initial oil saturation was calculated volumetrically from the amount of injected oil and produced water. Secondary recovery of the heavy oil was done by flooding the core with brine at a rate of 0.4 cm^3/min, until no more oil was produced. The residual crude heavy oil in the core was measured from the volume of oil produced.

For the core flooding experiment, the mother inoculum was prepared by 24 h-grown isolates in Lysogeny broth (LB) medium (in Luria Bertani broth) (OD$_{620}$ = 1.324; 1.06 × 10^9 CFU/mL for *B. firmus* and OD$_{620}$ = 1.672; 1.34 × 10^9 CFU/mL for *B. halodurans*). Freshly prepared sterile BH medium was added to the mother inoculum in a ratio of 1:4. One PV of the mixture was injected into the core and the system was shut in for 9 days at 50 °C. For evaluating the potential of the strains in extra heavy oil recovery, after the shut in period, the extra recovered oil was collected in graduated tubes by flooding with brine, and then measured. A control experiment was performed at same conditions, but without the injection of the isolates. The effluent collected during the tertiary recovery was tested for the presence of the isolates by MALDI Biotyper. The extra recovered oil was analyzed by GC-MS for determining the biotransformation of heavy crude oil. Scanning electron microscopy (SEM; JEOL, JSM-7600F Field Emission SEM, Tokyo, Japan) analysis of the core specimen from the outlet, middle and inlet portions was done after fixation using glutaraldehyde and osmium, dehydration using ethanol and critical point drying. The specimens were then mounted on stubs and were coated with gold using sputter coater for SEM analysis [20].

2.8. Statistical Analysis

All data analyses were done using the statistical software MINITAB 14 (Minitab, Ltd., Coventry, UK) with a maximal Type 1 error rate of 0.05. Kruskal-Wallis non parametric test was used where the assumptions of analysis of variance (ANOVA) were not met.

3. Results

3.1. Characterization of Soil and Oil Samples

The heavy crude oil-contaminated soil samples were collected and stored appropriately. The pH of the 10 soil samples were measured as 8.5 ± 0.5. The eTPH of the soil samples were ~4.2% and the moisture content of ~0.024 m^3/m^3. The mineral compositions of the 10 soil samples measured by XRD showed that all of the soil samples contained calcite and quartz; albite and palygorskite were present in 8 soil samples out of 10 samples tested. Other minerals observed were anorthite, dolomite, gypsum, halite, microcline, muscovite, rutile, suhailite and takanelite (Table 1). The heavy crude oil sample viscosity was determined as 650,000 mPa·s and as 4.57° API.

3.2. Isolation and Identification of Oil-Oxidizing Bacteria, Bacillus firmus and Bacillus halodurans

The isolates that were capable of utilizing heavy crude oil as carbon source were isolated based on their morphology. The isolates which showed maximum growth on agar plates in short period of time were selected for the study. The isolates were identified initially by MALDI-Biotyper as *Bacillus firmus* and *Bacillus halodurans* with a score value above 1.8. Phylogenetic analysis of the 16S rRNA genes of the isolates BG4 and BG5 revealed >97% similarity to the sequences of *Bacillus firmus* and *Bacillus halodurans*, respectively.

Table 1. Minerology of soil samples from XRD analysis.

Mineral	Soil Samples									
	SA	SB	SC	SD	SE	SF	SG	SH	SI	SJ
albite	1	1	0	1	0	1	1	1	1	1
anorthite	1	0	0	0	0	0	0	1	0	0
calcite	1	1	1	1	1	1	1	1	1	1
dolomite	0	1	0	0	0	0	0	0	0	0
gypsum	1	0	1	1	0	1	1	0	1	1
halite	0	1	0	1	0	0	1	1	0	1
microcline	0	0	0	0	0	0	0	1	1	0
muscovite	0	0	0	0	0	1	0	0	0	0
palygorskite	1	1	1	1	1	1	1	0	1	0
quartz	1	1	1	1	1	1	1	1	1	1
rutile	0	0	1	0	0	0	0	0	0	0
suhailite	0	1	0	0	0	0	0	0	0	0
takanelite	0	0	0	0	1	0	0	0	0	0

1 = present; 0 = absent.

3.3. Growth Characteristics of Bacteria during Crude Oil Degradation under Aerobic Conditions

The growth characteristic study of the two isolates showed that heavy crude oil concentration up to 7% (w/v) had no significant effect on pH for both the isolates, where the pH increased from ~7.5 to ~9.5, at all crude oil concentrations. In contrast, the OD_{620} values showed significant effects for 1 and 3% (w/v) heavy crude oil concentrations for *B. firmus* in BH medium, while no significant effect was found for *B. halodurans*. Statistical analysis was performed using MINITAB 14 for determining the effect of heavy crude oil on the growth of the isolates. The ANOVA *p*-value for OD_{620} for *B. firmus* was $p = 0.004 < 0.05$ and the post hoc analysis, Tukey test showed that the growth rate at 1% and 3% (w/v) was significantly different from 5% and 7% (w/v) (Figures 1 and 2).

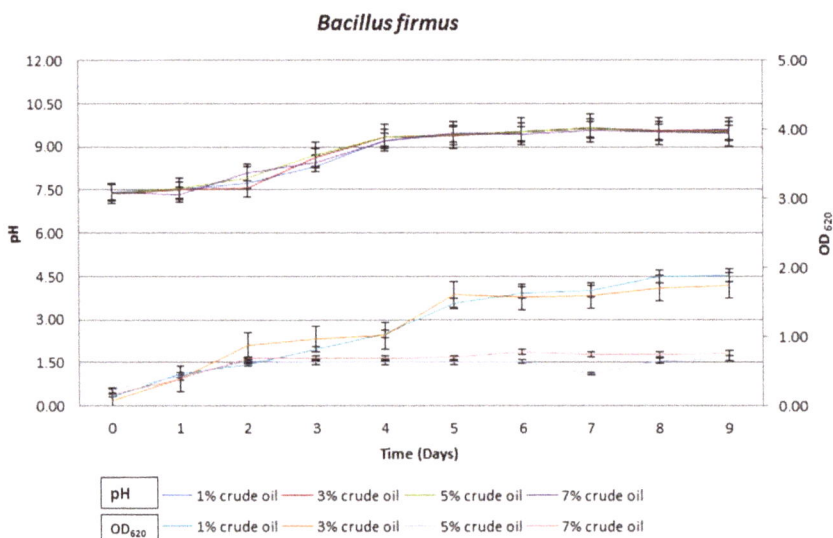

Figure 1. Growth profile of *B. firmus* in Bushnell-Haas (BH) medium. The isolate showed higher growth in presence of 1% and 3% heavy crude oil.

Bacillus halodurans

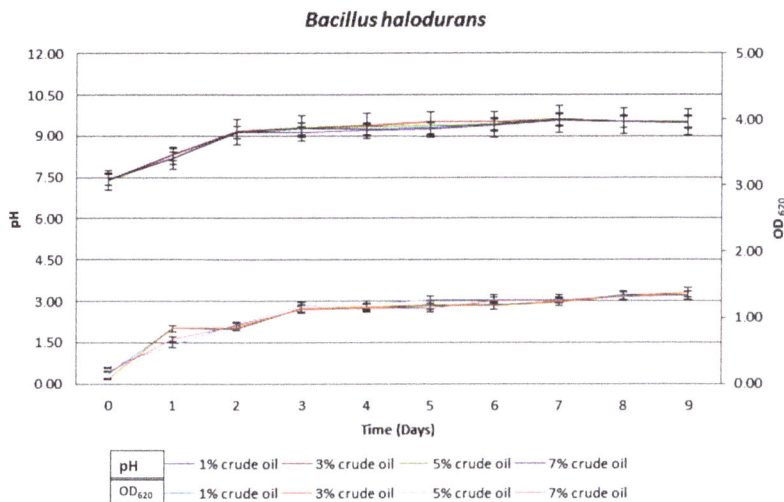

Figure 2. Growth profile of *B. halodurans* in BH medium. There was no effect of heavy crude oil concentrations up to 7% for the isolate.

3.4. Biotransformation Studies Using GC-MS

The biotransformed heavy crude oil incubated with *B. firmus* and *B. halodurans* for nine days in BH medium was extracted with DCM on the third, sixth and ninth days of incubation and was purified by passing through Silica G60 column. The fractions sequentially eluted with hexane and hexane:DCM (1:1) were analyzed using GC-MS. The analysis showed 81.36% biotransformation of heavy crude oil for *B. firmus* and 81.93% for *B. halodurans* compared to the abiogenic control. The total aromatic fractions reduced during the period of incubation were 70.80% for *B. firmus* and 47.77% for *B. halodurans* and aliphatics with 58.22 and 88.21% respectively. An increase in the concentration of aliphatic compounds was also observed (Figures 3–6).

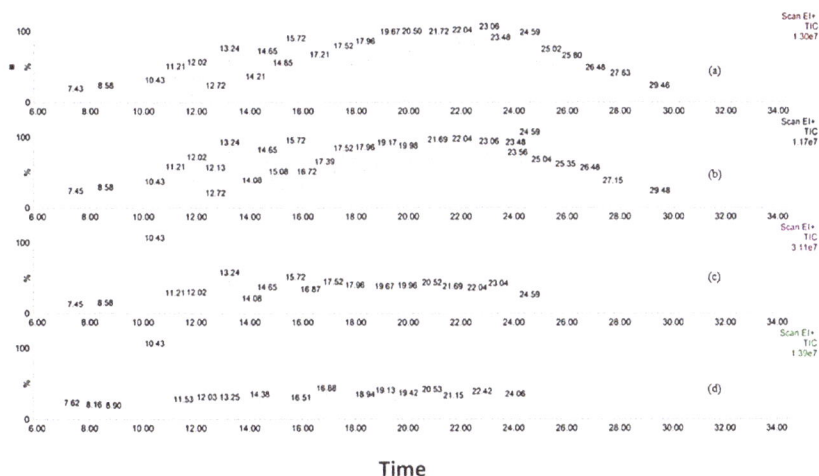

Figure 3. GC-MS chromatogram of heavy crude oil biotransformation by *B. firmus* on day 3 (**b**); day 6 (**c**); and day 9 (**d**); as compared to the control (**a**).

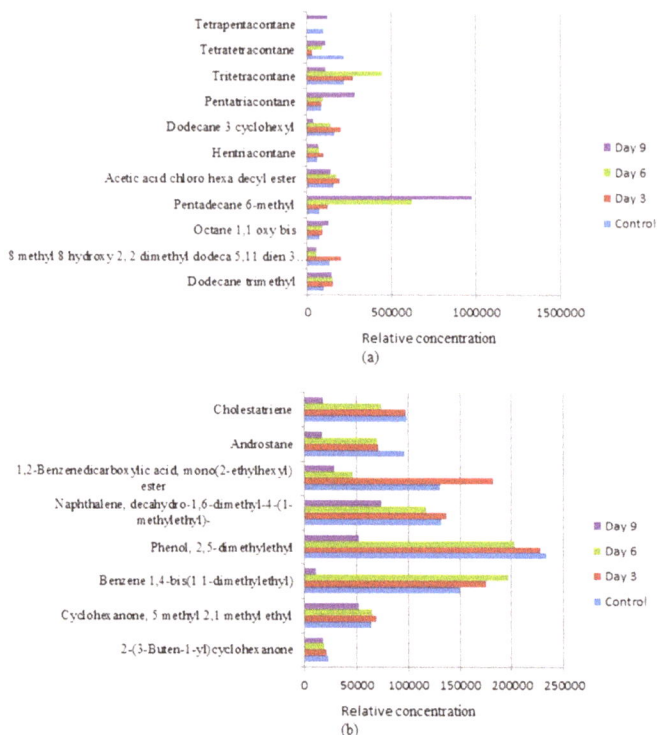

Figure 4. GC-MS analysis of bio-fractionated heavy crude oil by *B. firmus* (**a**) aliphatic compounds (**b**) aromatic compounds.

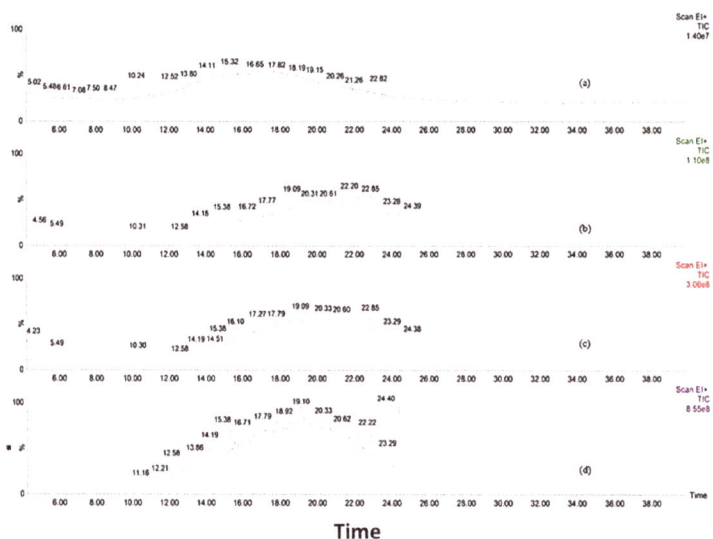

Figure 5. GC-MS chromatogram of heavy crude oil biotransformation by *B. halodurans* on day 3 (**c**); day 6 (**b**) and day 9 (**a**); as compared to the control (**d**).

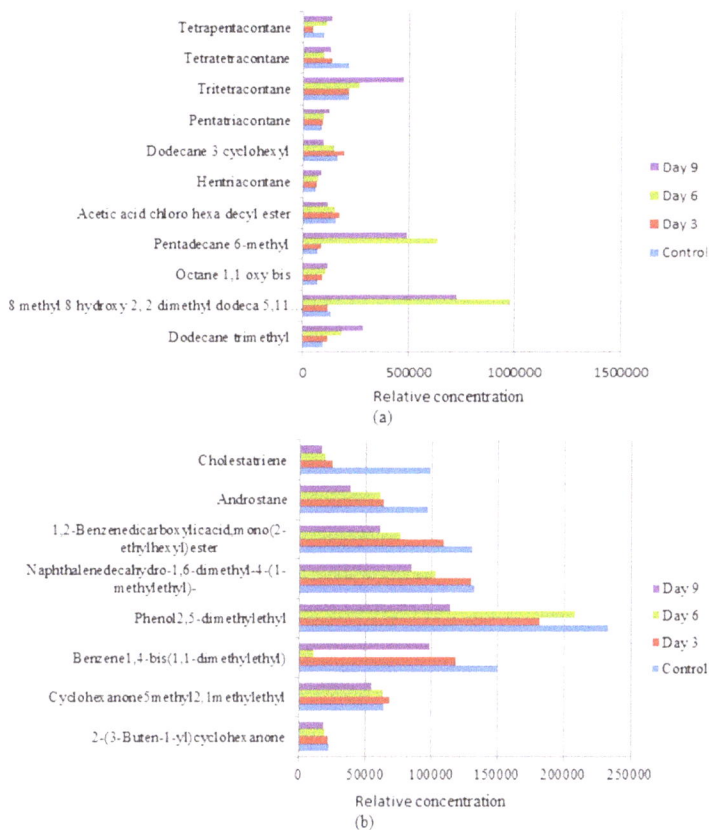

Figure 6. GC-MS analysis of bio-fractionated heavy crude oil by *B. halodurans* (**a**) aliphatic compounds (**b**) aromatic compounds.

3.5. Core Flooding Experiments

Berea sandstone cores were used to evaluate the potential of isolates in heavy oil recovery. The experiment was conducted under anaerobic condition mimicking the oil field conditions. Throughout the experiment, the temperature was maintained at 50 °C and a pressure of 1000 psi. The oil initially in place (OIIP) in the core for *B. firmus* BG4 was 13.2 mL (S_{oi} = 79.6%). The water flooding (5 PV) resulted in recovery of 50.75% of initial oil (OI) corresponding to 6.7 mL of the initial oil present in the core. After nine days incubation with the isolate *B. firmus* BG4, the system was again injected with 5 PV of brine that resulted in a total recovery of 61.22% (7.18 mL) of initial oil present in the core, in which 10.46% (0.68 mL) was contributed by the action of the isolate.

The OIIP in the core for *B. halodurans* BG5 was 14.2 mL, S_{oi} = 84.02%. The water flooding resulted in recovery of 54.92% (7.8 mL) of OIIP for *B. halodurans*. The tertiary recovery by *B. halodurans* after nine days shut-in period resulted in 7.8% (0.5 mL) extra recovery of residual oil by the biotransformation of heavy crude oil compared to the control experiment. The extra recovery measurements were based on the residual oil (RO) present in the core (Figure 7a,b). No pressure changes were observed during bacterial flooding. The effluent analyzed using MALDI Biotyper revealed the presence of the isolates.

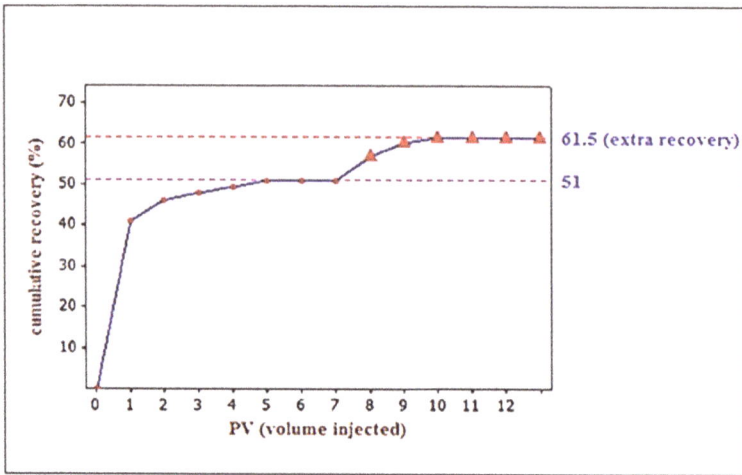

(a)

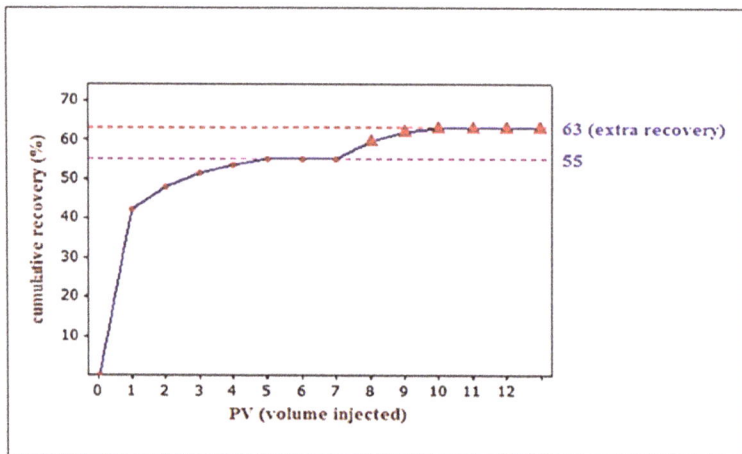

(b)

Figure 7. Cumulative oil recovery (**a**) by *B. firmus* and (**b**) by *B. halodurans*, after nine days shut-in period.

The migration of isolates inside the core was further determined by SEM analysis and the bioconversion of heavy crude oil was estimated by GC-MS analysis [3,40–42]. The extra recovered oil was analyzed using GC-MS which revealed that the biotransformation of heavy crude has occurred anaerobically. The percentage of aromatic compounds was reduced, and the concentration of lighter hydrocarbons has increased (Figure 8a–c; Tables 2 and 3). SEM analysis of the core indicated the presence of both isolates inside the core, which indicated their ability to grow anaerobically (Figures 9a,b and 10a,b).

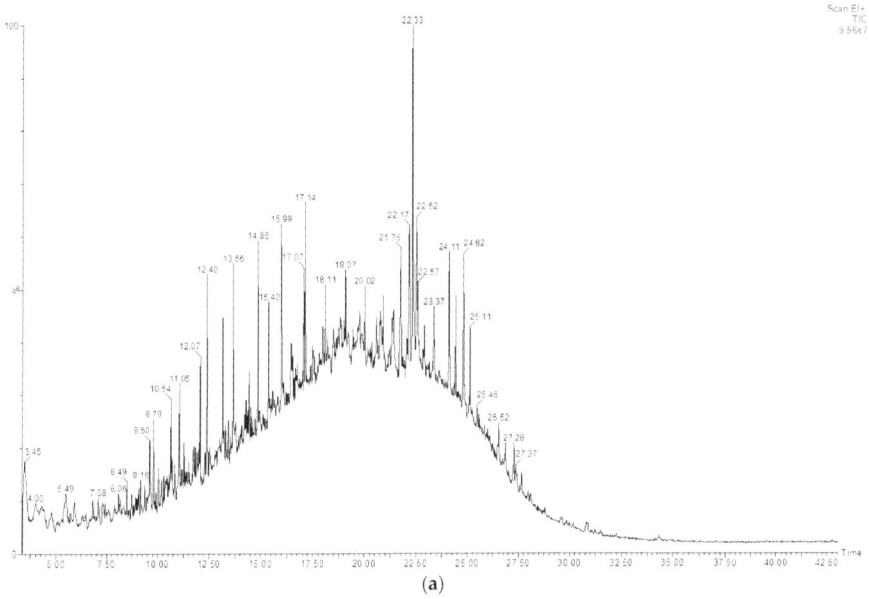

(a)

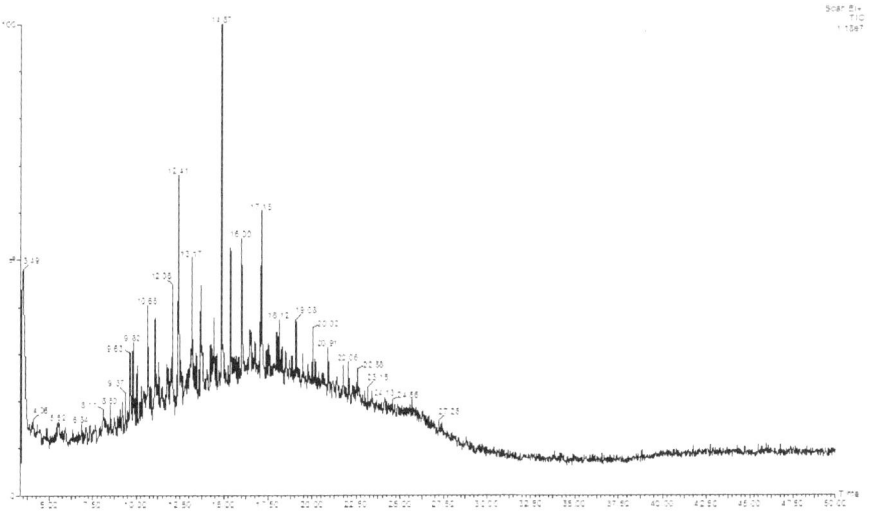

(b)

Figure 8. *Cont.*

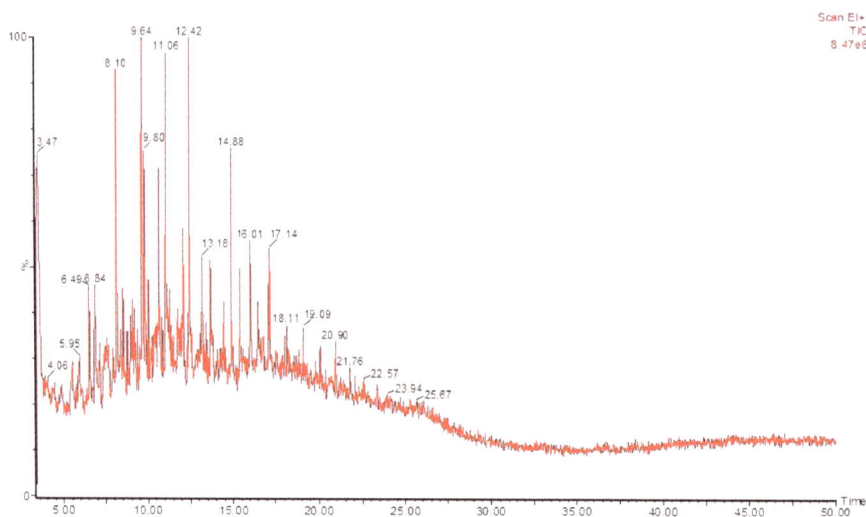

(c)

Figure 8. GC-MS analysis of (**a**) control heavy crude oil used for core flooding experiment (**b**) extra recovered oil by the action of *B. firmus* during core flooding experiment (**c**) extra recovered oil by the action of *B. halodurans* during core flooding experiment.

Table 2. GC-MS chromatogram analysis for the extra recovered oil due to biotransformation by *B. firmus*.

RT	Identified Compound	Carbon No.
9.63	2-methyl-1-pentanol	C6
9.82	cycloheptanol	C7
10.06	1,2-dibromo-octane	C8
10.65	2,4,4-trimethyl-1-hexene	C9
11.08	1,2-dibromo-2-methyl-undecane	C12
12.41	1,2-dibromododecane	C12
13.17	3,7,11-trimethyl-1-dodecanol	C15
13.67	1-nonadecanol	C19
14.87	hexadecanoic acid, (3-bromoprop-2-ynyl) ester	C19
15.39	1-bromoeicosane	C20
17.15	5,15-dimethylnonadecane	C21
18.12	2-nitro-1,3-bis(octyloxy)benzene	C22
19.08	7-hexyldocosane	C28
20.02	11-decyldocosane	C32
21.76	tritriacontane	C33
22.06	1-hexadecylheptadecylcyclohexane	C39
22.58	tetratetracontane	C44

Table 3. GC-MS chromatogram analysis for the extra recovered oil due to biotransformation by
B. halodurans.

RT	Identified Compound	Carbon No.
3.47	1,2-dibromo-2-methylundecane	C4
4.06	2-nitrocyclohexanone	C6
5.51	2,5-heptadecadione	C7
5.95	1,7-dichloroheptane	C7
6.49	2,2-dimethyl-3-pentanol	C7
6.84	1-chloro-heptane	C7
8.10	*N*-methylcyclohexanamine	C7
9.64	acetic acid, hexyl ester	C8
9.80	1,2-dibromo-octane	C8
11.06	1,2-dibromododecane	C12
12.42	1-chlorododecane	C12
13.18	1-nonadecanol	C19
14.88	1- eicosanol	C20
16.01	9-octadecenyl acetate	C20
17.14	dimethylnonadecane	C21
18.11	tetracosane	C24
19.09	7-hexyldocosane	C28
20.90	2-(1-decylundecyl)-1,4-dimethyl cyclohexane	C29
21.76	11-decyldocosane	C32
22.57	tritriacontane	C33

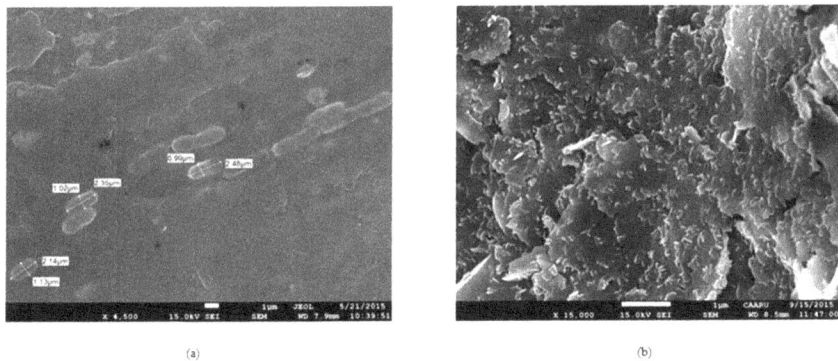

(a)　　　　　　　　　　(b)

Figure 9. Scanning electron microscope (SEM) image (**a**) *B. firmus* in fresh BH medium; and
(**b**) inside core.

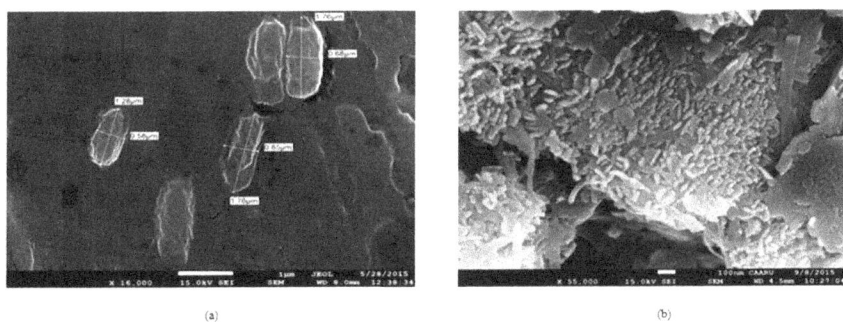

(a)　　　　　　　　　　(b)

Figure 10. SEM image (**a**) *B. halodurans* in fresh BH medium; and (**b**) inside core.

4. Discussion

Heavy crude oil, of significant economic value, poses difficulty in recovery because of its high viscosity and low flow characteristics. EOR methods were employed to overcome the difficulty. MEOR is a tertiary recovery method which can enhance the recovery of crude oil [43]. The soil sample pH was found to be 8.5 ± 0.5, which was slightly alkaline in nature. It has already been reported that the fractionation of hydrocarbons is higher under slightly alkaline conditions [44–46]. The eTPH was found to be ~4.2%. The mineralogy study of the 10 soil samples showed that minerals such as calcite, quartz, albite, palygorskite, anorthite, dolomite, gypsum, halite, microcline, muscovite, rutile, suhailite and takanelite are present in the soil samples. The heavy crude oil sample viscosity and API gravity were determined as 650,000 mPa·s and 4.57°, respectively.

In this study, two indigenous strains, *B. firmus* and *B. halodurans* having the potential of biotransforming heavy crude oil were isolated from heavy crude oil contaminated soil samples collected from one of the Oman oil fields. There are reports for the ability of native bacteria to mineralize crude oil hydrocarbons in oil contaminated sites [47,48]. Identification of the isolates were done using protein profiling by Bruker's MALDI Biotyper [36] and by 16S rRNA gene analysis showing >97% sequence identity with respective genes in the National Center for Biotechnology Information (NCBI) database [49].

Heavy crude oil is a complex mixture of organic compounds [50] and is somewhat resistant to microbial action. Only a few microbes can act on crude oil, most of the strains identified being able to act on a narrow range of substrates [51]. The community composition of indigenous bacteria in Gulf beach sands indicated the abundance of members of the Gammaproteobacteria and Alphaproteobacteria as the major players in oil degradation [52]. Polycyclic Aromatic Hydrocarbons (PAH)-degrading capabilities of *Arthrobacter*, *Burkholderia*, *Mycobacterium*, *Pseudomonas*, *Sphingomonas* and *Rhodococcus* were studied extensively [53]. *B. stearothermophilus* was reported to utilize only hydrocarbons of C_{15} to C_{17} [54], whereas *A. borkumensis* AP1, SK2, and SK7 could act only on alkanes ranging from C_6 to C_{16} [55]. The isolates *B. firmus* and *B. halodurans* are the first reports to biotransform heavy crude oil of 4.57° API gravity.

The crude oil utilization capability of the isolates were determined by the study of growth characteristics in heavy crude as the sole carbon source, the technique has been used in several studies to determine the oil degradation potential of *Pseudomonas* and *Bacillus* sp. [47,56]. Higher concentrations of hydrocarbons might inhibit biodegradation by limiting nutrient or oxygen supply or by its toxic effects [57]. *B. firmus* and *B. halodurans* were shown to have significant growth in BH medium with up to 7% heavy crude oil, which implicates the isolates' tolerance to higher concentrations of heavy crude oil. The pH of the medium turned became more alkaline during the growth period. It was already reported that a slightly alkaline pH may enhance the rate of biodegradation [44–46]. The findings suggest the isolates as being potential candidates for biotransformation of heavy crude oil.

Using GC-MS analysis showed 81.36% biotransformation of heavy crude oil for *B. firmus* and 81.93% for *B. halodurans* compared to the abiogenic control, which was about 8–10%. Spore forming consortia isolated from Oman oil fields which could biotransform heavy oil after 12–21 days of treatment has already been reported [20,26]. The ability of mixed bacterial consortia to degrade 28–51% of saturates and 0–18% of aromatics present in crude oil or up to 60% crude oil was also reported [58,59]. *B. stearothermophilus* isolated from Kuwait oil fields was able to degrade pure hydrocarbons of a chain length of C_{15} to C_{17}, but were not able to degrade crude oil. *A. borkumensis* AP1, SK2, and SK7 was reported capable of utilizing only alkanes ranging from C_6 to C_{16} [55]. This study showed that isolates, *B. firmus* and *B. halodurans* were mostly utilizing aromatic fractions in the crude oil and fractionation of which led to increase in the amount of aliphatic compounds. Several enzymes such as oxidoreductase (laccases and cytochrome-P450 mono-oxygenase), xylene monooxygenase, catechol 2,3-dioxygenase, benzoyl-CoA reductase and others, are reported to play an important role in bacterial biodegradation of crude oil and polycyclic aromatic hydrocarbons [5]. We are further analyzing these bacterial isolates for presence of genes encoding for such enzymes, which are responsible for heavy crude oil biotransformation.

Heavy oil that is trapped in oil reservoirs after primary and secondary recovery can be recovered by biotransforming the heavier fractions to lighter ones. *Bacillus* spp. that could degrade higher *n*-alkanes (>C_{27}) under anaerobic conditions were reported [60]. The most abundant compound present during the ninth day of incubation was hexadecanoic acid (RT 14.87) for *B. firmus* and dodecane (RT 12.42) for *B. halodurans* (Tables 2 and 3). Bacteria from the oil fields in Japan and China degrading *n*-alkane were reported [33,61]. *B. firmus* and *B. halodurans* in this study showed fractionation of higher *n*-alkanes having carbon numbers up to C_{54}. It was reported that *Thermus* sp. which was isolated from the reservoir of the Shengli oil field in East China, was capable of transforming crude oils [62]. MEOR studies using *Bacillus* spp. showed an extra recovery of 9.6% at 37 °C and 7.2% at 55 °C in core flood rig studies using crude oil of 26° API, due to the combined effect of biosurfactant and its biotransforming ability [63]. An extra recovery of 16% of 13.3°API crude oil was reported with *B. licheniformis* [64,65]. Youssef et al. [66] reported all the possibilities associated with microbial processes (both beneficial in EOR and detrimental) relevant to petroleum industry as in-depth analysis. In this study the extra recovered oil from tertiary recovery was 10.44% and 7.69%, respectively, for *B. firmus* and *B. halodurans*.

5. Conclusions

The ability of the isolates, *B. firmus* and *B. halodurans*, to grow at higher concentrations of heavy crude oil and their biotransformation ability by converting heavy fractions of crude oil to lighter ones, by utilizing mostly aromatic compounds indicated that the isolates showed promise for MEOR. The extra recovery of crude heavy oil in the core flood experiments and migration of bacteria in porous sand stone cores further confirms this. To the best of our knowledge, this is the first report of *B. firmus* and *B. halodurans* capable of biotransforming heavy crude oil of 4.57° API. All these findings have indicated that both isolates *B. firmus* and *B. halodurans* are promising candidates for MEOR applications and should be studied further.

Acknowledgments: We would like to acknowledge the help provided by Central Analytical and Applied Research Unit, Sultan Qaboos University, Oman, for SEM, MALDI-Biotyper, 16S rRNA sequencing, and GC analysis. No specific funding was received for this research work.

Author Contributions: B.S., S.N.B., Y.M.A.-W., A.E.E., and S.J.J. conceived and designed the experiments; B.S. performed the experiments; B.S., S.N.B., Y.M.A.-W., A.E.E., S.J.J., and A.S.A.-B. analyzed the data; S.N.B., Y.M.A.-W., A.E.E., S.J.J., and A.S.A.-B. contributed reagents/materials/analysis tools; B.S. wrote the paper; All the authors reviewed and edited the final manuscript.

Conflicts of Interest: The authors declare no conflict of interest.

References

1. Babadagli, T. Development of mature oil fields—A review. *J. Pet. Sci. Eng.* **2007**, *57*, 221–246. [CrossRef]
2. Bao, M.; Kong, X.; Jiang, G.; Wang, X.; Li, X. Laboratory study on activating indigenous microorganisms to enhance oil recovery in Shengli Oilfield. *J. Pet. Sci. Eng.* **2009**, *66*, 42–46. [CrossRef]
3. Shibulal, B.; Al-Bahry, S.N.; Al-Wahaibi, Y.M.; Elshafie, A.E.; Al-Bemani, A.S.; Joshi, S.J. Microbial Enhanced Heavy Oil Recovery by the Aid of Inhabitant Spore-Forming Bacteria: An Insight Review. *Sci. World J.* **2014**, *2014*. [CrossRef] [PubMed]
4. Huc, A.-Y. *Heavy Crude Oils: From Geology to Upgrading: An Overview*; Editions Technip: Paris, France, 2010.
5. Al-Sayegh, A.; Al-Wahaibi, Y.; Joshi, S.; Al-Bahry, S.; Elshafie, A.; Al-Bemani, A. Bioremediation of Heavy Crude Oil Contamination. *Open Biotechnol. J.* **2016**, *10*, 301–311. [CrossRef]
6. Temizel, C.; Rodriguez, D.; Saldierna, N.; Narinesingh, J. Stochastic Optimization of Steam flooding Heavy Oil Reservoirs. In Proceedings of the Society of Petroleum Engineers (SPE) Trinidad and Tobago Section Energy Resources Conference, Port of Spain, Trinidad and Tobago, 13–15 June 2016.
7. Joshi, S.J. Isolation and Characterization of Biosurfactant Producing Microorganisms and Their Possible Role in Microbial Enhanced Oil Recovery (MEOR). Ph.D. Thesis, Maharaja Sayajirao University of Baroda, Vadodara, India, 2008.

8. Sivasankar, P.; Kumar, G.S. Influence of pH on dynamics of microbial enhanced oil recovery processes using biosurfactant producing *Pseudomonas putida*: Mathematical modelling and numerical simulation. *Bioresour. Technol.* **2017**, *224*, 498–508. [CrossRef] [PubMed]
9. Banat, I.M. Biosurfactants production and possible uses in microbial enhanced oil recovery and oil pollution remediation: A review. *Bioresour. Technol.* **1995**, *51*, 1–12. [CrossRef]
10. Xu, T.; Chen, C.; Liu, C.; Zhang, S.; Wu, Y.; Zhang, P. A novel way to enhance the oil recovery ratio by *Streptococcus* sp. BT-003. *J. Basic Microbiol.* **2009**, *49*, 477–481. [CrossRef] [PubMed]
11. Brown, L.R. Microbial enhanced oil recovery (MEOR). *Curr. Opin. Microbiol.* **2010**, *13*, 316–320. [CrossRef] [PubMed]
12. Bryant, R.S.; Douglas, J. *Survival of MEOR Systems in Porous Media*; National Institute for Petroleum and Energy Research: Bartlesville, OK, USA, 1986; pp. 1–32.
13. Lazar, I.; Petrisor, I.; Yen, T. Microbial enhanced oil recovery (MEOR). *Pet. Sci. Technol.* **2007**, *25*, 1353–1366. [CrossRef]
14. Nielsen, S.M.; Shapiro, A.A.; Michelsen, M.L.; Stenby, E.H. 1D Simulations for Microbial Enhanced Oil Recovery with Metabolite Partitioning. *Transp. Porous Media* **2010**, *85*, 785–802. [CrossRef]
15. Almeida, P.; Moreira, R.; Almeida, R.; Guimaraes, A.; Carvalho, A.; Quintella, C.; Esperidia, M.; Taft, C. Selection and application of microorganisms to improve oil recovery. *Eng. Life Sci.* **2004**, *4*, 319–325. [CrossRef]
16. Jinfeng, L.; Lijun, M.; Bozhong, M.; Rulin, L.; Fangtian, N.; Jiaxi, Z. The field pilot of microbial enhanced oil recovery in a high temperature petroleum reservoir. *J. Pet. Sci. Eng.* **2005**, *48*, 265–271. [CrossRef]
17. Wentzel, A.; Ellingsen, T.E.; Kotlar, H.K.; Zotchev, S.B.; Throne-Holst, M. Bacterial metabolism of long-chain *n*-alkanes. *Appl. Microbiol. Biotechnol.* **2007**, *76*, 1209–1221. [CrossRef] [PubMed]
18. Grishchenkov, V.G.; Townsend, R.T.; McDonald, T.J.; Autenrieth, R.L.; Bonner, J.S.; Boronin, A.M. Degradation of petroleum hydrocarbons by facultative anaerobic bacteria under aerobic and anaerobic conditions. *Process Biochem.* **2000**, *35*, 889–896. [CrossRef]
19. Li, Q.; Kang, C.; Wang, H.; Liu, C.; Zhang, C. Application of microbial enhanced oil recovery technique to Daqing Oilfield. *Biochem. Eng. J.* **2002**, *11*, 197–199. [CrossRef]
20. Shibulal, B.; Al-Bahry, S.N.; Al-Wahaibi, Y.M.; Elshafie, A.E.; Al-Bemani, A.S.; Joshi, S.J. The potential of indigenous *Paenibacillus ehimensis* BS1 for recovering heavy crude oil by biotransformation to light fractions. *PLoS ONE* **2017**, *12*, e0171432. [CrossRef] [PubMed]
21. Etoumi, A.; Musrati, I.E.; Gammoudi, B.E.; Behlil, M.E. The reduction of wax precipitation in waxy crude oils by *Pseudomonas* species. *J. Ind. Microbiol. Biotechnol.* **2008**, *35*, 1241–1245. [CrossRef] [PubMed]
22. Binazadeh, M.; Karimi, I.A.; Li, Z. Fast biodegradation of long chain *n*-alkanes and crude oil at high concentration with *Rhodococcus* sp. Moj-3449. *Enzym. Microb. Technol.* **2009**, *45*, 195–202. [CrossRef]
23. Zhang, Y.; Xu, Z.; Ji, P.; Hou, W.; Dietrich, F. Microbial EOR laboratory studies and application results in Daqing oilfield. In Proceedings of the Society of Petroleum Engineers (SPE) Asia Pacific Oil and Gas Conference and Exhibition, Jakarta, Indonesia, 20–22 April 1999.
24. Felix, J.A.; Cooney, J.J. Response of spores and vegetative cells of *Bacillus* spp. in a hydrocarbon–water system. *J. Appl. Bacteriol.* **1971**, *34*, 411–416. [CrossRef] [PubMed]
25. Calvo, C.; Toledo, F.L.; González-López, J. Surfactant activity of a naphthalene degrading *Bacillus pumilus* strain isolated from oil sludge. *J. Biotechnol.* **2004**, *109*, 255–262. [CrossRef] [PubMed]
26. Al-Bahry, S.N.; Al-Wahaibi, Y.M.; Al-Hinai, B.; Joshi, S.J.; Elshafie, A.E.; Al-Bemani, A.S.; Al-Sabahi, J. Potential in heavy oil biodegradation via enrichment of spore forming bacterial consortia. *J. Pet. Explor. Prod. Technol.* **2016**, *6*, 787–799. [CrossRef]
27. Zhuang, W.Q.; Tay, J.H.; Maszenan, A.M.; Krumholz, L.R.; Tay, S.L. Importance of Gram-positive naphthalene-degrading bacteria in oil-contaminated tropical marine sediments. *Lett. Appl. Microbiol.* **2003**, *36*, 251–257. [CrossRef] [PubMed]
28. Ijah, U.; Ukpe, L. Biodegradation of crude oil by *Bacillus* strains 28A and 61B isolated from oil spilled soil. *Waste Manag.* **1992**, *12*, 55–60. [CrossRef]
29. Da Cunha, C.D.; Rosado, A.S.; Sebastián, G.V.; Seldin, L.; von der Weid, I. Oil biodegradation by *Bacillus* strains isolated from the rock of an oil reservoir located in a deep-water production basin in Brazil. *Appl. Microbiol. Biotechnol.* **2006**, *73*, 949–959. [CrossRef] [PubMed]

30. Cooper, D.G.; Macdonald, C.R.; Duff, S.J.B.; Kosaric, N. Enhanced production of surfactin from *Bacillus subtilis* by continuous product removal and metal cation additions. *Appl. Environ. Microbiol.* **1981**, *42*, 408–412. [PubMed]

31. Alinnor, I.J.; Nwachukwu, M.A. Determination of total petroleum hydrocarbon in soil and groundwater samples in some communities in Rivers State, Nigeria. *J. Environ. Chem. Ecotoxicol.* **2013**, *5*, 292–297.

32. EPA. gov: Method 3541 (SW-846): Automated Soxhlet Extraction. 1994. Available online: https://www.epa.gov/sites/production/files/2015-06/documents/epa-3541.pdf (accessed on 6 November 2016).

33. Kato, T.; Haruki, M.; Imanaka, T. Isolation and characterization of psychotrophic bacteria from oil-reservoir water and oil sands. *Appl. Microbiol. Biotechnol.* **2001**, *55*, 794–800. [CrossRef] [PubMed]

34. Hilyard, E.J.; Jones-Meehan, J.M.; Spargo, B.J.; Hill, R.T. Enrichment, Isolation, and Phylogenetic Identification of Polycyclic Aromatic Hydrocarbon-Degrading Bacteria from Elizabeth River Sediments. *Appl. Environ. Microbiol.* **2008**, *74*, 1176–1182. [CrossRef] [PubMed]

35. Kleinheinz, G.T.; Bagley, S.T. A filter-plate method for the recovery and cultivation of microorganisms utilizing volatile organic compounds. *J. Microbiol. Methods* **1997**, *29*, 139–144. [CrossRef]

36. Koubek, J.; Uhlik, O.; Jecna, K.; Junkova, P.; Vrkoslavova, J.; Lipov, J.; Kurzawova, V.; Macek, T.; Mackova, M. Whole-cell MALDI-TOF: Rapid screening method in environmental microbiology. *Int. Biodeterior. Biodegrad.* **2012**, *69*, 82–86. [CrossRef]

37. Wunschel, S.C.; Jarman, K.H.; Petersen, C.E.; Valentine, N.B.; Wahl, K.L.; Schauki, D.; Jackman, J.; Nelson, C.P.; White, E. Bacterial analysis by MALDI-TOF mass spectrometry: An inter-laboratory comparison. *J. Am. Soc. Mass Spectrom.* **2005**, *16*, 456–462. [CrossRef] [PubMed]

38. Minai-Tehrani, D.; Herfatmanesh, A. Biodegradation of Aliphatic and Aromatic Fractions of Heavy Crude Oil–Contaminated Soil: A Pilot Study. *Bioremediat. J.* **2007**, *11*, 71–76. [CrossRef]

39. United States Environmental Protection Agency (USEPA): Method 8270D. *Semivolatile Organic Compounds by Gas Chromatography/Mass Spectrophotometry*; Revision 5; (SW-846 Update V); USEPA: Washington, DC, USA, July 2014.

40. Raiders, R.A.; Knapp, R.M.; McInerney, M.J. Microbial selective plugging and enhanced oil recovery. *J. Ind. Microbial.* **1989**, *4*, 215–229. [CrossRef]

41. Davey, M.E.; Gevertz, D.; Wood, W.A.; Clark, J.B.; Jenneman, G.E. Microbial selective plugging of sandstone through stimulation of indigenous bacteria in a hypersaline oil reservoir. *Geomicrobiol. J.* **1998**, *15*, 335–352. [CrossRef]

42. Al-Hattali, R.; Al-Sulaimani, H.; Al-Wahaibi, Y.; Al-Bahry, S.; Elshafie, A.; Al-Bemani, A.; Joshi, S.J. Fractured carbonate reservoirs sweep efficiency improvement using microbial biomass. *J. Pet. Sci. Eng.* **2013**, *112*, 178–184. [CrossRef]

43. Beckman, J.W. The Action of Bacteria on Mineral Oil. *Ind. Eng. Chem. News Ed.* **1926**, *4*, 10.

44. Dibble, J.T.; Bartha, R. Effect of environmental parameters on the biodegradation of oil sludge. *Appl. Environ. Microbiol.* **1979**, *37*, 729–739. [PubMed]

45. Hambrick, G.A.; DeLaune, R.D.; Patrick, W.H. Effect of Estuarine Sediment pH and Oxidation-Reduction Potential on Microbial Hydrocarbon Degradation. *Appl. Environ. Microbiol.* **1980**, *40*, 365–369. [PubMed]

46. Rahman, K.S.; Thahira-Rahman, J.; Lakshmanaperumalsamy, P.; Banat, I.M. Towards efficient crude oil degradation by a mixed bacterial consortium. *Bioresour. Technol.* **2002**, *85*, 257–261. [CrossRef]

47. Sepahi, A.A.; Golpasha, I.D.; Emami, M.; Nakhoda, A. Isolation and Characterization of Crude Oil Degrading *Bacillus* Spp. *J. Environ. Health Sci. Eng.* **2008**, *5*, 149–154.

48. Ojo, O. Petroleum-hydrocarbon utilization by native bacterial population from a wastewater canal Southwest Nigeria. *Afr. J. Biotechnol.* **2006**, *5*, 333.

49. Stackebrandt, E.; Goebel, B. Taxonomic note: A place for DNA-DNA reassociation and 16S rRNA sequence analysis in the present species definition in bacteriology. *Int. J. Syst. Evol. Microbiol.* **1994**, *44*, 846–849. [CrossRef]

50. Head, I.M.; Jones, D.M.; Larter, S.R. Biological activity in the deep subsurface and the origin of heavy oil. *Nature* **2003**, *426*, 344–352. [CrossRef] [PubMed]

51. Liu, H.; Xu, J.; Liang, R.; Liu, J. Characterization of the Medium- and Long-Chain *n*-Alkanes Degrading *Pseudomonas aeruginosa* Strain SJTD-1 and Its Alkane Hydroxylase Genes. *PLoS ONE* **2014**, *9*, e105506. [CrossRef] [PubMed]

52. Kostka, J.E.; Prakash, O.; Overholt, W.A.; Green, S.J.; Freyer, G.; Canion, A.; Delgardio, J.; Norton, N.; Hazen, T.C.; Huettel, M. Hydrocarbon-Degrading Bacteria and the Bacterial Community Response in Gulf of Mexico Beach Sands Impacted by the Deepwater Horizon Oil Spill. *Appl. Environ. Microbiol.* **2011**, *77*, 7962–7974. [CrossRef] [PubMed]

53. Brooijmans, R.J.W.; Pastink, M.I.; Siezen, M.J. Hydrocarbon-degrading bacteria: The oil-spill clean-up crew. *Microb. Biotechnol.* **2009**, *2*, 587–594. [CrossRef] [PubMed]

54. Sorkhoh, N.; Ibrahim, A.; Ghannoum, M.A.; Radwan, S. High-temperature hydrocarbon degradation by *Bacillus stearothermophilus* from oil-polluted Kuwaiti desert. *Appl. Microbiol. Biotechnol.* **1993**, *39*, 123–126. [CrossRef]

55. Van Beilen, J.B.; Marin, M.M.; Smits, T.H.; Röthlisberger, M.; Franchini, A.G.; Witholt, B.; Rojo, F. Characterization of two alkane hydroxylase genes from the marine hydrocarbonoclastic bacterium *Alcanivorax borkumensis*. *Environ. Microbiol.* **2004**, *6*, 264–273. [CrossRef]

56. Mittal, A.; Singh, P. Isolation of hydrocarbon degrading bacteria from soils contaminated with crude oil spills. *Indian J. Exp. Boil.* **2009**, *47*, 760.

57. Palanisamy, N.; Ramya, J.; Kumar, S.; Vasanthi, N.S.; Chandran, P.; Khan, S. Diesel biodegradation capacities of indigenous bacterial species isolated from diesel contaminated soil. *J. Environ. Health Sci. Eng.* **2014**, *12*, 142. [CrossRef] [PubMed]

58. Chhatre, S.; Purohit, H.; Shanker, R.; Khanna, P. Bacterial consortia for crude oil spill remediation. *Water Sci. Technol.* **1996**, *34*, 187–193. [CrossRef]

59. Vasudevan, N.; Rajaram, P. Bioremediation of oil sludge-contaminated soil. *Environ. Int.* **2001**, *26*, 409–411. [CrossRef]

60. Gudiña, E.J.; Pereira, J.F.B.; Rodrigues, L.R.; Coutinho, J.A.P.; Teixeira, J.A. Isolation and study of microorganisms from oil samples for application in Microbial Enhanced Oil Recovery. *Int. Biodeterior. Biodegrad.* **2012**, *68*, 56–64. [CrossRef]

61. She, Y.-H.; Zhang, F.; Xia, J.-J.; Kong, S.-Q.; Wang, Z.-L.; Shu, F.-C.; Hu, J.-M. Investigation of biosurfactant-producing indigenous microorganisms that enhance residue oil recovery in an oil reservoir after polymer flooding. *Appl. Biochem. Biotechnol.* **2011**, *163*, 223–234. [CrossRef] [PubMed]

62. Hao, R.; Lu, A.; Wang, G. Crude-oil-degrading thermophilic bacterium isolated from an oil field. *Can. J. Microbiol.* **2004**, *50*, 175–182. [CrossRef] [PubMed]

63. Wu, L.; Yao, J.; Jain, A.K.; Radhika, C.; Xudong, D.; Hans, R.H. An efficient thermotolerant and halophilic biosurfactant-producing bacterium isolated from Dagang oil field for MEOR application. *Int. J. Curr. Microbiol. Appl. Sci.* **2014**, *3*, 586–599.

64. Al-Sayegh, A.; Al-Wahaibi, Y.; Al-Bahry, S.; Elshafie, A.; Al-Bemani, A.; Joshi, S. Microbial enhanced heavy crude oil recovery through biodegradation using bacterial isolates from an Omani oil field. *Microb. Cell Fact.* **2015**, *14*, 141. [CrossRef] [PubMed]

65. Al-Sayegh, A.; Al-Wahaibi, Y.; Al-Bahry, S.; Elshafie, A.; Al-Bemani, A.; Joshi, S. Enhanced oil recovery using biotransformation technique on heavy crude oil. *Int. J. GEOMATE* **2017**, *13*, 75–79. [CrossRef]

66. Youssef, N.; Elshahed, M.S.; McInerney, M.J. Microbial processes in oil fields: Culprits, problems, and opportunities. *Adv. Appl. Microbial.* **2009**, *66*, 141–251.

MDPI

St. Alban-Anlage 66

4052 Basel

Switzerland

Tel. +41 61 683 77 34

Fax +41 61 302 89 18

www.mdpi.com

Colloids Interfaces Editorial Office

E-mail: colloids@mdpi.com

www.mdpi.com/journal/colloids

www.ingramcontent.com/pod-product-compliance
Lightning Source LLC
Chambersburg PA
CBHW051838210326

41597CB00033B/5698